# Methods in Enzymology

## Volume 416
## GLYCOMICS

# METHODS IN ENZYMOLOGY

EDITORS-IN-CHIEF

## John N. Abelson    Melvin I. Simon

DIVISION OF BIOLOGY
CALIFORNIA INSTITUTE OF TECHNOLOGY
PASADENA, CALIFORNIA

### FOUNDING EDITORS

## Sidney P. Colowick and Nathan O. Kaplan

*Methods in Enzymology*

*Volume 416*

# Glycomics

EDITED BY

## Minoru Fukuda

GLYCOBIOLOGY PROGRAM
CANCER RESEARCH CENTER
THE BURNHAM INSTITUTE FOR MEDICAL RESEARCH
LA JOLLA, CALIFORNIA

ELSEVIER

AMSTERDAM • BOSTON • HEIDELBERG • LONDON
NEW YORK • OXFORD • PARIS • SAN DIEGO
SAN FRANCISCO • SINGAPORE • SYDNEY • TOKYO
Academic Press is an imprint of Elsevier

Academic Press is an imprint of Elsevier
525 B Street, Suite 1900, San Diego, California 92101-4495, USA
84 Theobald's Road, London WC1X 8RR, UK

This book is printed on acid-free paper.

For information on all Elsevier Academic Press publications
visit our Web site at www.books.elsevier.com

ISBN-13: 978-0-12-182821-9
ISBN-10: 0-12-182821-2

PRINTED IN THE UNITED STATES OF AMERICA
06  07  08  09  9  8  7  6  5  4  3  2  1

# Table of Contents

## Section I. Carbohydrate–Protein Interactions–Physical Measurements

## Section II. Phage Display Library

# Section III. Glycosyltransferase Expression Profiling

# Section IV. CHO Mutants in Glycan Biosynthesis

# Section V. Function of Sulfation

## Section VI. Functions Revealed by Forced Expression or Gene Knockout

# Contributors to Volume 416

Article numbers are in parentheses following the names of contributors.
Affiliations listed are current.

SCOTT T. ACTION (23), *Departments of Bio-medical Engineering and Electrical and Computer Engineering, University of Virginia, Charlottesville, Virginia*

TOMOYA O. AKAMA (20), *Glycobiology Program, Cancer Research Center, The Burnham Institute, La Jolla, California*

JESUS ANGULO (2), *Instituto de Investigaciones Químicas (CSIC-US), Sevilla, Spain*

HISASHI ASHIDA (12), *Department of Immunoregulation, Research Institute for Microbial Diseases, Osaka University, Osaka, Japan*

ANDREW J. BENIE (2), *Carlsberg Laboratory, Copenhagen, Denmark*

THORSTEN BIET (2), *Institute of Chemistry, University of Lubeck, Lubeck, Germany*

ASTRID BLUME (2), *Carlsberg Laboratory, Copenhagen, Denmark*

SHIHAO CHEN (24), *Glycobiology Program, Cancer Research Center, The Burnham Institute for Medical Research, La Jolla, California*

ASSOU EL-BATTARI (7), *Université de la Méditerranée, Marseille, France*

JEFFREY D. ESKO (13), *Department of Cellular and Molecular Medicine, Glycobiology Research and Training Center, University of California, San Diego, La Jolla, California*

BETH A. FRAZIER (13), *Washington University School of Medicine, St. Louis, Missouri*

MICHIKO N. FUKUDA (4, 20), *Glycobiology Program, Cancer Research Center, The Burnham Institute, La Jolla, California*

MINORU FUKUDA (8, 19, 22), *Glycobiology Program, Cancer Research Center, The Burnham Institute for Medical Research, La Jolla, California*

HIROKO HABUCHI (14), *Institute for Molecular Science of Medicine, Aichi Medical University, Nagakute, Aichi, Japan*

OSAMI HABUCHI (14), *Department of Chemistry, Aichi University of Education, Kariya, Aichi, Japan*

JENNIFER A. HAMMOND (10), *The Scripps Research Institute, La Jolla, California*

STEVEN R. HEAD (10), *The Scripps Research Institute, La Jolla, California*

GUIDO J. JENNISKENS (5), *Department of Biochemistry, Raboud University Nijmegen Medical Centre, Nijmegen Centre for Molecular Life Sciences, Nijmegen, The Netherlands*

HIROTO KAWASHIMA (16, 18), *Department of Microbiology, School of Pharmaceutical Sciences, University of Shizuoka, Shizuoka, Japan*

KOJI KIMATA (14), *Institute for Molecular Science of Medicine, Aichi Medical University, Nagakute, Aichi, Japan*

TAROH KINOSHITA (12), *Department of Immunoregulation, Research Institute for Microbial Diseases, Osaka University, Osaka, Japan*

MOTOHIRO KOBAYASHI (9), *Department of Pathology, Shinshu University School of Medicine, Matsumoto, Japan*

TETSUHITO KOJIMA (17), *Department of Medical Technology, Nagoya University School of Health Sciences, Nagoya, Japan*

ROGER LAWRENCE (13), *Department of Cellular and Molecular Medicine, Glycobiology Research and Training Center, University of California, San Diego, La Jolla, California*

JOOST F. M. LENSEN (5), *Department of Biochemistry, Raboud University Nijmegen Medical Centre, Nijmegen Centre for Molecular Life Sciences, Nijmegen, The Netherlands*

KLAUS LEY (23), *Department of Biomedical Engineering, Robert M. Berne Cardiovascular Research Center, University of Virginia, Charlottesville, Virginia*

YUSUKE MAEDA (12), *Department of Immunoregulation, Research Institute for Microbial Diseases, Osaka University, Osaka, Japan*

RODGER P. MCEVER (21), *Oklahoma Medical Research Foundation, Oklahoma City, Oklahoma*

JUNYA MITOMA (19), *Division of Glyco-Signal Research, Institute of Molecular Biomembrane and Glycobiology, Tohoku Pharmaceutical University, Sendai, Japan*

MEGUMI MORIMOTO-TOMITA (15), *Department of Anatomy, UCSF, San Francisco, California*

TAKASHI MURAMATSU (14, 17), *Department of Health Science, Faculty of Psychological and Physical Sciences, Aichi Gakuin University, Aichi, Japan*

HISAKO MURAMATSU (17), *Department of Biochemistry and Division of Animal Models, Center for Neurological Disease and Cancer, Nagoya University, Graduate School of Medicine, Nagoya, Japan*

JUN NAKAYAMA* (8), *Glycobiology Program, Cancer Research Center, The Burnham Institute for Medical Research, La Jolla, California*

HISASHI NARIMATSU (6), *Glycogene Function Team of Research Center for Glycoscience, National Institute of Advanced Industrial Science and Technology (AIST), Ibaraki, Japan*

MASAHIKO NEGISHI (1), *Pharmacogenetics Section, Laboratory of Reproductive and Developmental Toxicology, National Institute of Environmental Health Sciences, National Institutes of Health, Research Triangle Park, North Carolina*

JOHN E. PAK (3), *Department of Molecular and Medical Genetics, University of Toronto, Toronto, Ontario, Canada*

MONICA PALCIC (2), *Carlsberg Laboratory, Copenhagen, Denmark*

FRANCISCO PARRA (2), *Instituto de Biotechnología de Asturias, Departamento de Bioquímica y Biología Molecular, Universidad de Oviedo, Oviedo, Spain*

SANTOSH KUMAR PATNAIK (11), *Department of Cell Biology, Albert Einstein College of Medicine, Bronx, New York*

HANNELORE PETERS (2), *Institute of Chemistry, University of Lubeck, Lubeck, Germany*

THOMAS PETERS (2), *Institute of Chemistry, University of Lubeck, Lubeck, Germany*

JOHN PICKARD (23), *Department of Biomedical Engineering, University of Virginia, Charlottesville, Virginia*

*Current address: Department of Pathology, Shinshu University School of Medicine, Matsumoto, Japan.*

CHRISTOPH RADEMACHER (2), *Institute of Chemistry, University of Lubeck, Lubeck, Germany*

JAMES M. RINI (3), *Departments of Molecular and Medical Genetics and Biochemistry, University of Toronto, Toronto, Ontario, Canada*

STEVEN D. ROSEN (15), *Department of Anatomy, UCSF, San Francisco, California*

TAKASHI SATO (6), *Glycogene Function Team of Research Center for Glycoscience, National Institute of Advanced Industrial Science and Technology (AIST), Ibaraki, Japan*

NICOLE C. SMITS (5), *Department of Biochemistry, Raboud University Nijmegen Medical Centre, Nijmegen Centre for Molecular Life Sciences, Nijmegen, The Netherlands*

MACK SOBHANY (1), *Pharmacogenetics Section, Laboratory of Reproductive and Developmental Toxicology, National Institute of Environmental Health Sciences, National Institute of Health, Research Triangle Park, North Carolina*

MARKUS SPERANDIO (23), *Children's Hospital, Division of Neonatology, University of Heidelberg, Heidelberg, Germany*

PAMELA STANLEY (11), *Department of Cell Biology, Albert Einstein College of Medicine, Bronx, New York*

MASAMI SUZUKI (8), *Glycobiology Program, Cancer Research Program, The Burnham Institute for Medical Research, La Jolla, California*

MISA SUZUKI (8), *Glycobiology Program, Cancer Research Center, The Burnham Institute for Medical Research, La Jolla, California*

GERDY B. TEN DAM (5), *Department of Biochemistry, Raboud University Nijmegen Medical Centre, Nijmegen Centre for Molecular Life Sciences, Nijmegen, The Netherlands*

AKIRA TOGAYACHI (6), *Glycogene Function Team of Research Center for Glycoscience, National Institute of Advanced Industrial Science and Technology (AIST), Ibaraki, Japan*

KENJI UCHIMURA (14, 15), *Department of Anatomy, Program in Immunology, UCSF, San Francisco, California*

SUNIL UNNIKRISHNAN (23), *Department of Biomedical Engineering, University of Virginia, Charlottesville, Virginia*

TOIN H. VAN KUPPEVELT (5), *Department of Biochemistry, Raboud University Nijmegen Medical Centre, Nijmegen Centre for Molecular Life Sciences, Nijmegen, The Netherlands*

TESSA J. M. WIJNHOVEN (5), *Department of Biochemistry, Raboud University Nijmegen Medical Centre, Nijmegen Centre for Molecular Life Sciences, Nijmegen, The Netherlands*

LIJUN XIA (21), *Oklahoma Medical Research Foundation, Oklahoma City, Oklahoma*

LIJUAN ZHANG (13), *Washington University School of Medicine, St. Louis, Missouri*

# Preface

In the past decade, we have seen an explosion of progress in understanding the roles of carbohydrates in biological systems. This explosive progress was made with the efforts in determining the roles of carbohydrates in immunology, neurobiology, and many other disciplines, examining each unique system and employing new technology. Thanks to Academic Press editorial management, particularly to Ms. Cindy Minor, three books in the *Methods in Enzymology* series, namely *Glycobiology* (Volume 415), *Glycomics* (Volume 416), and *Functional Glycomics* (Volume 417), have been dedicated to disseminating information on methods in determining the biological roles of carbohydrates. These books are designed to provide an introduction of new methods to a wide variety of readers who would like to participate in and contribute to the advancement of glycobiology. The methods covered include structural analysis of carbohydrates, biological and chemical synthesis of carbohydrates, expression and determination of ligands for carbohydrate-binding proteins, gene expression profiling using microarrays, and generation of gene knockout mice and their phenotype analyses. The book also covers recent advances in special topics such as chaperones for glycosyltransferase, the roles of glycosylation in signal transduction, chemokine and cytokine binding, muscle development, glycolipids in neural development, and cell-cell interaction. I believe we have a collection of outstanding contributors who represent their respective expertise and field.

The Glycomics volume (416) covers methods on physical measurements of carbohydrate-protein interactions, phage display library, glycosyltransferase expression profiling including *in situ* hybridization and gene microarrays, CHO mutants in glycosyltransferases, GPI-anchor and proteoglycans, functional analysis by transfection, and gene knockout. I believe that this book will be useful to a wide variety of readers from graduate students and researchers in academics and industry, to those who would like to teach glycobiology at various levels. We hope this book will contribute to explosive progress in glycobiology.

MINORU FUKUDA
June 16th, 2006

# METHODS IN ENZYMOLOGY

# Section I

# Carbohydrate–Protein
# Interactions–Physical Measurements

# [1]  Characterization of Specific Donor Binding to α1,4-N-Acetylhexosaminyltransferase EXTL2 Using Isothermal Titration Calorimetry

By MACK SOBHANY and MASAHIKO NEGISHI

## Abstract

Glycosyltransferases encompass one of the largest families of enzymes found in nature. Their principle function is to catalyze the transfer of activated donor-sugar molecules to various acceptor substrates. The molecular basis that governs this specific transfer reaction, such as how a given transferase determines donor-sugar specificity, remains to be elucidated. Human α1,4-N-acetylhexosaminyltransferase (EXTL2) transfers N-acetylglucosamine and N-acetylgalactosamine but does not transfer glucose or galactose. Isothermal titration calorimetry (ITC) is a powerful technique used to characterize various binding reactions, including both protein–ligand and protein–protein interactions. ITC provides the binding stoichiometry, affinity, and the thermodynamic parameters free energy ($\Delta G$), enthalpy ($\Delta H$), and entropy ($\Delta S$) of these binding interactions. This chapter describes our ITC study demonstrating the two-step mechanism that regulates the specific binding of N-acetylhexosamines to EXTL2.

## Overview

EXTL2 is a member of the exostosin (EXT)-related family of enzymes and was originally characterized as the enzyme responsible for the completion of the specific linker region of the heparan chain (Kitagawa et al., 1999). X-ray crystal structures solved in the presence of UDP-N-acetylhexosamines reveal that EXTL2 does not form any direct interaction with the signature N-acetyl group, which resides in an open space of the substrate binding cleft (Negishi et al., 2003; Pedersen et al., 2003). This observation raises an interesting question as to how EXTL2 determines its donor substrate specificity towards the N-acetylhexosamines. Given the available structure information, ITC provides an excellent tool to explore the molecular mechanism underlying the specific donor–enzyme recognition (Sobhany et al., 2005).

ITC directly measures the heat evolved or absorbed as a result of a given binding process. An ITC instrument essentially consists of a sample cell, a reference cell, and an injecting syringe. In an ITC experiment, the

METHODS IN ENZYMOLOGY, VOL. 416                          0076-6879/06 $35.00
                                                        DOI: 10.1016/S0076-6879(06)16001-2

solution containing the protein or enzyme of interest is loaded into the sample cell of the instrument and the ligand solution is placed into the injecting syringe. A feedback control system provides thermal power constantly to maintain the exact same temperature in both the reference cell and the sample cell. When heat is generated or absorbed within the sample cell, the temperature within the cell will change and the feedback control system will alter the amount of power supplied to minimize the temperature imbalance (Velazquez-Campoy and Freire, 2005). The syringe injects precise amounts of the ligand into the sample cell; the syringe employs a spinning mechanism for subsequent mixing (Fig. 1). Temperature differences between the reference cell and sample cell are measured. Data analysis from a single ITC experiment yields the binding stoichiometry ($n$) and binding constant $K_a$. Analysis of the reaction heat as a function of concentration provides a complete thermodynamic characterization of a binding reaction: $\Delta H$, $\Delta S$, and $\Delta G$ (Perozzo et al., 2004).

In general, simple reversible associations between a protein $P$ and a ligand $L$, $P + L \leftrightarrow PL$ are characterized by the binding constant $K_a$ or the dissociation constant $K_d$:

$$K_a = [PL]/[P][L] = 1/K_d,$$

where $[P]$ and $[L]$ are the concentrations of the free reactants and $[PL]$ is the concentration of the complex generated by the reactants (Perozzo et al., 2004; Velazquez-Campoy et al., 2004). These values are related to the Gibbs free energy of binding ($\Delta G$) and can be expressed in terms of the changes of enthalpy ($\Delta H$) and entropy ($\Delta S$) in the process: $\Delta G = -RT \ln K_a = \Delta H - T\Delta S$. R is the universal gas law constant (1.9872 cal/K·mol) and T is the absolute temperature at which the reaction took place in units Kelvin (Velazquez-Campoy et al., 2004). The perceptible response in an ITC experiment is the change in heat correlated with each injection of ligand. For every given injection of ligand, the heat generated or absorbed is directly proportional to the total amount of formed complex and can be expressed as $q = V_0 \Delta H \Delta [PL]$; $q$ is the measured heat that is absorbed/generated associated with the change in complex concentration, $\Delta[PL]$, $V_0$ is the reaction volume of the sample cell, and $\Delta H$ is the molar enthalpy of binding (Perozzo et al., 2004). As an ITC experiment progresses, the concentration of unoccupied binding sites begins to decrease and the changes in heat decrease in magnitude correspondingly as ligand is added until saturation is reached. From this titration curve, the parameters $n$ (stoichiometry), binding constant $K_a$, and $\Delta H$ are directly determined. $\Delta S$ and $\Delta G$ are then calculated from these values. These

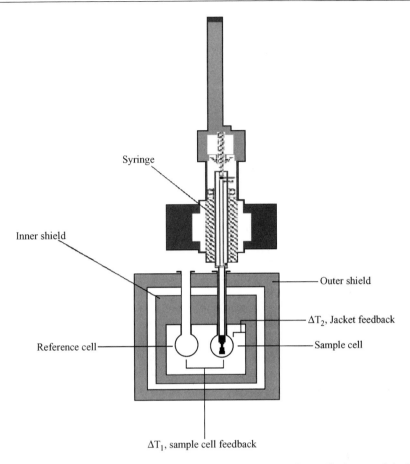

FIG. 1. Illustration of an isothermal titration calorimeter. Two cells in an adiabatic environment are connected to the outside through narrow access tubes. A thermoelectric device measures the temperature difference between the two cells. A second device measures the difference in temperature between the cells and the jacket. During the process of a chemical reaction occurring in the sample cell, heat is generated or absorbed. The temperature difference between the sample and reference cells ($\Delta T_1$) is kept at zero by the addition of heat to the sample or reference cell. The integral of the power required to maintain $\Delta T_1 = 0$ over time is a measure of total heat resulting from the process being studied. A spinning syringe is used for injecting and subsequent mixing of the reactants in the experiment (*VP-ITC MicroCalorimeter User's Manual*, MicroCal, LLC).

calculations are performed by the software that accompanies the ITC apparatus. The software also performs computer modeling to identify the number of binding sites possessed by the protein.

General Experimental Design

Optimization of protein and ligand concentrations and the injection volume is necessary to obtain high-quality data. A few preliminary experiments may need to be performed to establish these conditions. The protein concentration required for ITC depends on the given protein and its substrate binding affinity. For experiments that can be expected to produce a high substrate binding affinities, protein concentrations as low as 5 $\mu M$ can be used to successfully obtain good ITC data. However, protein–carbohydrate interactions are quite weaker than protein–protein and protein–nucleic acid interactions and higher concentrations of protein may be necessary to obtain observable changes in heat due to binding. For glycosyltransferases, protein concentrations of ~30 $\mu M$ are sufficient for most ITC experiments. Generally, the ligand concentration is 10–20 times higher than that of the protein because several injections of small aliquots into the sample cell must take place in order to generate a sufficient number of data points for analysis. Upon completion, a titration experiment should approach or reach complete saturation of the binding sites. The time between successive injections is also an important parameter; if protein–ligand interaction is rapid, thermal equilibrium will be achieved within a short period (generally 3–4 min is sufficient in such cases). Conversely, heat signals of a slow process require much more time to reach thermal equilibrium and the time between injections should be adjusted accordingly. Additionally, it is absolutely critical for the solutions of the ligand and protein to be pure and exactly identical with respect to pH, buffer capacity, and salt concentration; essentially, it is preferable for the protein and ligand to be dissolved in the same buffer. Formation of air bubbles should be avoided, as air bubbles in the syringe can cause variation in injection volumes, and bubbles in the sample cell interfere with the thermal contact of solution and cell wall. Therefore, it is very important to thoroughly degas all solutions before use in the experiment.

Donor Sugar Binding to EXTL2

Figure 2 shows the ITC profiles of the binding of UDP-GlcNAc to EXTL2, and Table I summarizes the binding characteristics and the thermodynamic parameters of various donor substrates. UDP, UDP-GlcNAc, and UDP-GalNAc are all found to bind to EXTL2 at a 1:1 ratio, thus agreeing with the X-ray crystal structure showing only one donor molecule per EXTL2 molecule (Pedersen et al., 2003). The binding affinities of the donor substrates are similar; the $K_d$ value of UDP binding is 23 $\mu M$ and

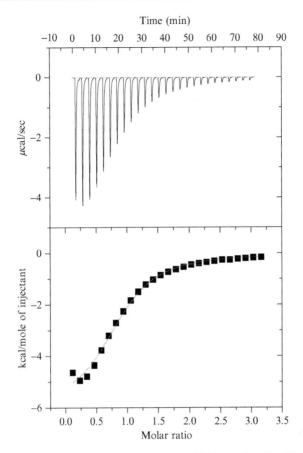

FIG. 2. Calorimetric profile of the binding of UDP-GlcNAc to EXTL2. The reaction cell contained a solution of EXTL2 (1.4 ml and 165.4 $\mu M$). The syringe contained 4 m$M$ UDP-GlcNAc dissolved in the same buffer as EXTL2. (Top) Raw calorimetric data obtained from the injection of 6-$\mu l$ aliquots of UDP-GalNAc at 3-min intervals. The lower plot shows the integrated binding isotherm with the experimental points (■) and best fit. The final fitting parameters for the best fit provide the values for N (number of sites), $\Delta$H (cal/mol), $\Delta$S (cal/mol/deg), and K (binding constant in $M^{-1}$).

those of UDP-GlcNAc and UDP-GalNAc are ~61 $\mu M$. UDP-Gal and UDP-Glc are found not to bind to EXTL2. These results provide direct evidence that despite having no direct interaction with the enzyme, the $N$-acetyl group determines the binding of UDP-$N$-acetylhexosamines to EXTL2.

TABLE I
THERMODYNAMIC PARAMETERS, $\mu$G (CAL/MOL), $\mu$H (CAL/MOL), $\mu$S (CAL/MOL/DEG), AND $N$ (NUMBER OF BINDING SITES) FOR DONOR BINDING WITH WILD-TYPE EXTL2 IN $H_2O$ SOLUTION AT 303° $K^a$

| Substrate | $\Delta$G | $\Delta$H | $\Delta$S | $K_d$ ($\mu$M) | $n$ |
|---|---|---|---|---|---|
| UDP | $-6443.4 \pm 193.7$ | $-6034 \pm 193.7$ | 1.351 | 22.62 | 0.869 |
| UDP-GalNAc | $-5839.1 \pm 72.14$ | $-2097 \pm 72.14$ | 12.35 | 61.46 | 0.968 |
| UDP-GlcNAc | $-5844.3 \pm 68.32$ | $-1975 \pm 68.32$ | 12.77 | 60.80 | 1.106 |
| UDP-Gal | n.d. | n.d. | n.d. | n.d. | n.d. |
| UDP-Glc | n.d. | n.d. | n.d. | n.d. | n.d. |

[a] Substrate and EXTL2 concentrations were 4 m$M$ and 166 $\mu$M, respectively. n.d., no detectable binding.

ITC has previously been employed with $\alpha$1,3-galactosyltransferase, which also displayed a $K_d$ of 60 $\mu$M for its binding to UDP-Gal (Boix *et al.*, 2002). Therefore, the interaction of glycosyltransferases with carbohydrates appeared to be around 100 $\mu$M, which is weaker than other specific interactions. Protein–protein interactions are generally quite strong; the association of a monoclonal antibody with cytochrome $c$ yields a binding affinity of 71.43 PM (Pierce *et al.*, 1999). Protein–nucleic acid interactions also yield robust binding affinities; for instance, the binding of *Thermus aquaticus* DNA polymerase to template DNA at 30° produces a $K_d$ of 6.2 nM (Datta and LiCata, 2003).

The binding reactions of UDP-GlcNAc and UDP-GalNAc are accompanied with an increase in $\Delta$S. A comparison of the $\Delta$H and T$\Delta$S values reveals that absolute values of the T$\Delta$S values are larger than those of $\Delta$H, indicating that the binding reactions are driven by changes in entropy. Increases in entropy in systems at constant temperature are known to correlate with the loss of previously oriented waters of hydration into the bulk solvent (Sievers *et al.*, 2004). Thus, the *N*-acetyl group may expel water from the area of enzyme–donor complex into the bulk solvent in order to position UDP-*N*-acetylhexosamine for its binding to EXTL2. To provide further evidence that water molecules may be involved in determining the specific donor binding, ITC experiments were performed in the denser and more viscous deuterium oxide ($D_2O$). Both UDP-GlcNAc and UDP-GalNAc maintain their ability to bind to EXTL2 in $D_2O$ solution with similar affinities and $\Delta S$ values to those observed in $H_2O$ solution. The importance of water in donor binding activity was substantiated by testing the binding of UDP-Gal and UDP-Glc in $D_2O$ solution. UDP-Gal and UDP-Glc were found to be capable of binding to EXTL2 in $D_2O$ solution, although their binding affinities to EXTL2 were still four to fivefold weaker than those of the UDP-$N$-acetylhexosamines. The bindings of UDP-Glc and UDP-Gal are accompanied with only a marginal increase in entropy, compared with the large increases in entropy observed with those of the UDP-$N$-acetylhexosamines in $D_2O$ solution. These data are consistent with the hypothesis that water molecules are critically involved in EXTL2 donor binding activity.

While EXTL2 forms no direct interaction with the $N$-acetyl group, its residues, such as Arg135, Asp246, and Arg293, interact directly with the hydroxyl groups of $N$-acetylhexosamine moiety (Pedersen $et\ al.$, 2003; Sobhany $et\ al.$, 2005). Alanine mutations of those residues result in an increase of donor binding affinity ($K_d$) and $K_m$ for donor hydrolysis (Sobhany $et\ al.$, 2005). Moreover, the $K_m$ and $K_d$ values are very similar in wild-type and in a given mutant of EXTL2. Thus, those residues determine the donor binding affinity and hydrolysis activity via their direct hydrogen bond interactions with the donor molecule, while the $N$-acetyl group determines the donor specificity by the entropy-driven positioning of UDP-$N$-acetylhexosamines in the active site. These experiments exemplify that when coupled with available structural data, ITC can be a powerful tool in elucidating donor substrate recognition and catalytic mechanisms of glycosyltransferases.

*Materials for Protein Expression and Purification*

1. BL21 (DE3) Competent Cells (Stratagene, La Jolla, CA) transformed with pMAL-2c EXTL2 expression vector (New England Biolabs, Beverly, MA).
2. Luria–Bertani Media: 10 g of Bacto-Tryptone, 5 g of Bacto-Yeast Extract, 10 g of NaCl, dissolved in distilled water, final volume 1 liter with pH adjusted to 7.5. Working media solution modified with 100 $\mu$g/ml ampicillin.
3. 2X YT Media: 16 g Bacto-Tryptone, 10 g of Bacto-Yeast Extract, 10 g of NaCl, dissolved in distilled water, final volume 1 liter with pH adjusted to 7.5. Working media solution modified with 100 $\mu$g/ml ampicillin.
4. DU 640 Spectrophotometer (Beckman Coulter, Fullerton, CA).
5. Isopropylthio-$\beta$-D-galactoside (IPTG) (Invitrogen, Carlsbad, CA).
6. FRENCH Pressure Cell and Press (Thermo IEC, Needham Heights, MA).
7. J6-MI Centrifuge and Optima L-90K ultracentrifuge (Beckman Coulter, Fullerton, CA).
8. Innova 44 Shaking Incubator (New Brunswick Scientific, Edison, NJ).
9. Amylose Resin (New England Biolabs, Beverly, MA).

10. Molecularporous membrane tubing (Spectrum Laboratories, Inc., Rancho Dominguez, CA).
11. Bio-Rad Protein Assay (Bio-Rad, Hercules, CA).

*Solutions*

1. Lysis Buffer: 30 m$M$ Tris–Cl (pH 7.5), 350 m$M$ NaCl, 5 m$M$ dithiothreitol (DTT) (ICN Biomedicals, Inc., Aurora, OH).
2. Elution Buffer: 30 m$M$ Tris–Cl, 350 m$M$ NaCl, 5 m$M$ DTT modified with 20 m$M$ Maltose.
3. Reaction Buffer: 25 m$M$ HEPES (pH 7.5), 100 m$M$ NaCl, 20 m$M$ MnCl$_2$, 1 m$M$.

*Method for Protein Expression and Purification*

1. 1 ml of BL21 (DE3) cells transformed with pMAL-2c EXTL2 expression vector was added to 100 ml of Luria–Bertani media modified with 100 $\mu$g/ml ampicillin and placed on a shaker at 210 rpm at 37° for an overnight culture.
2. 10 ml of pre-culture each was added to ten 2-liter flasks containing 1 liter of 2X YT media modified with 100 $\mu$g/ml ampicillin.
3. These flasks were placed in the Innova 44 Shaking Incubator set to shake at 210 rpm at 37°. When A$_{550}$ reached 0.8, isopropylthio-$\beta$-D-galactoside was added to each flask, final concentration of 0.2 m$M$.
4. The temperature of the incubator was then set to 23° and the flasks were allowed to shake overnight.
5. The cells were then harvested by spinning at 4000 rpm for 25 min and then resuspended in lysis buffer.
6. The cells were disrupted in a French Press at room temperature under pressures varying between 8000 and 24,000 psi.
7. Following disruption, the cells were spun down at 38,000 rpm for 45 min.
8. The soluble fraction obtained was then allowed to bind to amylose resin at 4° for 2 h with gentle agitation.
9. The resin was then spun down at 4000 rpm for 5 min. The supernatant was poured off and the resin was washed with the lysis buffer. Resin was pelleted and washed a total of three times.
10. Protein was then eluted with the elution buffer.
11. Following elution, the protein solution was placed into Molecularporous membrane tubing and dialyzed overnight into reaction buffer. Following dialysis, protein concentration was determined using Bio-Rad Protein Assay.

*Materials for Isothermal Titration Calorimetry*

1. Uridine Diphosphate *N*-acetylglucosamine (Sigma-Aldrich, St. Louis, MO).
2. ThermoVac Sample Degasser and Thermostat (MicroCal, LLC, Northampton, MA).
3. VP-ITC MicroCalorimeter (MicroCal, LLC, Northampton, MA).
4. Origin Scientific Graphing and Analysis Software (OriginLab, Northampton, MA).

*Solutions*

1. Reaction Buffer: 25 mM HEPES (pH 7.5), 100 m$M$ NaCl, 20 m$M$ MnCl$_2$, 1 m$M$ CaCl$_2$.

Method for Isothermal Titration Calorimetry

1. Degas all protein samples, buffers, and donor substrate solutions thoroughly before testing.
2. Load EXTL2 sample into the sample cell of the VP-ITC Micro Calorimeter.
3. Load reaction buffer into the reference cell.
4. Dissolve Uridine Diphosphate *N*-acetylglucosamine (UDP-GlcNAc) in the reaction buffer at a concentration 10–20 times that of the EXTL2. Then load the UDP-GlcNAc solution into the syringe.
5. Set the VP-ITC MicroCalorimeter to perform 26 injections of 6 $\mu$l at 180-s intervals. Run program; data acquisition and analysis will be completed by the Origin software package.

Acknowledgment

This research was supported by the Intramural Research Program of the National Institutes of Health (NIH), and the National Institutes of Environmental Health Sciences (NIEHS).

References

Boix, E., Zhang, Y., Swaminathan, J., Brew, K., and Achaya, K. R. (2002). Structural basis of ordered binding of donor and acceptor substrates to the retaining glycosyltransferase, alpha-1,3-galactosyltransferase. *J. Biol. Chem.* **277,** 28310–28318.
Datta, K., and LiCata, V. J. (2003). Thermodynamics of the binding of *Thermus aquaticus* DNA polymerase to primed-template DNA. *Nucleic Acids Res.* **31,** 5590–5597.

Kitagawa, H., Shimakawa, H., and Sugahara, K. (1999). The tumor suppressor EXT-like gene *EXTL2* encodes an $\alpha$1,4-*N*-acetylhexosaminyltransferase that transfers *N*-acetylgalactosamine and *N*-acetylglucosamine to the common glycosaminoglycan-protein linkage region. *J. Biol. Chem.* **274,** 13933–13937.

Negishi, M., Dong, J., Darden, T. A., Pedersen, L. G., and Pedersen, L. C. (2003). Glucosaminylglycan biosynthesis: What we can learn from the X-ray crystal structures of glycosyltransferases GlcAT1 and EXTL2. *Biochem. Biophys. Res. Commu.* **303,** 393–398.

Pedersen, L. C., Dong, J., Taniguchi, F., Kitagawa, H., Krahn, J. M., Pedersen, L. G., Sugahara, K., and Negishi, M. (2003). Crystal structure of an alpha 1,4-*N*-acetylhexosaminyltransferase (EXTL2), a member of the exostosin gene family involved in heparan sulfate biosynthesis. *J. Biol. Chem.* **278,** 14420–14428.

Perozzo, R., Folkers, G., and Scapozza, L. (2004). Thermodynamics of protein–ligand interactions: History, presence, and future aspects. *J. Recept. Sig. Transd.* **24,** 1–52.

Pierce, M. M., Raman, C. S., and Nall, B. T. (1999). Isothermal titration calorimetry of protein-protein interactions. *Methods* **19,** 213–221.

Sievers, A., Beringer, M., Rodina, M. V., and Wolfenden, R. (2004). The ribosome as an entropy trap. *Proc. Natl. Acad. Sci. USA* **101,** 7897–7901.

Sobhany, M., Dong, J., and Negishi, M. (2005). Two-step mechanism that determines the donor binding specificity of human UDP-*N*-acetylhexosaminyltransferase. *J. Biol. Chem.* **280,** 23441–23445.

Velazquez-Campoy, A., Leavitt, S. A., and Freire, E. (2004). Characterization of protein-protein interactions by isothermal titration calorimetry. *Methods Mol. Biol.* **261,** 35–54.

Velazquez-Campoy, A., and Freire, E. (2005). ITC in the post-genomic era...? Priceless. *Biophys. Chem.* **115,** 115–124.

"VP-ITC MicroCalorimeter User's Manual." MicroCal, LLC, Northampton, MA.

# [2]    NMR Analysis of Carbohydrate–Protein Interactions

*By* Jesus Angulo, Christoph Rademacher, Thorsten Biet,
Andrew J. Benie, Astrid Blume, Hannelore Peters,
Monica Palcic, Francisco Parra, and Thomas Peters

## Abstract

Carbohydrate–protein interactions are frequently characterized by dissociation constants in the $\mu M$ to m$M$ range. This is normally associated with fast dissociation rates of the corresponding complexes, in turn leading to fast exchange on the nuclear magnetic resonance (NMR) chemical shift time scale and on the NMR relaxation time scale. Therefore, NMR experiments that take advantage of fast exchange are well suited to study carbohydrate–protein interactions. In general, it is possible to analyze

METHODS IN ENZYMOLOGY, VOL. 416
Copyright 2006, Elsevier Inc. All rights reserved.

0076-6879/06 $35.00
DOI: 10.1016/S0076-6879(06)16002-4

ligand binding by observing either protein signals or ligand resonances. Because most receptor proteins to which carbohydrates bind are rather large with molecular weights significantly exceeding 30 kDa, the analysis of the corresponding protein spectra is not trivial, and only very few studies have been addressing this issue so far. We, therefore, focus on NMR experiments that employ observation of free ligand, that is, carbohydrate signals to analyze the bound state. Two types of NMR experiments have been extremely valuable to analyze carbohydrate–protein interactions at atomic resolution. Whereas transferred nuclear Overhauser effect (NOE) experiments deliver bioactive conformations of carbohydrates binding to proteins, saturation transfer difference (STD) NMR spectra provide binding epitopes and valuable information about the binding thermodynamics and kinetics. We demonstrate the power of a combined transfer NOE/STD NMR approach for the analysis of carbohydrate–protein complexes using selected examples.

Introduction

Molecular recognition of carbohydrate structures is involved in a number of important biological processes such as recognition of antigenic carbohydrates on the surface of bacteria and viruses by antibodies, initiation of inflammatory processes by selectin–carbohydrate interactions, and the processing of glycan chains by highly specific glycosyltransferases (Bertozzi and Kiessling, 2001; Dube and Bertozzi, 2005; Hakomori, 2001; Raman *et al.*, 2005; Unligil and Rini, 2000). Therefore, the investigation of carbohydrate–protein interactions at an atomic level is of significant scientific interest. To fully understand the binding process, one would ideally have a comprehensive picture of the conformational properties of carbohydrate chains in aqueous solution on the one hand and of the protein-bound state on the other hand. The inherent flexibility of carbohydrate chains has posed a number of complications on their conformational analysis in aqueous solution, as this has been realized many years ago (for reviews, see Duus *et al.*, 2000; Imberty and Perez, 2000). NMR in conjunction with computer simulations has contributed significantly to an understanding of the conformational properties of carbohydrates. The investigation of carbohydrate–protein complexes has the advantage that upon binding, usually one so-called *bioactive conformation* is selected and problems of complex conformational equilibria do not exist. Instead, other difficulties arise because of the usually low affinity of carbohydrate–protein interactions. In comparison to, say, DNA–protein interactions, dissociation constants $K_D$ are at least three orders of magnitude larger ranging from $\mu M$ to m$M$ values. In addition, labeling of carbohydrates with NMR active isotopes such as $^{15}N$ and $^{13}C$ is not straightforward because carbohydrates are secondary gene products. Synthetic routes towards labeled sugars are rather cost intensive and have not been used frequently (Homans et al., 1998; Milton et al., 1998). A potential perspective is the production of isotope-labeled carbohydrates as constituents of bacterial cell walls by growing the bacteria on suitable isotope-enriched media (Martin-Pastor and Bush, 1999; Xu and Bush, 1996, 1998). Yet, this will not facilitate access to mammalian glycan structures. Therefore, NMR techniques common for the observation of high-affinity ligand–protein interactions (Hajduk et al., 1997; Petros et al., 2006) are usually not applicable to carbohydrate ligands.

A common feature of NMR techniques that are used to study carbohydrate–protein interactions is that they use the resonance signals of the free ligand to recruit information about the bound state. To achieve this goal, the dissociation rate of the ligand has to be fast on the NMR time scale (Lian et al., 1994). In general, one has to distinguish between the NMR chemical shift time scale, and the NMR relaxation time scale. If exchange is fast on the NMR chemical shift time scale, one observes one average resonance signal for a signal of the bound and of the free ligand. A process that is fast on the NMR chemical shift time scale is also fast on the NMR relaxation time scale. If a process is slow on the NMR chemical shift time scale, two separate signals are observed for a signal of the free and of the bound ligand. Nevertheless, the process may still be fast on the NMR relaxation time scale and one may use the resonance signals of the free ligand to obtain information about the bound state. For carbohydrates, the majority of cases are fast on the NMR chemical shift time scale and on the NMR relaxation time scale.

To facilitate the observation of bound state properties on signals of the free ligand, there must be some mechanism that enhances the "information" of the bound state in order to make it detectable on an excess of free ligand. This is necessary because at low carbohydrate ligand to protein ratios, the fraction of bound protein is rather low, and only a significant excess of ligand over protein guarantees that most binding sites are saturated. The transferred NOE experiment (Balaram et al., 1972a,b) and variants thereof are ideally suited for this purpose, but one has to keep in mind that there is an optimal carbohydrate ligand to protein ratio for the observation of transferred NOEs (Clore and Gronenborn, 1983; Neuhaus and Williamson, 2000; Ni, 1994). The immense value of saturation transfer experiments (Cayley et al., 1979; Hyde et al., 1980; Keller and Wuthrich, 1978; Wuthrich et al., 1978) for the observation of bound ligand properties has been demonstrated using so-called STD NMR experiments (Mayer and Meyer, 1999, 2001). In the following sections, we describe examples in which these techniques have been applied to analyze bioactive conformations of carbohydrates, as well as binding epitopes at atomic resolution.

Experimental Principles

*Transferred NOE Experiments*

The theory of the transferred NOE has been described in numerous publications (Balaram *et al.*, 1972a,b; Clore and Gronenborn, 1983; Moseley *et al.*, 1995) and several excellent reviews have been written (Neuhaus and Williamson, 2000; Ni, 1994; Post, 2003). The observation of transferred NOE is possible for ligand–protein complexes that are characterized by dissociation constants $K_D$ that are in the $\mu M$ to m$M$ range. Too tight binding places the systems outside the range of fast exchange on the NMR time scale and no transferred NOEs are observed. The majority of carbohydrate–protein interactions fall into the experimental window of the transferred NOE experiment, and consequently numerous transferred NOE studies have been published on carbohydrate–protein systems (for reviews, see Jimenez-Barbero *et al.*, 1999; Peters and Pinto, 1996).

In a transferred nuclear Overhauser and exchange spectroscopy (NOESY) experiment, for example, during the mixing time, there is exchange between the free and the bound form of the carbohydrate and the resulting observed ("transferred") NOEs are due to averaged cross-relaxation rates $\sigma$ of protons for the free and the bound state of the ligand. Because proteins to which carbohydrates bind are much larger, transferred NOEs are always negative. To observe transferred NOEs, a simple inequality [Eq. (1)] has to be fulfilled, where $\sigma$ is the cross-relaxation rate and N the number of molecules in the free ($\sigma_f$, $N_f$) and the bound ($\sigma_b$, $N_b$) state:

$$|N_b \times \sigma_b| \gg |N_f \times \sigma_f| \tag{1}$$

The setup of transferred NOE experiments is identical to the setup of "normal" NOE experiments. The only difference is the preparation of the sample. The intensity of transferred NOEs strongly depends on the excess of ligand over protein, and there is an optimum ratio at which transferred NOEs are of maximal size. In many cases, this optimum is at a molar ratio of carbohydrate to protein between 10 and 50 to 1. Depending on the size of the carbohydrate ligand, three regimes may be distinguished:

   a. The molecular weight of the carbohydrate ligand leads to correlation times ranging in the order of tens to hundreds of picoseconds, and therefore NOEs of the free ligand are positive. At 500 MHz, this is usually the case up to the size of trisaccharides. If charges are present as, for example, in sialic acid residues, the tumbling of the molecule is slower and one may observe negative NOEs already for a trisaccharide.

b. If the molecular weight is such that the correlation time approaches zero crossing conditions, no NOEs will be observable. For uncharged carbohydrates at 500 MHz, this is usually the case for tetrasaccharides and pentasaccharides.
c. Larger carbohydrates have correlation times of several nanoseconds and, therefore, display negative NOEs at frequencies of 500 MHz and higher.

In cases (a) and (b), the discrimination of transferred NOEs from free ligand NOEs is straightforward because at carbohydrate to protein ratios in which inequality [Eq. (1)] is fulfilled, the sign of the NOE changes upon binding from positive to negative. At the same time, the mixing time at which a maximum NOE is observed is reduced and in the range of 200 ms, as compared to 600–1000 ms for the free ligand. Because of this change in sign, the experiment has also been used to identify binding in mixtures of low molecular weight compounds (Meyer *et al.*, 1997). Figure 1 illustrates this dependence on the mixing time and the differences between positive NOEs and negative transferred NOEs.

In case (c), discrimination is less straightforward and usually requires the acquisition of NOESY experiments with different mixing times.

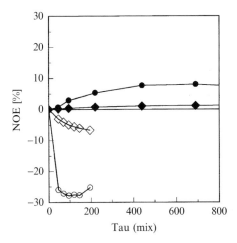

FIG. 1. Nuclear Overhauser effects (NOEs) for free $\alpha$-L-Fuc-$(1\rightarrow6)$-$\beta$-D-GlcNAc-OMe (filled symbols) and trNOEs for $\alpha$-L-Fuc-$(1\rightarrow6)$-$\beta$-D-GlcNAc-OMe in the presence of *Aleuria aurantia* agglutinin (open symbols), measured at 600 MHz as a function of the mixing time $\tau_{mix}$. Circles and diamonds refer to proton pairs H6proR$^{GlcNAc}$-H6proS$^{GlcNAc}$ and H1$^{Fuc}$-H6proS$^{GlcNAc}$, respectively. Maximum NOEs for the free disaccharides are found around mixing times $\tau_{mix}$ of 1000 ms, whereas maximum trNOEs are reached at mixing times lower than 300 ms.

*Saturation Transfer Difference NMR*

Although saturation transfer experiments have been known for quite a while, only after STD NMR had been introduced to identify binding of ligands to receptor proteins (Mayer and Meyer, 1999) was this technique widely used to quickly identify binding and binding epitopes at the same time (for a review, see Meyer and Peters, 2003). Although STD NMR has been applied to a huge variety of ligand–protein systems, as it covers a wider range of $K_D$ values than the transferred NOE experiment, namely from n$M$ to high m$M$ values, it is extremely well suited for the investigation of carbohydrate–protein interactions, which were also the first systems to be studied by STD NMR (Haselhorst *et al.*, 1999, 2001; Mayer and Meyer, 2001). Developments have extended the method to the investigation of ligand binding to whole cells and native viruses, rendering STD NMR a valuable tool to study membrane-bound receptor proteins (Claasen *et al.*, 2005).

In contrast to transferred NOE experiments STD NMR experiments do not critically depend on the ligand-to-protein ratio. As signal intensity that does not result from the saturation transfer process from the protein to the ligand is eliminated by different spectroscopy signal intensities in STD, spectra always reflect purely the bound state. In practice, one is working with carbohydrate-to-protein ratios ~50:1 at protein concentrations in the micromolar range. A standard pulse sequence for STD NMR is shown in Fig. 2. This pulse sequence can be combined with any other pulse sequence such as HSQC or TOCSY. The shaped pulses that are used for saturation

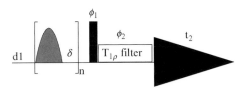

FIG. 2. Pulse scheme for saturation transfer difference (STD) nuclear magnetic resonance (NMR). Pulse sequence for the one-dimensional (1D) STD NMR spectra. The subtraction is performed after every scan via phase cycling. The on and off resonance frequency of the selective pulse is, therefore, switched after every scan. Phases are $\phi_1 = (x, -x, -x, x, y, -y, -y, y, -x, x, x, -x, -y, y, y, -y)$; $\phi_2 = 2\ (y, -y),\ 2\ (-x, x)$; and $\phi_{rec} = 2\ (x),\ 2\ (-x),\ 2\ (y), 2\ (-y),\ 2(-x),\ 2(x),\ 2\ (-y),\ 2(y)$. The length of the selective pulse is 50 ms, the delay $\delta$ between the pulses is 1 ms, the duration of the presaturation period is adjusted by the number of pulses n (typically n = 40), and d1 is an additional short relaxation delay. The field strength $\omega_1 = \gamma \times B_1$ of the selective saturation Gauss pulse is ~100 Hz.

of the protein usually are gaussian pulses. To achieve efficient saturation of the protein and thus optimal STD effects, one has to adjust the pulse length. Usually, 720° pulses lead to very satisfying results. Shaped pulses such as E-BURP may also be used for band-selective excitation, but their application is less robust than that of simple shapes such as the gaussian shape.

In practice, the advent of cryogenic probe technology has greatly enhanced the scope of STD NMR. It is now possible to even identify and characterize binding of ligands to proteins if the protein concentration is in the nanomolar range.

## Binding of UDP Sugars to Glycosyltransferases

The atomic details of binding of activated sugars to glycosyltransferases has been difficult to investigate for several reasons, one being that most glycosyltransferases are membrane-bound proteins. Although a number of crystal structures are now available, there is still an ongoing debate about the mechanisms that are used by glycosyltransferases. In most cases, the bioactive conformation of the donor substrate has not been experimentally determined. In this example, we demonstrate the benefit of transferred NOE experiments and STD NMR for the investigation of complexes between glycosyltransferases and donor substrates.

### Binding of Donor Substrates to β-(1,4)-Galactosyltransferase T1

The first example addresses the binding of UDP-Gal and UDP-Glc to the inverting $\beta$-(1,4)-galactosyltransferase T1 [$\beta$-(1,4)-GalT1] (Biet and Peters, 2001; Jayalakshmi *et al.*, 2004). Using STD NMR experiments, Biet and Peters (2001) identified the binding epitopes of UDP-Gal and UDP-Glc when bound to the enzyme, as shown in Fig. 3. It is evident that the enzyme discriminates between the two ligands, which is in accordance with the fact that UDP-Gal, but not UDP-Glc, is a donor substrate. As seen in Fig. 3, the enzyme does not "recognize" the glucose residue of UDP-Glc. When crystal structural data of the complex became available (Ramakrishnan *et al.*, 2002), it turned out that the binding epitope very well fits the STD NMR data on a qualitative basis.

Based on the crystal structure, it was possible to perform a quantitative analysis of the UDP-Gal/$\beta$-(1,4)-GalT1 complex, essentially leading to a refined solution structure. For this structural refinement, we used a full relaxation and exchange matrix approach employing the program CORCEMA-ST (Jayalakshmi *et al.*, 2004). It turned out that the crystal structure and the solution structure were rather similar.

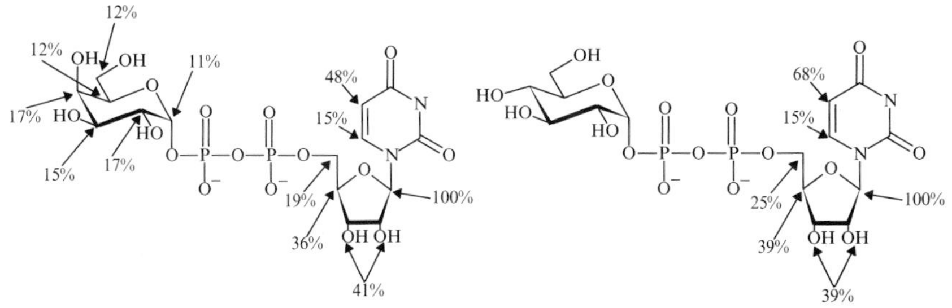

FIG. 3. Relative saturation transfer difference (STD) effects for UDP-Gal (left) and UDP-Glc (right) bound to $\beta$4-Gal-T1 at 500 MHz. The STD spectra were measured at 293-K with 2-k scans. The reference spectra were measured with 1-k scans. Saturation transfer was achieved by using 40 selective gaussian pulses with a duration of 50 ms and a spacing of 1 ms. The protein envelope was irradiated at 0 ppm (on-resonance) and 40 ppm (off-resonance). Subsequent substraction was achieved via phase cycling. Values were calculated by determining individual signal intensities in the STD spectrum ($I_{STD}$), and in the reference one-dimensional (1D) NMR spectrum ($I_0$). The ratios of the intensities $I_{STD}/I_0$ were normalized using the largest STD effect (anomeric proton H1' of the ribose unit, 100%) as a reference. Arrows indicate the position of the proton experiencing an STD effect. The value of 41% corresponds to the cumulative saturation transfer for H2' and H3'. Likewise, 12% denotes the sum effect for both protons H6 of the galactose residue. It is obvious that the galactose residue of UDP-Gal, but not the glucose residue of UDP-Glc, is recognized by the enzyme.

## Binding of Donor Substrates to Human Blood
## Group B Galactosyltransferase

In contrast to $\beta$-(1,4)-GalT1, human blood group B galactosyltransferase (GTB) is a retaining glycosyltransferase, and the precise mechanism is still a matter of debate; for example, it is not experimentally verified whether a covalent intermediate in this process of double inversion at the anomeric center of the galactose residue occurs or whether the reaction has the character of an $S_N i$ reaction. Even though a crystal structure of GTB exists, the bioactive conformation of UDP-Gal and its precise location in the active site have been unknown (Patenaude *et al.*, 2002).

This example demonstrates that the combined use of transferred NOE experiments and STD NMR allows one to deduce the bioactive conformation and the binding kinetics of the donor substrate UDP-Gal (Angulo *et al.*, submitted for publication), and that the presence of bivalent metal cations switches the recognition of the galactose residue of UDP-Gal. Using a fragment-based approach, STD NMR also allows the definition of minimal binding requirements for donor-substrate analogs (Blume *et al.*, submitted for publication).

The analysis of the conformation of UDP-Gal and its analog UDP-Glc when bound to GTB leads to a set of transferred NOEs that is not observed for the free ligands. It should be noted that to observe transferred NOEs in this case, we had to move to rather low and thus unusual carbohydrate ligand-to-protein ratios of 2:1 because the off-rate of UDP-Gal is slow (see later discussion). At higher ratios, inequality [Eq. (1)] is no longer fulfilled and one observes only effects from the free ligand. To unambiguously observe transferred NOEs against the protein signal background at such low ligand-to-protein ratios, we used perdeuterated GTB for these experiments, which had the additional advantage of removing effects stemming from spin diffusion. The "extra" NOEs in the bound form are across the diphosphate bridge and unambiguously indicate folded bound conformations of UDP-Gal and UDP-Glc, as illustrated in Fig. 4.

This folded bound conformation has been the starting point for a complete structural analysis of the UDP-Gal/GTB complex based on

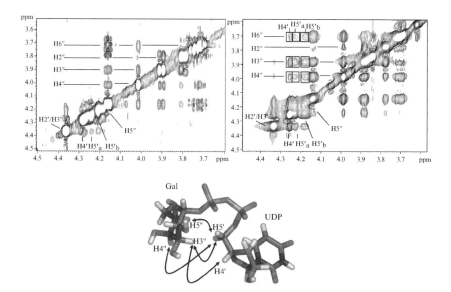

Fig. 4. Conformational change of the sugar nucleotides upon binding to group B galactosyltransferase (GTB). NOESY spectra of UDP-Gal in the presence (right panel) and in the absence (left panel) of GTB. Spectra were recorded at 700 and 500 MHz, respectively. NOEs across the pyrophosphate bridge (black squares in right panel) are only observed for the bound states. (Bottom) Conformation of bound UDP-Gal that gives the best R-NOE factor (0.27) for the STD data, and that is in agreement with the observed trNOEs across the diphosphate bridge (arrows in bottom panel). (Used, with permission, from Angulo *et al.*, in press.)

transferred NOEs and STD effects. Interestingly, and in contrast to the inverting $\beta$-(1,4)-GalT1, UDP-Glc binds to the enzyme in an essentially identical manner as UDP-Gal, as concluded from almost identical binding epitopes and transferred NOEs. Nevertheless, virtually no transfer of the glucose residue to acceptor substrates is observed. We, therefore, hypothesize that after UDP-Gal has been bound to GTB, the binding of an acceptor substrate induces a hitherto unknown conformational change of the ternary complex that in turn leads to a glycosyl transfer reaction but only in the case of UDP-Gal.

It is well established that the presence of bivalent metal cations such as $Mg^{2+}$ or $Mn^{2+}$ significantly enhances the rate of galactosyl transfer. We, therefore, performed STD NMR experiments and transferred NOE experiments of GTB complexes with UDP-Gal or UDP-Glc in the presence and in the absence of $Mg^{2+}$. We observed that binding still occurs in the absence of $Mg^{2+}$, but GTB no longer properly "recognizes" the galactose or the glucose residue. In accordance with this observation, transferred NOE experiments give no indication of a folded bound conformation in the absence of $Mg^{2+}$ (data not shown). For an illustration, Fig. 5 shows STD spectra of UDP-Gal in the presence and in the absence of $Mg^{2+}$.

As this has been shown for the example of the bifunctional enzyme UDP-GlcNAc 2-epimerase/ManNAc kinase, STD NMR is extremely well suited to reveal the minimal structural requirements of a ligand or a substrate for binding to an active site (Blume et al., 2004). In the case of UDP-GlcNAc 2-epimerase/ManNAc kinase, it was shown that UMP is the smallest fragment that is recognized by this enzyme that has some relation to glycosyltransferases in that it also has one binding pocket that recognizes an UDP-activated sugar, in this case UDP-GlcNAc. For GTB, the fragment-based approach clearly shows that in this case even the base alone binds to the enzyme's donor substrate pocket (Blume et al., submitted for publication). Such differences are the basis for the development of selective inhibitors of such enzymes. From competitive STD NMR titration experiments (Mayer and Meyer, 2001), one obtains a qualitative affinity ranking of UDP-based substrate analogs (Fig. 6).

Based on the bound conformation of UDP-Gal bound to GTB, as inferred from transferred NOE experiments (Fig. 4), it is possible to perform a structural refinement of the UDP-Gal/GTB complex based on the crystal structure and STD buildup curves with CORCEMA-ST (Jayalakshmi and Krishna, 2002). As part of this structural refinement process, the program also delivers details about the binding kinetics of the ligand. Assuming a $K_D$ value of 17 $\mu M$ for UDP-Gal binding to GTB, it is found that the off-rate constant $k_{off}$ for UDP-Gal is 10 Hz. A similar analysis was performed for the acceptor H-disaccharide, leading to off-rates that are at least a factor

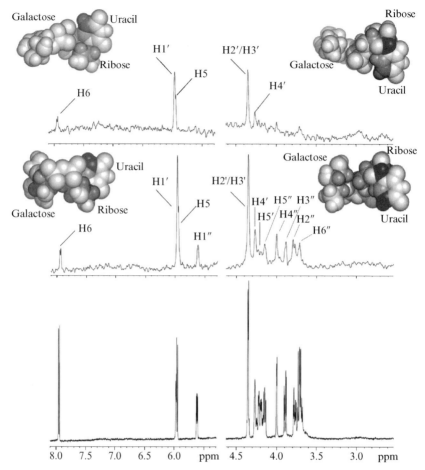

FIG. 5. Mg$^{2+}$ ions trigger the recognition of the hexopyranose ring of UDP-Gal. Saturation transfer difference (STD) nuclear magnetic resonance (NMR) spectra of UDP-Gal in the presence of GTB and in the absence (top) or presence (middle) of Mg$^{2+}$. (Bottom) Reference spectrum of UDP-Gal. It is obvious that protons of the galactose ring give an STD response only if Mg$^{2+}$ is present, indicating that only in this case the pyranose unit is properly "recognized" by the enzyme. The models shown as inserts reflect the binding epitope as determined from relative STD values (grey scale code: dark grey 80–100%, medium grey 60–80%, grey 40–60%, light grey 20–40%, white 0–20%). The models also reflect the bound conformation with and without Mg$^{2+}$. (Used, with permission, from Angulo et al., in press.)

of 10 higher. Therefore, the binding of the activated sugar donor is much slower than the binding of acceptor substrates. It is obvious that this has important implications for the elucidation of suitable mechanistic models describing the galactosyl transfer reaction.

## Protein NMR Investigations into Glycosyltransferases

As mentioned earlier, glycosyltransferases are usually membrane-bound proteins, and even if soluble functional fragments are available, their high molecular weight places them into that class of proteins where a structural analysis with NMR still is very demanding. On the other hand,

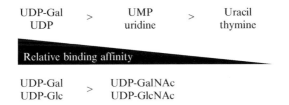

FIG. 6. Relative binding affinities of UDP-Gal and derivatives. Competitive saturation transfer difference (STD) nuclear magnetic resonance (NMR) titrations show that all these ligands compete with each other and, therefore, bind to the same binding site. (Used, with permission, from Blume et al., in press.)

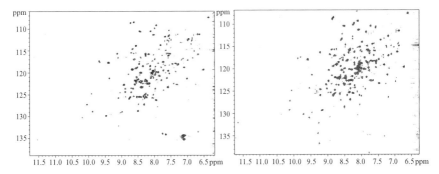

FIG. 7. $^1$H,$^{15}$N-TROSY spectra of uniformly $^2$H,$^{15}$N-labeled group B galactosyltransferase (DN-GTB) at 25° (left) and 50° (right) and 700 MHz (cryoprobe). The $^1$H,$^{15}$N-TROSY spectrum of a uniformly $^{15}$N-labeled sample of GTB (N-GTB) is shown for comparison. The protein concentration was 800 $\mu M$ for DN-GTB and 600 $\mu M$ for N-GTB Probe in 25-m$M$ Bis–Tris buffer at pH 6.5.

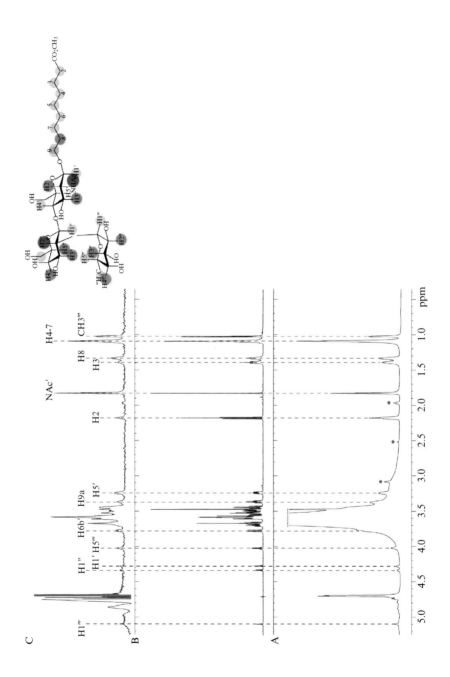

crystal structure analysis delivers structural details that subsequently may be used as a starting point for further in-depth binding studies using NMR experiments or for the analysis of the protein dynamics upon binding with NMR. Such experiments are well established and may be applied to glycosyltransferases once isotope-labeled material is available. GTB is a suitable test candidate because this enzyme is recombinantly available from *Escherichia coli* in sufficient quantities. We have established experimental protocols for uniform and amino acid–selective labeling of GTB. As a first example, we have prepared uniformly $^2$H- and $^{15}$N-labeled GTB that was then subjected to $^1$H,$^{15}$N-TROSY (Pervushin *et al.*, 1997) experiments (unpublished data). These experiments revealed that the enzyme gives spectra with the dispersion of signals improving at higher temperatures (e.g., 50°). Although this constitutes a first step to study protein dynamics and conformation upon binding of donor or acceptor substrates, it will now be necessary to find optimal experimental conditions for the data acquisition. Figure 7 shows the first $^1$H,$^{15}$N-TROSY spectra of a sample of uniformly $^2$H,$^{15}$N-labeled GTB recorded at 700 MHz at temperatures of 25 and 50° reflecting a folded state of the enzyme.

The TROSY spectrum in Fig. 7 also shows that deuteration is essential, because it significantly improves the quality of the TROSY spectrum. The complexity of the spectrum also underlines the importance of amino acid–selective labeling, and future binding studies will certainly have to focus on selectively labeled GTB.

FIG. 8. (A) Reference spectrum (with off-resonance irradiation at 57 ppm) for 0.5-m$M$ H-antigen type II ($\alpha$-L-Fuc(1,2)-$\beta$-D-Gal-(1,4)-$\beta$-D-GlcNAc-O-(CH$_2$)$_8$-CO$_2$CH$_3$ [see formula in the inset]) in the presence of 0.11 $\mu M$ rabbit hemorrhagic disease virus (RHDV) virus-like particles (VLPs) (20-$\mu M$ binding sites) in 100 mM d$_{19}$-Bis–Tris in D$_2$O at pH 6.0. Asterisks mark impurities. The large signal at 3.6 ppm results from an unknown impurity of the VLP preparation. (B) $^1$H NMR of 0.5-m$M$ H-antigen type II in D$_2$O. This spectrum was used as a reference for the evaluation of qualitative saturation transfer difference (STD) effects, as shown in the inset. Because of severe overlap resulting from the impurity at 3.6 ppm, the methyl group of the $N$-acetyl group was used to normalize the STD values. (C) STD spectrum for H-antigen type II in the presence of VLPs with on-resonance irradiation at $-2.3$ ppm and a saturation time of 2 s. In a control experiment, no direct irradiation of the sugar was observed (data not shown). All spectra were recorded with 2048 scans at 288 K on a Bruker Avance DRX 700-MHz spectrometer equipped with a cryoprobe employing a W5 WATERGATE sequence for solvent suppression. The gray scale code in the formula is as follows: dark gray, strong STD, close proximity to protons of the virus binding pocket; medium gray, medium STD; light gray, weak STD.

Binding of Carbohydrates to Native Viruses

The last example presented addresses the binding of carbohydrates to viruses (Huang *et al.*, 2003; Lindesmith *et al.*, 2003; Smith and Helenius, 2004). It has been shown that it is possible to study the binding of low molecular weight ligands and even of larger receptor fragments to native viruses using STD NMR (Benie *et al.*, 2003; Moser *et al.*, 2005). This is possible because the lines of the viral coat proteins are rather broad and extend into spectral regions where one can safely saturate the virus signals without touching resonances of peptide fragments even of the size of a few kilodaltons. The first system that we studied was a human rhinovirus sero-type 2 (HRV-2) (Benie *et al.*, 2003). We succeeded to obtain high-quality STD spectra that revealed the binding epitopes of low molecular weight compounds that bind into the canyon binding site on the viral surface. In a second step, we showed that it is possible to also study the binding epitope of receptor fragments that bind to the viral surface, albeit the fact that these fragments had a molecular weight of ~7 kDa (Moser *et al.*, 2005). These promising results led us to speculate that one should also be able to target the binding of carbohydrates to viral surfaces employing STD NMR. It is well established that influenza viruses attach to target cells via interactions between hemagglutinin on the viral surface and sialic acid residues on the surface of host cells. It is less well known that a number of other viruses such as the Norwalk virus (NV), the rabbit hemorrhagic disease virus (RHDV), or the hepatitis A virus (HAV) specifically adhere to carbohydrate struc-tures on the surface of host cells. In the case of NV and RHDV, these structures are blood group determinants, whereas in the case of HAV, mucin-related structures are discussed (Sanchez *et al.*, 2004). In addition, it has been shown that HAV adhesion is sialic acid dependent. Although the biological significance of these interactions is not yet fully understood, it appears that STD has a great potential for the investigations of these inter-actions. In a pilot experiment, we, have, therefore investigated the binding of an H-type II blood group antigen to recombinant RHDV virus-like particles (VLPs). The STD spectrum (shown in Fig. 8) clearly indicates binding of the carbohydrate antigen to the virus.

For a precise epitope mapping, the spectrum is not yet suitable because it contains impurities of the virus preparation that lead to artifact signals in the reference spectrum rendering a quantitative analysis of STD effects impossible. Nevertheless, the qualitative analysis of STDs was possible and yielded the preliminary binding epitope shown in Fig. 8. Future work will focus on the refinement of this binding epitope. We believe that STD NMR will significantly assist the design of corresponding viral entry inhibitors in general by precisely defining the pharmacophoric groups that are essential for the interaction with the viral coat.

## Outlook

It is obvious that NMR offers valuable tools for the investigation of carbohydrate–protein interactions. Techniques that are based on transferred NOEs or on STD NMR will give us a detailed picture of the binding processes from the perspective of the ligand. The examples in this chapter demonstrate that these experiments may be performed even on rather complex systems in near-physiological environments. In combination with other techniques such as crystal structure analysis, NMR has especially helped and will help us to understand the atomic details of carbohydrate–protein interactions.

## Acknowledgments

T. P. thanks the *Deutsche Forschungsgemeinschaft* for support within the special research area SFB 470, project B3. The 700-MHz NMR facility has also been funded by the *Deutsche Forschungsgemeinschaft* (Me 1830/1). J. A. thanks the European Union for a Marie-Curie Intra-European Fellowship (MEIF-CT-2003–500861). We also thank the VW Foundation (center project grant "conformational control of biomolecular functions") for generous financial support. F. P. has been funded by FEDER and Spanish Ministerio de Educación y Ciencia (grant BIO2003–04237).

## References

Balaram, P., Bothner-By, A. A., and Dadok, J. (1972a). Negative nuclear Overhauser effects as probes of macromolecular structure. *J. Am. Chem. Soc.* **94,** 4015–4017.

Balaram, P., Bothner-By, A. A., and Breslow, E. (1972b). Localization of tyrosine at the binding site of neurophysin II by negative nuclear Overhauser effects. *J. Am. Chem. Soc.* **94,** 4017–4018.

Benie, A. J., Moser, R., Bauml, E., Blaas, D., and Peters, T. (2003). Virus–ligand interactions: Identification and characterization of ligand binding by NMR spectroscopy. *J. Am. Chem. Soc.* **125,** 14–15.

Bertozzi, C. R., and Kiessling, L. L. (2001). Chemical glycobiology. *Science* **291,** 2357–2364.

Biet, T., and Peters, T. (2001). Molecular recognition of UDP-Gal by $\beta$-1,4-galactosyl-transferase T1. *Angew Chem. Int. Ed.* **40,** 4189–4192.

Blume, A., Benie, A. J., Stolz, F., Schmidt, R. R., Reutter, W., Hinderlich, S., and Peters, T. (2004). Characterization of ligand binding to the bifunctional key enzyme in the sialic acid biosynthesis by NMR: I. Investigation of the UDP-GlcNAc 2-epimerase functionality. *J. Biol. Chem.* **279,** 55715–55721.

Cayley, P. J., Albrand, J. P., Feeney, J., Roberts, G. C., Piper, E. A., and Burgen, A. S. (1979). Nuclear magnetic resonance studies of the binding of trimethoprim to dihydrofolate reductase. *Biochemistry* **18,** 3886–3895.

Claasen, B., Axmann, M., Meinecke, R., and Meyer, B. (2005). Direct observation of ligand binding to membrane proteins in living cells by a saturation transfer double difference (STDD) NMR spectroscopy method shows a significantly higher affinity of integrin alpha (IIb)beta3 in native platelets than in liposomes. *J. Am. Chem. Soc.* **127,** 916–919.

Clore, G. M., and Gronenborn, A. M. (1983). Theory of the time-dependent transferred nuclear Overhauser effect: Applications to structural analysis of ligand protein complexes in solution. *J. Magn. Reson.* **53**, 423–442.

Dube, D. H., and Bertozzi, C. R. (2005). Glycans in cancer and inflammation–potential for therapeutics and diagnostics. *Nat. Rev. Drug Discov.* **4**, 477–488.

Duus, J., Gotfredsen, C. H., and Bock, K. (2000). Carbohydrate structural determination by NMR spectroscopy: Modern methods and limitations. *Chem. Rev.* **100**, 4589–4614.

Hajduk, P. J., Meadows, R. P., and Fesik, S. W. (1997). Discovering high-affinity ligands for proteins. *Science* **278**, 497, 499.

Hakomori, S. (2001). Tumor-associated carbohydrate antigens defining tumor malignancy: Basis for development of anti-cancer vaccines. *Adv. Exp. Med. Biol.* **491**, 369–402.

Haselhorst, T., Espinosa, J. F., Jimenez-Barbero, J., Sokolowski, T., Kosma, P., Brade, H., Brade, L., and Peters, T. (1999). NMR experiments reveal distinct antibody-bound conformations of a synthetic disaccharide representing a general structural element of bacterial lipopolysaccharide epitopes. *Biochemistry* **38**, 6449–6459.

Haselhorst, T., Weimar, T., and Peters, T. (2001). Molecular recognition of sialyl Lewis(x) and related saccharides by two lectins. *J. Am. Chem. Soc.* **123**, 10705–10714.

Homans, S. W., Field, R. A., Milton, M. J., Probert, M., and Richardson, J. M. (1998). Probing carbohydrate–protein interactions by high-resolution NMR spectroscopy. *Adv. Exp. Med. Biol.* **435**, 29–38.

Huang, P., Farkas, T., Marionneau, S., Zhong, W., Ruvoen-Clouet, N., Morrow, A. L., Altaye, M., Pickering, L. K., Newburg, D. S., LePendu, J., and Jiang, X. (2003). Noroviruses bind to human ABO, Lewis, and secretor histo-blood group antigens: Identification of 4 distinct strain-specific patterns. *J. Infect. Dis.* **188**, 19–31.

Hyde, E. I., Birdsall, B., Roberts, G. C., Feeney, J., and Burgen, A. S. (1980). Proton nuclear magnetic resonance saturation transfer studies of coenzyme binding to *Lactobacillus casei* dihydrofolate reductase. *Biochemistry* **19**, 3738–3746.

Imberty, A., and Perez, S. (2000). Structure, conformation, and dynamics of bioactive oligosaccharides: Theoretical approaches and experimental validations. *Chem. Rev.* **100**, 4567–4588.

Jayalakshmi, V., Biet, T., Peters, T., and Krishna, N. R. (2004). Refinement of the conformation of UDP-galactose bound to galactosyltransferase using the STD NMR intensity-restrained CORCEMA optimization. *J. Am. Chem. Soc.* **126**, 8610–8611.

Jayalakshmi, V., and Krishna, N. R. (2002). Complete relaxation and conformational exchange matrix (CORCEMA) analysis of intermolecular saturation transfer effects in reversibly forming ligand-receptor complexes. *J. Magn. Reson.* **155**, 106–118.

Jimenez-Barbero, J., Asensio, J. L., Canada, F. J., and Poveda, A. (1999). Free and protein-bound carbohydrate structures. *Curr. Opin. Struct. Biol.* **9**, 549–555.

Keller, R. M., and Wuthrich, K. (1978). Assignment of the heme c resonances in the 360 MHz H NMR spectra of cytochrome *c*. *Biochim. Biophys. Acta* **533**, 195–208.

Lian, L. Y., Barsukov, I. L., Sutcliffe, M. J., Sze, K. H., and Roberts, G. C. (1994). Protein–ligand interactions: Exchange processes and determination of ligand conformation and protein–ligand contacts. *Methods Enzymol.* **239**, 657–700.

Lindesmith, L., Moe, C., Marionneau, S., Ruvoen, N., Jiang, X., Lindblad, L., Stewart, P., LePendu, J., and Baric, R. (2003). Human susceptibility and resistance to Norwalk virus infection. *Nat. Med.* **9**, 548–553.

Martin-Pastor, M., and Bush, C. A. (1999). New strategy for the conformational analysis of carbohydrates based on NOE and 13C NMR coupling constants. Application to the flexible polysaccharide of *Streptococcus mitis* J22. *Biochemistry* **38**, 8045–8055.

Mayer, M., and Meyer, B. (1999). Characterization of ligand binding by saturation transfer difference NMR spectroscopy. *Angew Chem. Int. Ed.* **38,** 1784–1788.

Mayer, M., and Meyer, B. (2001). Group epitope mapping by saturation transfer difference NMR to identify segments of a ligand in direct contact with a protein receptor. *J. Am. Chem. Soc.* **123,** 6108–6117.

Meyer, B., and Peters, T. (2003). NMR spectroscopy techniques for screening and identifying ligand binding to protein receptors. *Angew Chem. Int. Ed. Engl.* **42,** 864–890.

Meyer, B., Weimar, T., and Peters, T. (1997). Screening mixtures for biological activity by NMR. *Eur. J. Biochem.* **246,** 705–709.

Milton, M. J., Harris, R., Probert, M. A., Field, R. A., and Homans, S. W. (1998). New conformational constraints in isotopically ($^{13}$C) enriched oligosaccharides. *Glycobiology* **8,** 147–153.

Moseley, H. N., Curto, E. V., and Krishna, N. R. (1995). Complete relaxation and conformational exchange matrix (CORCEMA) analysis of NOESY spectra of interacting systems; two-dimensional transferred NOESY. *J. Magn. Reson. B* **108,** 243–261.

Moser, R., Snyers, L., Wruss, J., Angulo, J., Peters, H., Peters, T., and Blaas, D. (2005). Neutralization of a common cold virus by concatemers of the third ligand binding module of the VLDL-receptor strongly depends on the number of modules. *Virology* **338,** 259–269.

Neuhaus, D., and Williamson, M. P. (2000). "The Nuclear Overhauser Effect in Structural and Conformational Analysis." Wiley-VCH, New York, Chichester, Weinheim, Brisbane, Singapore, Toronto.

Ni, F. (1994). Recent developments in transferred NOE methods. *Prog. NMR Spectrosc.* **26,** 517–606.

Patenaude, S. I., Seto, N. O., Borisova, S. N., Szpacenko, A., Marcus, S. L., Palcic, M. M., and Evans, S. V. (2002). The structural basis for specificity in human ABO(H) blood group biosynthesis. *Nat. Struct. Biol.* **9,** 685–690.

Petros, A. M., Dinges, J., Augeri, D. J., Baumeister, S. A., Betebenner, D. A., Bures, M. G., Elmore, S. W., Hajduk, P. J., Joseph, M. K., Landis, S. K., Nettesheim, D. G., Rosenberg, S. H., Shen, W., Thomas, S., Wang, X., Zanze, I., Zhang, H., and Fesik, S. W. (2006). Discovery of a potent inhibitor of the antiapoptotic protein Bcl-xL from NMR and parallel synthesis. *J. Med. Chem.* **49,** 656–663.

Peters, T., and Pinto, B. M. (1996). Structure and dynamics of oligosaccharides: NMR and modeling studies. *Curr. Opin. Struct. Biol.* **6,** 710–720.

Pervushin, K., Riek, R., Wider, G., and Wuthrich, K. (1997). Attenuated T2 relaxation by mutual cancellation of dipole–dipole coupling and chemical shift anisotropy indicates an avenue to NMR structures of very large biological macromolecules in solution. *Proc. Natl. Acad. Sci. USA* **94,** 12366–12371.

Post, C. B. (2003). Exchange-transferred NOE spectroscopy and bound ligand structure determination. *Curr. Opin. Struct. Biol.* **13,** 581–588.

Raman, R., Raguram, S., Venkataraman, G., Paulson, J. C., and Sasisekharan, R. (2005). Glycomics: An integrated systems approach to structure–function relationships of glycans. *Nat. Methods* **2,** 817–824.

Ramakrishnan, B., Balaji, P. V., and Qasba, P. K. (2002). Crystal structure of beta1, 4-galactosyltransferase complex with UDP-Gal reveals an oligosaccharide acceptor binding site. *J. Mol. Biol.* **318,** 491–502.

Sanchez, G., Aragones, L., Costafreda, M. I., Ribes, E., Bosch, A., and Pinto, R. M. (2004). Capsid region involved in hepatitis A virus binding to glycophorin A of the erythrocyte membrane. *J. Virol.* **78,** 9807–9813.

Smith, A. E., and Helenius, A. (2004). How viruses enter animal cells. *Science* **304,** 237–242.

Unligil, U. M., and Rini, J. M. (2000). Glycosyltransferase structure and mechanism. *Curr. Opin. Struct. Biol.* **10,** 510–517.

Wuthrich, K., Wagner, G., Richarz, R., and Perkins, S. J. (1978). Individual assignments of the methyl resonances in the 1H nuclear magnetic resonance spectrum of the basic pancreatic trypsin inhibitor. *Biochemistry* **17,** 2253–2263.

Xu, Q., and Bush, C. A. (1998). Measurement of long-range carbon–carbon coupling constants in a uniformly enriched complex polysaccharide. *Carbohydr. Res.* **306,** 335–339.

Xu, Q., and Bush, C. A. (1996). Dynamics of uniformly $^{13}$C-enriched cell wall polysaccharide of Streptococcus mitis J22 studied by $^{13}$C relaxation rates. *Biochemistry* **35,** 14512–14520.

# [3]   X-ray Crystal Structure Determination of Mammalian Glycosyltransferases

*By* JOHN E. PAK and JAMES M. RINI

## Abstract

The vast majority of mammalian glycosyltransferases are endoplasmic reticulum (ER) and Golgi resident type II membrane proteins. As such, producing large quantities of properly folded and active enzymes for X-ray crystallographic analysis is a challenge. Described here are the methods that we have developed to facilitate the structural characterization of these enzymes. The approach involves the production of a soluble Protein A–tagged form of the catalytic domain in a mammalian cell expression system. Production is scaled up in a perfusion-fed bioreactor with media flow rates of 3–5 liters/day. Expression levels are typically in the 1- to 4-mg/liter range and a simple and efficient purification method based on immunoglobulin G (IgG)–Sepharose affinity chromatography has been developed. Our approach to delimiting the catalytic domain and deglycosylating it when necessary is also discussed. Finally, we describe the selenomethionine labeling protocol used in our X-ray crystal structure determination of leukocyte-type Core 2 $\beta$1,6-$N$-acetylglucosaminyltransferase.

## Introduction

X-ray crystallographic analysis has provided fundamental insights into glycosyltransferase structure, substrate binding, and catalytic mechanism (Coutinho *et al.*, 2003; Lairson and Withers, 2004). Nevertheless, representative structures for only 23 of the 84 glycosyltransferase families

METHODS IN ENZYMOLOGY, VOL. 416
0076-6879/06 $35.00
DOI: 10.1016/S0076-6879(06)16003-6

(*http://afmb.cnrs-mrs.fr/CAZY/*) (Coutinho *et al.*, 2003) identified have been determined and there is much interest in extending the analysis to include enzymes with different linkage and substrate specificities and perhaps even new folds. The main impediment to progress stems from the fact that producing large quantities of glycosyltransferases for crystallization trials has proven to be difficult in many cases. The study of eukaryotic glycosyltransferases is particularly challenging, as these enzymes are typically ER or Golgi resident glycosylated membrane proteins. Although the expression and crystallization of these single-pass membrane proteins can be greatly simplified by working with a soluble form of the catalytic domain, these domains are themselves disulfide bond containing glycoproteins. As such, they would be best expressed in a system possessing the appropriate glycosylation machinery, chaperones, disulfide exchange enzymes, and quality-control mechanisms required to ensure that the secreted target proteins are properly folded and enzymatically active (Helenius and Aebi, 2004). Indeed, yeast (Fritz *et al.*, 2004; Lobsanov *et al.*, 2004), baculovirus/insect (Unligil *et al.*, 2000), and mammalian cell (Gastinel *et al.*, 1999; Gordon *et al.*, 2006; Pak *et al.*, 2006) expression systems have been used in the structure determination of a number of eukaryotic glycosyltransferases, and we anticipate that these approaches will play an increasingly important role in future efforts. Our interests are focused primarily on the study of mammalian glycosyltransferases and described here are the methods that we have developed for their large-scale production in a bioreactor-based mammalian cell expression system. These methods build in part on experience gained in the X-ray crystal structure determinations of rabbit *N*-acetylglucosaminyltransferase I (GnT I) (Gordon *et al.*, 2006; Unligil *et al.*, 2000) and mouse leukocyte-type Core 2 $\beta$1,6-*N*-acetylglucosaminyltransferase (C2GnT-L) (Pak *et al.*, 2006).

## Delimiting the Catalytic Domain

Mammalian glycosyltransferases are typically type II membrane glycoproteins possessing a short N-terminal cytoplasmic region, a single-pass transmembrane region, a neck region, and a C-terminal catalytic domain oriented into the ER or Golgi lumen (Paulson and Colley, 1989). In general, it is desirable to express the "minimum" fragment containing the catalytic domain, because unstructured regions, other domains connected by flexible linkers, or regions like the neck that are often highly glycosylated are not usually conducive to obtaining crystals (Derewenda, 2004). Given that the 84 glycosyltransferase families are defined as distinct amino acid–based sequence families, establishing relationships between them is not straightforward. As such, defining the limits of the catalytic domain in an uncharacterized glycosyltransferase or glycosyltransferase family may require considerable effort even though it appears that glycosyltransferases are restricted to only a few fold types (Breton *et al.*, 2006).

*Sequence Analysis, Secondary Structure Prediction and*
  *Homology Modeling*

Multiple sequence alignments and secondary structure predictions are important tools used in delimiting a soluble fragment for crystallization trials. When the sequences from widely divergent species are available, the catalytic domain of a given glycosyltransferase can often be identified by that region which is significantly more conserved than that of the rest of the protein. As shown in Fig. 1 for GnT I sequences, residue 107 (rabbit numbering) clearly represents a boundary between the relatively unconserved N-terminal region and the remainder of the protein. Moreover, residue 107 also corresponds to the start of the first predicted $\beta$-strand (see Fig. 1) of an approximately 300-residue region composed of alternating $\alpha$- and $\beta$-structural elements characteristic of the GT-A fold glycosyltransferases. This approach was used in defining the catalytic domain of GnT I (Sarkar *et al.*, 1998), and as subsequently shown by the X-ray crystal structure (Unligil *et al.*, 2000), residue 107 corresponds to the start of the GT-A fold catalytic domain. In the same way, including all members of a given glycosyltransferase family, or a subset of it, can provide an estimate of the limits of the catalytic domain as shown for family 14 (Wilson, 2002). Because the X-ray crystal structure of C2GnT-L (a family 14 member) shows that it possesses ~70 structured residues N-terminal to this point, it must be cautioned that comparisons including different glycosyltransferases may not provide a good starting point for a given sequence.

A high-quality homology model should provide reliable estimates of the limits of a fragment for study and should be generated whenever possible. Once one member of a family has been determined, this structure can be used to model the others (Heissigerova *et al.*, 2003). Homology modeling crossing glycosyltransferase families, however, is more challenging because of the absence of significant sequence similarity between families. Nevertheless, models for the human family 42 $\alpha$2,3 sialyltransferases have recently been generated based on the bacterial family 29 $\alpha$2,3/8 sialyltransferase, CstII (Sujatha and Balaji, 2006).

*Truncation Mutagenesis*

Truncation mutagenesis in conjunction with Western blot analysis (see later discussion) is used to generate or test trial expression fragments. By measuring the secreted expression levels of a fusion protein containing the trial fragment, relative to that of the fusion tag alone (Protein A in the

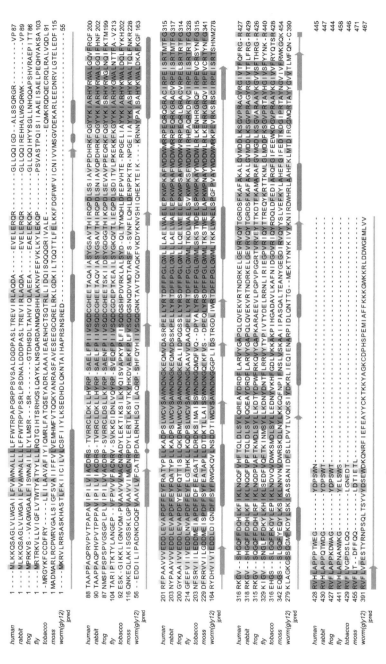

FIG. 1. Multiple sequence alignment and secondary structure prediction for *N*-acetylglucosaminyltransferase I (GnT I) sequences. Residues identical in at least four of the seven sequences are shaded. Sequences were aligned using ClustalW (Chenna *et al.*, 2003) and graphical output was produced by Jalview (Clamp *et al.*, 2004). Secondary structure prediction was performed using JNet Secondary Structure Prediction (Cuff and Barton, 1999), as implemented in Jalview (Clamp and Barton, 1999). Rods and arrows correspond to α-helices and β-strands, respectively.

work described here), an assessment of the "integrity" of the fragment can be made. Fusion proteins that do not include a complete and properly folded catalytic domain will not be well secreted. Typically both the intracellular and the extracellular levels of both the fusion protein and the fusion tag alone are measured. We have also found it valuable to use the vector containing the fusion tag alone as an internal control to normalize for differences in transfection efficiency when comparing the expression levels of trial fusion proteins. In this system, a well-secreted fusion protein is one whose levels approach that of the fusion tag alone.

*Limited Protease Digestion*

A powerful approach to delimiting domains for crystallization trials is the use of limited proteolysis (Derewenda, 2004). This method takes advantage of the fact that compact globular domains are relatively resistant to proteolytic cleavage. With membrane proteins of this type, the approach would typically be used in conjunction with truncation mutagenesis. Once a soluble fragment has been produced, proteolysis can be used to further reduce the size of the fragment. Figure 2 shows the results of a protease screen performed on an N-terminal Protein A fusion containing a mouse C2GnT-L fragment (residues 38–428) (Zeng *et al.*, 1997). As shown, clostripain, chymotrypsin, trypsin, V8, and elastase all generate a similar protease-resistant fragment. N-terminal sequencing and mass spectrometry can then be used to determine the limits of the new fragment. With glycoproteins such as these, analysis of the mass spectrometry data is greatly simplified if the molecules are fully deglycosylated using peptide:N-glycosidase F (PNGase F) before analysis (see later discussion). The limits of the protease-resistant fragment can then be used in the redesign of a new stable cell line, or as in the C2GnT-L case, the resistant fragment generated by proteolysis can be used directly.

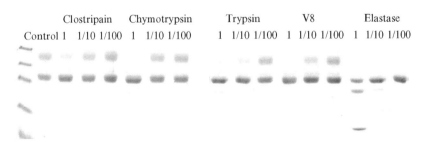

Fig. 2. Limited proteolysis screen of a Protein A–tagged fragment of C2GnT-L. Five proteases were tested at three protease concentrations (labeled 1, 1/10, 1/100). The digests were sampled after 2 h at 30°.

*Proteolysis Test Digest*

> 8 $\mu$l of protein (1.5 mg/ml)
> 2 $\mu$l of 10× buffer (as recommended by vendor)
> 4 $\mu$l of protease[1]
> 6 $\mu$l $H_2O$

1. Incubate at 30°
2. Sample 5 $\mu$l at 0, 2, and 6 h, as well as overnight and analyze by sodium dodecylsulfate (SDS)–polyacrylamide gel electrophoresis (PAGE)

### Vector Design and Construction

Building on our success with the Protein A–tagged C2GnT-L fusion protein, we have developed a mammalian cell expression vector designed to facilitate the production and purification of similarly tagged glycosyltransferase fragments. The expression vector pPA-TEV (Fig. 3) was derived from pIRESpuro3 (Clontech) but modified to incorporate the transin leader and N-terminal Protein A affinity tag (Sanchez-Lopez *et al.*, 1988) and a TEV protease recognition site. The puromycin acetyltransferase selectable marker is under the control of an internal ribosomal entry site (IRES) (Kaufman, 2000). This arrangement places the fusion protein and the drug resistance marker on the same messenger RNA (mRNA), increasing the likelihood that cells resistant to puromycin will also produce high levels of the fusion protein.

Although the vector contains a number of restriction sites in the multiple cloning site, we typically subclone the fragment of interest into the Fse

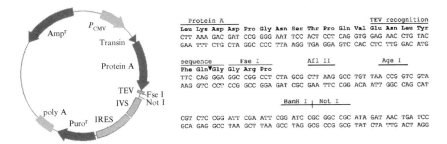

FIG. 3. pPA-TEV expression vector map and region surrounding the multiple cloning site. Protein A labels the last 4 residues of the 254 residue (53–306) Protein A tag.

---

[1] Three protease dilutions are used (1, 1/10, 1/100 in buffer). Stock protease concentrations are as follows: trypsin, 0.1 mg/ml; V8, 0.5 mg/ml; chymotrypsin, 0.5 mg/ml; elastase, 0.5 mg/ml; and clostripain, 0.1 mg/ml.

I/Not I pair. Both of these enzymes have eight base pair restriction sites and, as such, are unlikely to be found in most trial fragments. Moreover, the Fse I site has been positioned so that only four amino acids (Gly-Gly-Arg-Pro) are added to the N-terminus of the fragment after TEV protease digestion.

### Subcloning into pPA-TEV

1. The desired fragment is amplified by polymerase chain reaction (PCR) using the appropriate primers incorporating an in-frame Fse I site on the 5′ primer and on the 3′ primer, a stop codon to end translation 3′ to the Not I site.

   Example:

   5′ primer: 5′*NNNNNN*GGCCGGCCCNNN... 3′ (where *NNNNNN* are arbitrary nucleotides to ensure efficient Fse I cleavage and NNN corresponds to the codon of the first amino acid of the N-terminus of the required fragment.

   3′ primer: 5′*NNNNNN*GCGGCCGCCTANNN... 3′ (where *NNNN-NN* are arbitrary nucleotides to ensure efficient Not I cleavage, CTA is the inverse complement of a stop codon, and NNN is the inverse complement of the C-terminal amino acid of the required fragment.

2. Gel purify the PCR fragment. Digest ~0.5 $\mu$g of DNA with 3 U Fse I in a reaction volume of 25 $\mu$l. Incubate at 37° for 3 h.

3. Add 3 $\mu$l of 1 $M$ NaCl and 3 U of Not I. Incubate at 37° for 3 h. Gel purify.

4. For the vector, digest 1 $\mu$g of DNA with 3 U Fse I in a reaction volume of 15 $\mu$l. Incubate at 37° for 2 h.

5. Add 2 $\mu$l of 1 $M$ NaCl and 3 U of Not I. Incubate at 37° for 2 h.

6. Add 0.5 U calf intestinal alkaline phosphatase (CIP) to the digested vector. Incubate at 37° for 30 min. Gel-purify the fragment.

7. Perform ligation using a molar ratio of vector:insert of 1:3 for 5 min using Quick T4 DNA Ligase (NEB).

## Transient Expression and the Production of Stable Cell Lines

### Transient Expression Protocol

HEK293T cells are grown in Iscove's modified Dulbecco medium (IMDM) supplemented with 10% fetal bovine serum (FBS), 1X penicillin–streptomycin, and 1 mg/liter aprotinin in 6-cm tissue culture dishes to ~40% cell confluency. Three hours before transfection, the media is replaced by a fresh 5-ml aliquot of the same media. Ten micrograms of uncut circular DNA plasmid is then transfected into the cells using the calcium phosphate method. Ten microliters of media is sampled each day and stored at $-20°$ for Western blot analysis (Fig. 4). Each 10-$\mu$l sample of medium is run on a 12.5% SDS gel and transferred to nitrocellulose or PVDF membranes using standard protocols. The membranes are immunoblotted for 2 h at room temperature using a 1/5000 dilution of a monoclonal anti-Protein A antibody (Sigma, P2921). Following washing and overnight blocking with powdered skim milk, the membrane is incubated for 35 min at room temperature with a 1/10,000 dilution of HRP-conjugated goat antimouse IgG (Jackson ImmunoResearch, 115–035–003) and detected using Western Lighting Chemiluminescence Reagent Plus (Perkin-Elmer, NEL105).

*Stable Cell Line Production*

The transfection protocol described earlier is used in the generation of stable cell lines except that the plasmid DNA is linearized and gel-purified before transfection to facilitate optimal integration. Three days after transfection, puromycin drug selection is started by replacing the media with fresh media supplemented with 1–5 $\mu$g/ml puromycin. This results in the rapid (3–4 days) killing of un-transfected cells. The media is exchanged with fresh puromycin containing media every 2 days. Approximately 10 days into drug selection, small foci of cells appear; these cells are isolated and expanded into six-well tissue culture plates. Once cultures are established, secreted

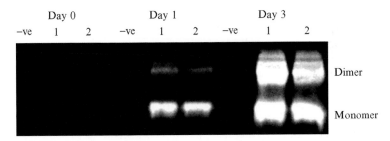

Fig. 4. Western blot analysis of media from transiently transfected HEK-293T cells. Cells were transfected with circular pPA-TEV expressing a soluble Protein A–tagged fragment of C2GnT-L at day 0 in duplicate (labeled "1" and "2"). Mock transfection (labeled "–ve") was performed in the absence of DNA. Ten microliters of sampled media was run on a non-reducing SDS-PAGE gel. Protein was detected using a mouse anti-Protein A antibody followed by an HRP-conjugated anti-mouse immunoglobulin G (IgG).

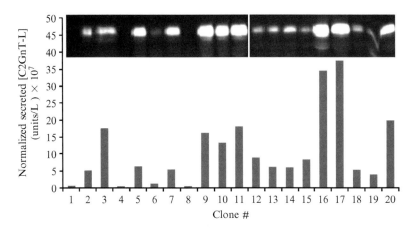

FIG. 5. Protein expression and glycosyltransferase activity from 20 stable clonal cell lines expressing a Protein A–tagged fragment of C2GnT-L using pPA-TEV. For each clone, the amount of secreted fusion protein (assessed via Western blot analysis) and the activity measured are normalized to the total number of cells. One unit is defined as the amount of enzyme that transfers 1 $\mu$mol of GlcNAc from UDP-GlcNAc onto the Gal$\beta$1,3GalNAc-pNP acceptor substrate in 1 min at 37°.

protein is detected by Western blot analysis, as described earlier for the transient transfection protocol. Figure 5 shows that protein production, measured by Western blot analysis, is well correlated with the level of enzyme activity for 20 C2GnT-L clones isolated in this way. Although the amount of protein produced from bulk cultures may be sufficient for some applications, Fig. 5 shows that there is significant variation from clone to clone and that the best expressers produce three to four times that of the average observed for the bulk culture.

Large-Scale Production in Bioreactors

Protein production is scaled up from the stably transformed mammalian cells in a Celligen Plus 2.2 liter bioreactor (1.2 liter working volume; New Brunswick) using the basket impeller (Fig. 6). In this configuration, the adherent cells attach to and grow on FibraCel (New Brunswick, M1176–9984), polyester disks ($\sim$6 mm diameter) contained within the basket. Because cells remain attached to the disks and, therefore, within the reactor throughout the entire run, media can be continuously pumped into and out of the reactor (i.e., perfusion fed culture). The bioreactor controls dissolved oxygen ($dO_2$), pH, and temperature, as well as the media perfusion rate. $dO_2$/pH control is achieved by a four-gas controller (air, $O_2$, $N_2$, and $CO_2$) in conjunction with $dO_2$ and pH probes and a base pump (5% sodium bicarbonate solution).

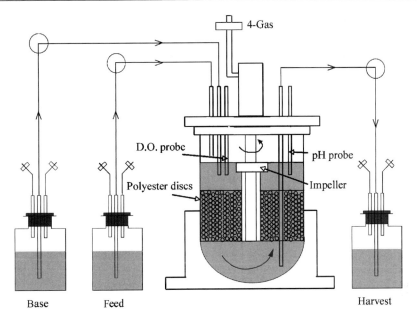

FIG. 6. Schematic representation of a bioreactor configured for perfusion-fed cell culture (not to scale). The FibraCel disks contained in the basket impeller serve as insoluble supports for the attachment and growth of adherent cell lines. Four-gas input (air, $O_2$, $N_2$, and $CO_2$), $dO_2$ and pH probes, and the base pump (5% sodium bicarbonate solution) constitute the basis for $dO_2$/pH control.

The combination of perfusion feeding and $dO_2$/pH control permits the maintenance of high-density cultures over extended periods (3–4 wks).

*Preparation of Cells for Seeding*

To generate the cells required to seed a reactor containing 30 g of FibraCel disks, cells are expanded into ten T175 tissue culture flasks in growth media: CHO-S-SFM-II media (Gibco, 12052–098) supplemented with 3% FBS, 2.25 g/liter glucose, 1X nonessential amino acids (Gibco, 11140–050), 1X penicillin–streptomycin (Gibco, 15140–148), 1 mg/liter aprotinin (BioShop, 9087–70–1), and 1–5 mg/liter puromycin. Cells are trypsinized, counted, and washed with fresh growth media and then resuspended to give $3 \times 10^8$ cells in 1.3 liters of room temperature growth media.

*Bioreactor Inoculation and Production*

The reactor is assembled and sterilized as outlined in the operating manual. After calibrating the $dO_2$ probe, the phosphate-buffered saline (PBS) solution used in the sterilization/calibration steps is pumped out of the reactor. The turbid 1.3 liter cell suspension is then pumped into the reactor and the impeller started at 50 rpm. The media clarifies after ~2 h as the cells attach to the FibraCel disks. After an additional 2 h, the impeller speed is increased to 70 rpm where it is fixed for the remainder of the run. The $dO_2$ and pH set points are fixed at 50% and 7.2, respectively, for the entire run. Fresh medium is not pumped into the reactor for the first 3–4 days while the cells grow on the FibraCel disks. Each day, a small amount of medium is removed from the reactor using the harvest pump and the glucose level is measured using a blood glucose strip/meter available at a local pharmacy. When the glucose level begins to drop, media perfusion is begun (e.g., 1 liter/day on day 4, 2 liters/day on day 5, and 3 liters/day on day 6). The perfusion rate for the remainder of the run is determined by the rate required to achieve a glucose level of 1 g/liter in the reactor (~3–5 liters/day with 30-g FibraCel disks). Figure 7 shows a representative bioreactor run for the production of the Protein A C2GnT-L fusion protein.

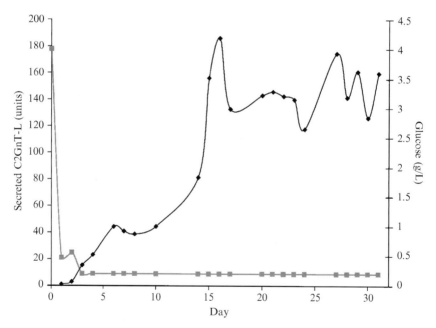

FIG. 7. Daily cell culture profile from a C2GnT-L bioreactor run. Total secreted fusion protein (diamonds) and glucose concentration in the harvested media (squares) are indicated for each day. Glycosyltransferase activity is defined as in Fig. 5.

Isolation and Purification from the Culture Media

Typically, the harvested medium is processed in 6- to 10-liter batches. The media is first centrifuged at low speed (1000×g) for 10 min in 1-liter buckets to pellet small amounts of debris. The medium is then concentrated 10-fold on a Prep/Scale spiral wound cross-flow ultrafiltration unit (10K regenerated cellulose, Millipore, SK1P003W4). This step has been found to significantly increase the yield in the subsequent affinity chromatography step (to 90–95% recovery) while reducing volumes to more manageable levels.

*Affinity Chromatography*

1. 10 ml of IgG-Sepharose beads (GE Healthcare, 17–0969–01) are added to each liter of the 10× concentrated media. Using roller bottles and a roller apparatus, the mixture is rolled at slow speed (~2 rpm) for 2 h at 4° to ensure gentle mixing of the IgG-Sepharose beads.
2. Pour the media/bead mixture into a 120-ml Econo-Column (Bio-Rad Laboratories) and wash with 30 bed volumes of wash buffer (50 m$M$ Tris pH = 7.6, 150 m$M$ NaCl).
3. For on-column TEV cleavage, add 2 bed volumes of TEV solution (0.05 mg/ml in wash buffer) and resuspend beads with gentle pipetting; incubate for 14 h at 4°. Collect protein by washing with 10 column volumes of wash buffer. For acid elution, wash with 5 bed volumes of 5 m$M$ glycine, pH 5.0, and elute with 7 × 1 bed volume fractions of 250 m$M$ glycine, pH 3.4. Neutralize each fraction immediately with 2 bed volumes of 250 m$M$ MES, pH 7.5. The fusion tag is then removed by TEV cleavage.

The sample is usually greater than 90% pure after the affinity chromatography step, and standard chromatographic techniques are then used to further purify it for crystallization trials. Typically, ion exchange and hydrophobic interaction chromatography are employed as required. The sample is then concentrated to 5–10 mg/ml and subject to a final polishing step on a Superdex 200 26/60 gel filtration column (GE Healthcare). Often donor substrate (e.g., UDP-GlcNAc) is added at this stage (in the millimolar concentration range) to help prevent aggregation that might otherwise occur during the concentration step (Unligil *et al.*, 2000). Our purified yields range from 1 to 4 mg/liter, and with perfusion rates of 3–5 liters/day, we can easily produce the tens of milligrams often required in the determination of an X-ray crystal structure.

## Deglycosylation and Crystallization

Mammalian expression systems produce N-linked oligosaccharide structures of the complex type. These branched structures possess a variable number of "arms" or "antennae" elaborated typically with N-acetylglucosamine, galactose, and sialic acid. Because of their size, heterogeneity, and inherent flexibility, oligosaccharides of this type can be an impediment to crystallization (Kwong *et al.*, 1999). Nevertheless, we initially attempt to crystallize the intact sample because the removal of carbohydrate often leads to a considerable loss in the solubility of the sample. If crystals are not obtained, or if attempts to improve crystals of poor quality are not successful, we then consider one of several deglycosylation strategies. The first attempts to strike a balance between carbohydrate removal and solubility. In this case, the arms of the N-linked glycans are pared down using a sialidase, a $\beta$-galactosidase and an N-acetylglucosaminidase. In principle, these structures can be reduced to the $(Man)_3(GlcNAc)_2$-Asn core common to all N-linked oligosaccharide structures. Although the removal of sialic acid and galactose is quantitative, our experience with glycoproteins secreted from CHO and HEK293 cells is that it is difficult to completely remove all of the branching GlcNAc moieties.

To effect a complete removal of N-linked oligosaccharides, PNGase F is employed. This enzyme cleaves the GlcNAc-Asn linkage while converting the asparagine residue to an aspartic acid. The advantage of PNGase F is that it removes all N-linked oligosaccharide types. The drawback with this enzyme is that the GlcNAc-Asn linkage is not always accessible in the folded glycoprotein, so it can be difficult to achieve quantitative deglycosylation under native conditions. Typically a digest of the sample in both the native and the denatured state is compared to assess whether this poses a problem in any given case. Alternately, endo-$\beta$-N-acetylglucosaminidases F2 and F3 (Endo F2 and Endo F3) can be considered. These enzymes cleave the glycosidic linkage between the two GlcNAc moieties of the chitobiose core, leaving a single GlcNAc linked to the protein (GlcNAc-Asn). The advantage of these enzymes is that the linkage cleaved is more accessible and, therefore, more easily cleaved under native conditions. The drawback with them is that unlike PNGase F, they show specificity for only certain branched structures and even a mixture of the two enzymes does not always lead to complete deglycosylation with material produced in CHO and HEK293 cells. In all cases, enzyme concentration, reaction time, and temperature should be varied to optimize removal. For the most part, these digests can be monitored by SDS-PAGE where the width and migration position of the sample bands provides an indication of the extent of reaction. As mentioned earlier, a PNGase digest under denaturing conditions is performed and this serves to establish the band width and position in the completely deglycosylated state.

*Deglycosylation Protocols*

*Partial Deglycosylation Test Digest*

18 $\mu$l glycoprotein (0.5 mg/ml)
5 $\mu$l 5× reaction buffer[2]
0.5 $\mu$l glycosidase: sialidase A (5 U/ml) or $\beta$-galactosidase (3 U/ml) or
$\beta$-N-acetylglucosaminidase (40 U/ml)
dH$_2$O to a final total volume of 25 $\mu$l

*PNGase F Deglycosylation Test Digest*

18 $\mu$l glycoprotein (0.5 mg/ml)[3]
2 $\mu$l 0.5 $M$ sodium phosphate, pH 7.5
2 $\mu$l PNGase-F (2.5 U/ml)

*Endo F2/F3 Deglycosylation Test Digest*

19 $\mu$l glycoprotein (0.5 mg/ml)
5 $\mu$l 5× reaction buffer (250 m$M$ sodium acetate, pH 4.5)
1 $\mu$l Endo F2/F3 (5 U/ml)

FOR ALL DEGLYCOSYLATION REACTIONS

1. Incubate at 37°
2. Sample 4 $\mu$l at 0 h, 2 h, and overnight and analyze by SDS-PAGE.


Selenomethionine Labeling

The MAD phasing method, in conjunction with the ability to biosynthetically label proteins with selenomethionine, has transformed our approach to solving the phase problem in protein crystallography (Ealick, 2000). As such, protocols have been developed to permit selenomethionine labeling in bacterial (Doublie, 1997; Hendrickson *et al.*, 1990), yeast (Bushnell *et al.*, 2001; Larsson *et al.*, 2002), insect (Bellizzi *et al.*, 1999), and mammalian (Wu *et al.*, 1994) cell expression systems. However, because selenomethionine is toxic, adding it to the culture media typically leads to a significant decrease

---

[2] All enzymes are from ProZyme (San Leandro, CA); buffer as described by vendor; for multiple enzyme digests, use buffer closest to neutral pH.
[3] For denatured control, add 2 $\mu$l 10× denaturing buffer (5% SDS, 10% $\beta$-mercaptoethanol) and incubate at 100° for 10 min before adding buffer and glycosidase.

in protein production levels and rapid cell death. As production levels are already relatively modest in a mammalian cell expression system, further decreases in yield make it difficult to produce sufficient quantities for structure determination. To take advantage of the inherently high capacity for production in our bioreactor-based system, we set out to develop a protocol that would allow us to selenomethionine label protein produced in this way. We reasoned that by establishing a high-density cell culture in the reactor before introducing selenomethionine, large amounts of protein could be produced before the cells are killed by its toxic effects. To test this idea, a bioreactor run was initiated using the supplemented CHO-S-SFM II–based growth media exactly as described earlier. At day 30, the bioreactor was drained and one reactor volume of a medium lacking methionine but containing selenomethionine was added. The selenomethionine-containing medium was based on a serum-free CHO cell formulation described previously (Sinacore *et al.*, 2000). Only the inorganic and amino acid components— with the exception of methionine and sodium selenite—of that formulation were used (Table I). These components were supplemented with serum (3% FBS), 1 mg/liter puromycin, 4.0 g/liter glucose, 30 mg/liter seleno-L-methionine (Sigma, S3132), 1X Insulin-Transferrin-Selenium-G Supplement (Gibco, 41400–045), 1X Basal Media Eagle Vitamins (Gibco, 21040–050), 1X penicillin–streptomycin–glutamine (Gibco, 10378–016), 1X Chemically defined lipid concentrate (Gibco, 11905–31), and 1 mg/liter aprotinin (BioShop, 9087–70–1) to give the selenomethionine-containing medium. After 24 h, the selenomethionine-containing medium in the bioreactor was discarded and a second reactor volume of fresh selenomethionine-containing medium added. Perfusion with the labeled medium was then resumed at a rate of ~1.5 liters/day for 8 more days. Although we have yet to optimize this procedure, more than 10 mg of selenomethionine-labeled protein was made and used in solving the X-ray crystal structure of C2GnT-L by the MAD phasing method (Pak *et al.*, 2006). Given that mammalian cells cannot biosynthesize methionine, selenomethionine incorporation will be quantitative once the available methionine pools are consumed. Figure 8 shows an anomalous difference Fourier electron density map that clearly shows 9 of the 11 selenium atoms.

## Perspective/Future Directions

The desire to study eukaryotic secreted and membrane proteins has driven the development of expression systems possessing the biosynthetic machinery required to reliably produce them. The resultant proteins are naturally glycosylated, however, which can present a problem from the crystallization standpoint. The protein is often deglycosylated as part of

TABLE I
INORGANIC AND AMINO ACID COMPONENTS OF THE LABELING MEDIA[a]

| Salts and amino acids | Molecular weight (g/mol) | Concentration ($\mu$g/ml) |
|---|---|---|
| Sodium chloride | 58.44 | 4400 |
| Potassium chloride | 74.55 | 310 |
| Calcium chloride, anhydrous | 111 | 58 |
| Sodium phosphate, monobasic, hydrate | 138 | 130 |
| Magnesium sulfate, anhydrous | 120.4 | 84 |
| Cupric sulfate, pentahydrate | 249.7 | 0.00279 |
| Ferrous sulfate, heptahydrate | 278 | 1.665 |
| Zinc sulfate, heptahydrate | 287.5 | 0.92 |
| Sodium bicarbonate | 84.01 | 2400 |
| L-alanine | 89.09 | 71 |
| L-arginine | 174.2 | 760 |
| L-asparagine monohydrate | 150.1 | 540 |
| L-aspartic acid | 133.1 | 270 |
| L-cysteine hydrochloride monohydrate | 175.6 | 700 |
| L-glutamic acid | 147.1 | 120 |
| L-glutamine | 146.1 | 1200 |
| Glycine | 75.07 | 60 |
| L-histidine hydrochloride monohydrate | 209.6 | 290 |
| L-isoleucine | 131.2 | 470 |
| L-leucine | 131.2 | 680 |
| L-lysine hydrochloride | 182.7 | 730 |
| L-phenylalanine | 165.2 | 330 |
| L-proline | 115.1 | 280 |
| L-serine | 105.1 | 630 |
| L-threonine | 119.1 | 380 |
| L-tryptophan | 204.2 | 130 |
| L-tyrosine disodium dihydrate | 261.24 | 420 |
| L-valine | 117.1 | 370 |

[a] Adapted from Sinacore et al. (2000).

the purification/crystallization process, and regardless of the expression system, it is often difficult to remove carbohydrate quantitatively. To address this issue, we have begun to explore methods for separating fully digested material from that which has not been completely processed. Both hydrophobic interaction chromatography and lectin affinity chromatography are being used; the former exploits solubility differences, and the latter can be used to specifically remove uncleaved material. Lectin-resistant cell lines producing only high mannose type N-linked oligosaccharides (Crispin et al., 2006; Davis et al., 1993; Reeves et al., 2002; Vischer and Hughes, 1981) have also been suggested as a means of producing more easily

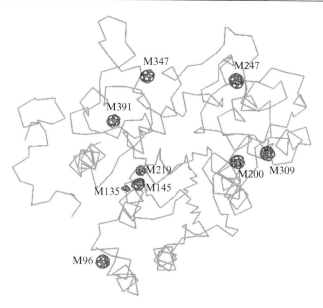

FIG. 8. Anomalous difference Fourier map of selenomethionine-labeled C2GnT-L contoured at $7\sigma$ showing 9 of 11 selenium positions observed at this contour level.

deglycosylated protein from a mammalian cell expression system. This approach would permit the use of the endoglycosidase H for deglycosylation, and we plan to evaluate these cell lines for scale up in our perfusion-fed cultures. Lastly, we intend to optimize our selenomethionine labeling protocol. The strength of our bioreactor-based approach for this purpose stems from the fact that large amounts of protein can be produced from high-density cultures before the toxic effects of selenomethionine kill the cells. To facilitate this process, we will first adapt our cell lines to the labeling media (supplemented with methionine). Once the cells have been adapted, we will develop a protocol for growing them in the reactor using labeling media at all steps. By reducing the shock associated with a change in media type when selenomethionine is finally introduced, we anticipate increasing both the production levels and the duration of the labeling phase.

## Acknowledgments

This work was supported by the Canadian Institutes of Health Research and the Protein Engineering Network of Centres of Excellence.

## References

Bellizzi, J. J., Widom, J., Kemp, C. W., and Clardy, J. (1999). Producing selenomethionine-labeled proteins with a baculovirus expression vector system. *Structure* **7**, R263–R267.

Breton, C., Snajdrova, L., Jeanneau, C., Koca, J., and Imberty, A. (2006). Structures and mechanisms of glycosyltransferases. *Glycobiology* **16**, 29R–37R.

Bushnell, D. A., Cramer, P., and Kornberg, R. D. (2001). Selenomethionine incorporation in *Saccharomyces cerevisiae* RNA polymerase II. *Structure* **9**, R11–R14.

Chenna, R., Sugawara, H., Koike, T., Lopez, R., Gibson, T. J., Higgins, D. G., and Thompson, J. D. (2003). Multiple sequence alignment with the Clustal series of programs. *Nucleic Acids Res.* **31**, 3497–3500.

Clamp, M., Cuff, J., Searle, S. M., and Barton, G. J. The Jalview Java alignment editor. *Bioinformatics* .

Coutinho, P. M., Deleury, E., Davies, G. J., and Henrissat, B. (2003). An evolving hierarchical family classification for glycosyltransferases. *J. Mol. Biol.* **328**, 307–317.

Crispin, M., Harvey, D. J., Chang, V. T., Yu, C., Aricescu, A. R., Jones, E. Y., Davis, S. J., Dwek, R. A., and Rudd, P. M. (2006). Inhibition of hybrid and complex-type glycosylation reveals the presence of the GlcNAc transferase I-independent fucosylation pathway. *Glycobiology* **8**, 748–756.

Cuff, J. A., and Barton, G. J. (1999). Evaluation and improvement of multiple sequence methods for protein secondary structure prediction. *Proteins* **34**, 508–519.

Davis, S. J., Puklavec, M. J., Ashford, D. A., Harlos, K., Jones, E. Y., Stuart, D. I., and Williams, A. F. (1993). Expression of soluble recombinant glycoproteins with predefined glycosylation: Application to the crystallization of the T-cell glycoprotein CD2. *Protein Eng.* **6**, 229–232.

Derewenda, Z. S. (2004). The use of recombinant methods and molecular engineering in protein crystallization. *Methods* **34**, 354–363.

Doublie, S. (1997). Preparation of selenomethionyl proteins for phase determination. *Methods Enzymol.* **276**, 523–530.

Ealick, S. E. (2000). Advances in multiple wavelength anomalous diffraction crystallography. *Curr. Opin. Chem. Biol.* **4**, 495–499.

Fritz, T. A., Hurley, J. H., Trinh, L. B., Shiloach, J., and Tabak, L. A. (2004). The beginnings of mucin biosynthesis: The crystal structure of UDP-GalNAc:polypeptide alpha-N-acetylgalactosaminyltransferase-T1. *Proc. Natl. Acad. Sci. USA* **101**, 15307–15312.

Gastinel, L. N., Cambillau, C., and Bourne, Y. (1999). Crystal structures of the bovine beta4galactosyltransferase catalytic domain and its complex with uridine diphosphogalactose. *EMBO J.* **18**, 3546–3557.

Gordon, R. D., Sivarajah, P., Satkunarajah, M., Ma, D., Tarling, C. A., Vizitiu, D., Withers, S. G., and Rini, J. M. (2006). X-ray crystal structures of rabbit N-acetylglucosaminyltransferase I (GnT I) in Complex with Donor Substrate Analogues. *J. Mol. Biol.* **360**, 67–79.

Heissigerova, H., Breton, C., Moravcova, J., and Imberty, A. (2003). Molecular modeling of glycosyltransferases involved in the biosynthesis of blood group A, blood group B, Forssman, and iGb3 antigens and their interaction with substrates. *Glycobiology* **13**, 377–386.

Helenius, A., and Aebi, M. (2004). Roles of N-linked glycans in the endoplasmic reticulum. *Annu. Rev. Biochem.* **73**, 1019–1049.

Hendrickson, W. A., Horton, J. R., and LeMaster, D. M. (1990). Selenomethionyl proteins produced for analysis by multiwavelength anomalous diffraction (MAD): A vehicle for direct determination of three-dimensional structure. *EMBO J.* **9**, 1665–1672.

Kaufman, R. J. (2000). Overview of vector design for mammalian gene expression. *Mol. Biotechnol.* **16**, 151–160.

Kwong, P. D., Wyatt, R., Desjardins, E., Robinson, J., Culp, J. S., Hellmig, B. D., Sweet, R. W., Sodroski, J., and Hendrickson, W. A. (1999). Probability analysis of variational crystallization and its application to gp120, the exterior envelope glycoprotein of type 1 human immunodeficiency virus (HIV-1). *J. Biol. Chem.* **274**, 4115–4123.

Lairson, L. L., and Withers, S. G. (2004). Mechanistic analogies amongst carbohydrate modifying enzymes. *Chem. Commun. (Camb.)* 2243–2248.

Larsson, A. M., Stahlberg, J., and Jones, T. A. (2002). Preparation and crystallization of selenomethionyl dextranase from Penicillium minioluteum expressed in *Pichia pastoris*. *Acta Crystallogr. D Biol. Crystallogr.* **58**, 346–348.

Lobsanov, Y. D., Romero, P. A., Sleno, B., Yu, B., Yip, P., Herscovics, A., and Howell, P. L. (2004). Structure of Kre2p/Mnt1p: A yeast alpha1,2-mannosyltransferase involved in mannoprotein biosynthesis. *J. Biol. Chem.* **279**, 17921–17931.

Pak, J. E., Arnoux, P., Zhou, S., Sivarajah, P., Satkunarajah, M., Xing, X., and Rini, J. M. (2006). X-ray crystal structure of leukocyte type Core 2 $\beta$1,6-N-acetylglucosaminyltransferase: Evidence for a convergence of metal ion independent glycosyltransferase mechanism. *J. Biol. Chem.*, in press.

Paulson, J. C., and Colley, K. J. (1989). Glycosyltransferases. Structure, localization, and control of cell type-specific glycosylation. *J. Biol. Chem.* **264**, 17615–17618.

Reeves, P. J., Callewaert, N., Contreras, R., and Khorana, H. G. (2002). Structure and function in rhodopsin: High-level expression of rhodopsin with restricted and homogeneous N-glycosylation by a tetracycline-inducible N-acetylglucosaminyltransferase I-negative HEK293S stable mammalian cell line. *Proc. Natl. Acad. Sci. USA* **99**, 13419–13424.

Sanchez-Lopez, R., Nicholson, R., Gesnel, M. C., Matrisian, L. M., and Breathnach, R. (1988). Structure-function relationships in the collagenase family member transin. *J. Biol. Chem.* **263**, 11892–11899.

Sarkar, M., Pagny, S., Unligil, U., Joziasse, D., Mucha, J., Glossl, J., and Schachter, H. (1998). Removal of 106 amino acids from the N-terminus of UDP-GlcNAc: Alpha-3-D-mannoside beta-1,2-N-acetylglucosaminyltransferase I does not inactivate the enzyme. *Glycoconj. J.* **15**, 193–197.

Sinacore, M. S., Drapeau, D., and Adamson, S. R. (2000). Adaptation of mammalian cells to growth in serum-free media. *Mol. Biotechnol.* **15**, 249–257.

Sujatha, M. S., and Balaji, P. V. (2006). Fold recognition and comparative modeling of human alpha2,3-sialyltransferases reveal their sequence and structural similarities to CstII from Campylobacter jejuni. *BMC Struct. Biol.* **6**, 9.

Unligil, U. M., Zhou, S., Yuwaraj, S., Sarkar, M., Schachter, H., and Rini, J. M. (2000). X-ray crystal structure of rabbit N-acetylglucosaminyltransferase I: Catalytic mechanism and a new protein superfamily. *EMBO J.* **19**, 5269–5280.

Vischer, P., and Hughes, R. C. (1981). Glycosyl transferases of baby-hamster-kidney (BHK) cells and ricin-resistant mutants. N-glycan biosynthesis. *Eur. J. Biochem.* **117**, 275–284.

Wilson, I. B. (2002). Functional characterization of *Drosophila melanogaster* peptide O-xylosyltransferase, the key enzyme for proteoglycan chain initiation and member of the core 2/I N-acetylglucosaminyltransferase family. *J. Biol. Chem.* **277**, 21207–21212.

Wu, H., Lustbader, J. W., Liu, Y., Canfield, R. E., and Hendrickson, W. A. (1994). Structure of human chorionic gonadotropin at 2.6 A resolution from MAD analysis of the selenomethionyl protein. *Structure* **2**, 545–558.

Zeng, S., Dinter, A., Eisenkratzer, D., Biselli, M., Wandrey, C., and Berger, E. G. (1997). Pilot scale expression and purification of soluble protein A tagged beta 1,6N-acetylglucosaminyltransferase in CHO cells. *Biochem. Biophys. Res. Commun.* **237**, 653–658.

# Section II

# Phage Display Library

# [4]   Screening of Peptide-Displaying Phage Libraries to Identify Short Peptides Mimicking Carbohydrates

*By* Michiko N. Fukuda

## Abstract

Peptide-displaying phage technology has numerous applications. Using a specificity-defined monoclonal anti-carbohydrate antibody, we can identify a series of short peptides that mimic the binding specificity of a specific carbohydrate. This chapter introduces pioneering work applying phage display technology to the glycobiology field, presents a step-by-step protocol for phage library screening, and provides useful hints for evaluating results including false positives, all of which should contribute to successful cloning. Thus, biopanning using a monoclonal antibody as the target is described in detail. Because peptides are useful alternatives to carbohydrate ligands, their potential use as structural or functional mimics of carbohydrate-binding proteins is discussed.

## Overview

Peptide-displaying phage technologies have provided a method to identify short peptide sequences specific to a target. This technology may also provide a methodology appropriate for producing recombinant carbohydrates, as unlimited phage-displaying carbohydrate mimics can be amplified. Because chemical synthesis of complex carbohydrates is still not automated, a specific peptide that functions in carbohydrate mimicry provides us with a practical tool for glycobiology.

Bacteriophage consists of DNA and several coat proteins. When a phage infects a host bacterium, the phage proliferates by copying its DNA, producing coat proteins, and assembling new phage particles. Amplified phage particles leak or burst from host bacterial cells in a remarkably efficient process. Importantly, each phage plaque or colony appearing on an agar plate represents a genetically homogenous clone.

Peptide-displaying phage was invented by Smith in 1985. He modified the genome of the filamentous fd phage so that one coat protein displayed a foreign gene product as a fusion protein on the phage surface. In fd phage, the foreign gene product is displayed at the N-terminal of the protein, and the C-terminal is fused to the pIII coat protein. Each fd phage displays five copies of the peptide.

METHODS IN ENZYMOLOGY, VOL. 416                                  0076-6879/06 $35.00
           DOI: 10.1016/S0076-6879(06)16004-8

T7-based phage has also been used for phage display technology (Hoffman *et al.*, 2002; Yamamoto *et al.*, 1999). In this system, 200–400 peptide copies are displayed on the phage surface as C-terminal peptides. Because of the high copy number, T7 phage enables identification of peptides with weak affinity.

### Screening a Phage Library

Peptide display phage screens have two basic components: the target macromolecule and the phage. Target macromolecules are commonly proteins such as antibodies, but they can be nucleic acids, carbohydrates, phospholipids, or even chemically undefined molecules. Undefined targets include a presumptive cell surface receptor expressed on cultured cells or presumptive carbohydrate-binding molecules in live animals. It is possible to identify a mimic peptide ligand to such a target by *ex vivo* and *in vivo* phage library screening. Phage technology can identify a specific peptide that binds to the target in the absence of biochemical or cell biological information (Kolonin *et al.*, 2004; Laakkonen *et al.*, 2002; Pasqualini and Ruoslahti, 1996).

The most commonly used method for *in vitro* library screening is biopanning (Fig. 1). The first identification of a peptide sequence mimicking a carbohydrate was made by Oldenburg *et al.* (1992) and Scott *et al.* (1992). They screened peptide-displaying phage libraries using the lectin concanavalin A (ConA) as a target and identified peptides that mimic mannosyl oligosaccharide. The identified peptides contained the consensus sequence Tyr-Pro-Tyr and bound ConA with a high affinity comparable to that of methyl $\alpha$-D-mannopyranoside. They proposed that, given the complexity of oligosaccharide synthesis, the prospect of finding peptides that competitively inhibit carbohydrate-specific receptors would simplify development of new therapeutic agents. Subsequently, Hoess *et al.* (1993) and Lou and Pastan (1999) identified Lewis Y mimicking peptide using monoclonal anti-Lewis Y antibody as a target. Many peptides were identified using monoclonal antibodies each specific to the carbohydrate epitope of glycolipids (Ishikawa *et al.*, 1998; Matsubara *et al.*, 1999; Taki *et al.*, 1997).

Successful identification by multiple laboratories of short peptides using monoclonal anti-carbohydrate antibodies indicates that monoclonal antibodies are excellent targets for phage library screening (Fukuda *et al.*, 2000; Harris *et al.*, 1997; O *et al.*, 2000; Taki *et al.*, 1997). By contrast, attempts to identify ligand mimic peptides using carbohydrate-binding proteins such as selectin have met with difficulties (Martens *et al.*, 1995), probably because the affinity of a carbohydrate-binding protein for its ligand is relatively weak compared to that of an antibody to an antigen. The basic protocol of *in vitro* biopanning of fd phage library using a monoclonal anti-carbohydrate

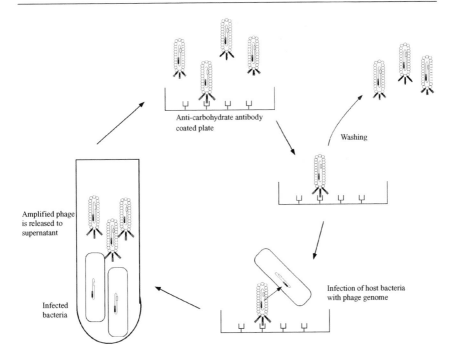

FIG. 1. Procedure for screening a peptide-displaying phage library by biopanning. The bottom of a plastic plate is coated with the target, such as an anti-carbohydrate antibody. As many as $10^{9-10}$ library clones are added to the plate (top), unbound phage is washed away (right), and clones binding the target are rescued by transformation of competent bacteria (bottom). Phage-infected host bacteria survive in media containing the appropriate antibiotics, and phage clones are amplified (left) for a second round of biopanning (top). After three to four cycles of library screening, the insert DNA is sequenced to determine the peptide sequence.

antibody as the target is provided in the following section. For T7-based phage library, see the comments at the end of protocol.

## Phage Library Screening: *In Vitro* Panning Protocol

### *Materials*

1. Peptide-displaying phage libraries are commercially available from New England BioLabs. A library construction kit is available from Novagen.
2. *Escherichia coli* K91kan strain, which is resistant to kanamycin (100 $\mu$g/ml) but sensitive to tetracycline (20 $\mu$g/ml) is available from

Invitrogen. Before starting the experiment, confirm that K91 cells to be used grow on LB agar containing kanamycin but fail to grow on kanamycin/tetracycline-containing LB agar.

3. 5X PEG/NaCl: Add water to 30 g polyethylene glycol (PEG) 8000 and 9.34 g NaCl to a total volume 100 ml. Store this reagent at room temperature after autoclaving.

4. Sequencing primer (fuse primer): 5′-CCCTCATAGTTAGCGTAA-CG-3′.

5. Reagents for DNA sequencing.

6. Phage-contaminate the equipment and spread easily. To avoid contamination, leave Pipetman pipettes upside down under an ultraviolet (UV) lamp every night while phage experiments are being performed.

*Procedures*

1. Add 100 $\mu$l water containing 10 $\mu$g purified monoclonal antibody to one well of a 96-well enzyme-linked immunosorbent assay (ELISA) plate. Leave the plate at 4° overnight. Because phage particles are very small, one well of a 96-well plate is enough to screen $10^{9-11}$ clones. Any plastic plate used for immunoassays (ELISA) can be used. Many commercially available monoclonal antibodies are pure immunoglobulin (IgG) or IgM. However, if a crude preparation such as ascites or serum is used, coat a well with purified second antibody such as goat anti-mouse IgG first and then add crude target antibody.

2. On the next day, remove the antibody solution from the well and wash the well three times with phosphate-buffered saline (PBS) containing 0.05% Tween 20 (PBST).

3. Fill the well with 3% bovine serum albumin (BSA) dissolved in PBST to block the well. Do not use nonfat milk, as it is rich in carbohydrates.

4. Add 10–20 $\mu$l of a phage library ($1 \times 10^{10}$ clones) and mouse serum (10 $\mu$l) in 3% BSA/PBST.

5. Incubate the library at room temperature for 30 min. Addition of mouse serum reduces false positives.

6. Remove unbound phage by aspiration, and wash the well 10 times with PBST.

7. Freshly cultured K91kan bacteria (100 $\mu$l) is then added to the well and left at room temperature for 1 h.

8. Bacteria are transferred to 10 ml LB tet/kan, LB containing tetracycline (20 $\mu$g/ml) and kanamycin (100 $\mu$g/ml). Elution of phage from the plastic plate is not required. K91 must be fresh and grown on the

day of panning. One colony can be inoculated 2 h before starting the procedure. Then grow it to OD600 = 2.0 (OD600 = 0.2 at 1/10 dilution).

9. Take triplicate aliquots of 0.1, 1.0, 10.0, and 100.0 $\mu$l from the 10-ml LB tet/kan, and spread each on four LB tet/kan agar plates (12 plates total).

10. Place LB tet/kan plates in a 37° incubator upside-down to avoid disruption of colonies by moisture. Culture the remaining transformed bacteria in 10 ml LB tet/kan at 37° in a shaker incubator.

11. Transformants will appear the next day on the agar plates. Count the number of colonies.

12. Centrifuge the amplified phage in 10-ml liquid culture at 8000 rpm for 15 min and transfer the supernatant to a new tube.

13. Add 2.5 ml 5× PEG/NaCl, incubate 3–4 h on ice, and centrifuge at 12,000 rpm to pellet phage.

14. Resuspend concentrated phage in 1 ml PBS.

### Second and Third Screens

Prepare two wells with or without target. Using two wells, select antibody-binding phage in the same manner as the first screen and determine colony number by plating 0.1, 1.0, 10.0, and 100.0 $\mu$l from the 10 ml of LB containing transformants. Colony numbers will differ between the control well and the target-coated well. As shown in Table I, numbers of colonies may increase in the second screen and further increase in the third. It is common to see colony number increased by a 1000-fold at the end of the third screen. Fourth and fifth screens may or may not be necessary. Repeated biopanning often results in amplification of dominant false positives or a loss in the variety of positive clones. It is important to stop biopanning at appropriate cycles or when more than 1000-fold increase has been achieved.

TABLE I

TOTAL NUMBERS OF COLONIES IN A 100-$\mu$L ALIQUOT; THE TARGET USED WAS
ANTI-LEWIS X (PMN6) ANTIBODY (MOUSE IgM)

| | No. of phage clones bound to control well | No. of phage clones bound to antibody-coated well |
| --- | --- | --- |
| First screen | NA | 79 |
| Second screen | 19 | 332 |
| Third screen | 13 | 1340 |

*Identification of Positive Clones*

1. Transfer each colony from the agar plate used for colony counting after the third screen to a tube, and culture bacteria in 5 ml LB tet/kan at 37° in a shaker incubator overnight.
2. Centrifuge the culture at 8000 rpm for 20 min and transfer the supernatant to a new tube.
3. Label each tube by clone number, and add 1.25 ml 5X PEG/NaCl, mix, and keep the mixture at 4° overnight.
4. Centrifuge the phage/PEG solution at 8000 rpm for 10 min.
5. Discard the supernatant and centrifuge at 15,000 rpm for 4 min.
6. Remove the remaining supernatant with a Pipetman pipette.
7. Add 1 ml Tris-buffered saline (TBS), resuspend phage with a Pipetman pipette, and transfer 200 $\mu$l to a numbered microcentrifuge tube.
8. Centrifuge the phage/TBS solution at 15,000 rpm for 5 min and transfer the supernatant to a new tube.
9. Save the numbered phage solution at 4° or −20°.
10. Wash StrataClean beads (Stratagene) with TBS and add 10 $\mu$l (of a 50% suspension) of beads to 200 $\mu$l of phage solution, then vortex vigorously for 30 s.
11. Centrifuge at 4000 rpm for 5 min and transfer 150 $\mu$l of the supernatant (DNA) to a labeled new microcentrifuge tube.
12. DNA is precipitated by adding 250 $\mu$l TE, 40 $\mu$l 3 $M$ sodium acetate, 1 $\mu$l glycogen, and 1 ml 100% ethanol, and incubating at −20° overnight.
13. On the next day, centrifuge DNA at 14,000 rpm for 10 min. A small white pellet (DNA) can now be seen.
14. Drain the liquid carefully. Wash the DNA pellet with 75% ethanol and air-dry. Resuspend the DNA in 12 $\mu$l sterile water.
15. Use 2 $\mu$l of DNA for sequencing by the dye termination reaction using the FUSE primer. An example of DNA sequencing is shown in Fig. 2. The cloned phage can be stored at −20° for several years.

The previously described procedures can be applied on T7-based phage library screening. As T7 phage produces plaque within 3 h, library screening goes fast. Reagents including sequencing primer and protocols are available from Novagen.

Troubleshooting and Validation of Identified Sequences

Successful library screening should reveal a consensus sequence (Table II). If no common sequence is seen among 20–30 clones, inspect those sequences for other shared characteristics. For example, if sequence data indicate that

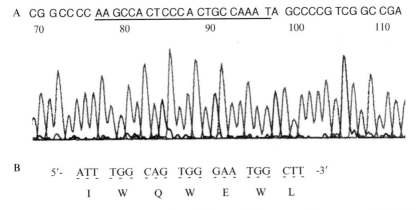

Fig. 2. Nucleotide sequence of cloned phage and insert peptide sequence. (A) DNA sequencing of cloned phage using anti-Lewis X antibody, MMA. Insert DNA sequence is underlined. (B) Complementary sequence of insert and translated peptide sequence.

TABLE II

Peptide Sequences Selected by Three Monoclonal Anti-Lewis X Antibodies, MMA, SSEA-1, and Pmn6 Clones

| MMA | SSEA1 | Pmn6 |
|---|---|---|
| IWAADWV | LWKFMGR | IWLQRRH |
| IWQWEWL[a] | IWRWAYF | LYLQRLR |
| VWSVEWV | | MWLQSVH |
| PWMGPEW | | YWLQTVR |
| | | LWLMKAR |

[a] See Fig. 2 for sequencing details of this clone.

*Note:* Consensus amino acid residues are underlined.

clones are hydrophobic peptides, it is possible that these phage clones are bound to the plastic. In such cases, inclusion of a stringent detergent in the wash buffer is advised. If phage clones encode peptides consisting primarily of charged amino acids, washing with PBS containing 0.5–1.0 $M$ NaCl may reduce such background. If false positives are acidic or basic peptides, washing at low or high pH levels may reduce nonspecific clones. However, if the phage clones show largely the same DNA sequence, such a result suggests that one clone is dominant by the end of the third screen. In such cases, select clones from an earlier (i.e., second) screen. Clones from an earlier screen should show more variability.

Binding of an identified peptide sequence to a carbohydrate-binding protein, such as selectin, can be done at this stage (Fukuda *et al.*, 2000). Thus, the carbohydrate-binding protein is coated on an ELISA plate, the cloned phage is added to the well, and the numbers of phage-binding colonies are counted. This assay should identify the best binder for the carbohydrate-binding protein; the numbers of colonies correlate with high-affinity binding to the protein of interest.

Further evaluation of an identified peptide should be done using a chemically synthesized peptide. Synthetic peptides can be linear, circular, or multivalent. Multivalent peptides (MAPS) are branched derivatives containing lysines, and many companies are capable of synthesizing them. Commercial companies can also make modified peptides (e.g., an N-terminal fatty acid conjugate that can mimic a glycolipid). A fatty acid–conjugated peptide can be incorporated into liposomes, which can be used to encapsulate drugs for targeted delivery (Oku *et al.*, 2002).

Addition of extra amino acid residues such as the KLAKLAK peptide makes a peptide with apoptosis-inducing capacity (Ellerby *et al.*, 1999; Gerlag *et al.*, 2001). By adding cysteine at either the N- or the C-terminus, carbohydrate-mimicry peptides can be further modified through a sulfhydryl cross-linker, maleimide. Thus, the peptide can be immobilized to maleimide-activated agarose beads or to macromolecules.

## Use of Carbohydrate-Mimicry Peptides

Groups identifying peptides binding to anti-carbohydrate antibodies and lectins assume that the identified peptides are carbohydrate mimetics or that they behave like carbohydrates. It is not clear whether peptides represent the structure of the carbohydrate or bind to the binding site of an antibody or lectin in a manner similar to the carbohydrate ligand.

Harris *et al.* (1997) addressed this question and identified a series of peptides using closely related monoclonal antibodies directed against the cell wall polysaccharide of group A *Streptococcus*. All of the identified peptides bind at or near the carbohydrate-binding site. Their results show that each monoclonal antibody binds a specific consensus peptide sequence, but they found no common peptide sequence among those selected by multiple monoclonal antibodies. We screened a phage library using three monoclonal antibodies directed to Lewis X antigen and found that each monoclonal antibody selected each consensus peptide sequence (Table II), of which conclusion is similar to that described earlier (Harris *et al.*, 1997).

On the other hand, Taki *et al.* (1997) identified a consensus 9-mer peptide recognized by two monoclonal antibodies directed to lactotetraosylceramide and neolactotetraosylceramide, indicating that 9-mer peptides are sufficient to mimic the galactosyl epitope structure. These studies suggest that peptides function in carbohydrate mimicry by interacting with carbohydrate-binding proteins at their epitope recognition site.

These results suggest that the identified peptide may not necessarily constitute the structural mimic of the carbohydrate, particularly when an epitope is part of a complex structure (Table II). Nonetheless, the peptide can be used as a convenient reagent to study carbohydrate-binding proteins, as exemplified by I-peptide, which inhibits cancer metastasis *in vivo* in the mouse (Fukuda *et al.*, 2000). I-peptide may be a powerful tool for studying carbohydrate-mediated metastasis (Fukuda *et al.*, 2000; Zhang *et al.*, 2002). A similar example is seen in the peptide identified as a mimic of ganglioside $G_{D1a}$, which inhibits adhesion of mouse lymphoma cells and hepatic endothelial cells (Ishikawa *et al.*, 1998; Takikawa *et al.*, 2000), indicating the involvement of $G_{D1a}$ in metastasis. The availability of unlimited amounts of peptide-displaying phage and readily available synthetic peptides should enable us to further analyze the role of carbohydrates in cancer.

## Acknowledgments

The author thanks Dr. Erkki Ruoslahti for helpful suggestions and offering his peptide-displaying phage libraries to our studies of carbohydrate-dependent cancer metastasis. The author thanks Dr. Elise Lamar for her assistance in editing the manuscript. This study has been supported by the National Institutes of Health (NIH CA71932), Department of Defense, Breast Cancer Idea grant DAMD17–02–0311, and The Susan G. Komen Breast Cancer Foundation grant BCTR0504175.

## References

Ellerby, H. M., Arap, W., Ellerby, L. M., Kain, R., Andrusiak, R., Rio, G. D., Krajewski, S., Lombardo, C. R., Rao, R., Ruoslahti, E., Bredesen, D. E., and Pasqualini, R. (1999). Anticancer activity of targeted pro-apoptotic peptides. *Nat. Med.* **5,** 1032–1038.

Fukuda, M. N., Ohyama, C., Lowitz, K., Matsuo, O., Pasqualini, R., Ruoslahti, E., and Fukuda, M. (2000). A peptide mimic of E-selectin ligand inhibits sialyl Lewis X-dependent lung colonization of tumor cells. *Cancer Res.* **60,** 450–456.

Gerlag, D. M., Borges, E., Tak, P. P., Ellerby, H. M., Bredesen, D. E., Pasqualini, R., Ruoslahti, E., and Firestein, G. S. (2001). Suppression of murine collagen-induced arthritis by targeted apoptosis of synovial neovasculature. *Arthritis Res.* **3,** 357–361.

Harris, S. L., Craig, L., Mehroke, J. S., Rashed, M., Zwick, M. B., Kenar, K., Toone, E. J., Greenspan, N., Auzanneau, F. I., Marino-Albernas, J. R., Pinto, B. M., and Scott, J. K. (1997). Exploring the basis of peptide-carbohydrate crossreactivity: Evidence for discrimination by peptides between closely related anti-carbohydrate antibodies. *Proc. Natl. Acad. Sci. USA* **94,** 2454–2459.

Hoess, R., Brinkmann, U., Handel, T., and Pastan, I. (1993). Identification of a peptide which binds to the carbohydrate-specific monoclonal antibody B3. *Gene* **128,** 43–49.

Hoffman, J. A., Laakkonen, P., Porkka, K., and Ruoslahti, E. (2002). "Phage Display: A Practical Approach." Oxford University Press, Oxford.

Ishikawa, D., Kikkawa, H., Ogino, K., Hirabayashi, Y., Oku, N., and Taki, T. (1998). GD1alpha-replica peptides functionally mimic GD1alpha, an adhesion molecule of metastatic tumor cells, and suppress the tumor metastasis. *FEBS Lett.* **441,** 20–24.

Kolonin, M. G., Saha, P. K., Chan, L., Pasqualini, R., and Arap, W. (2004). Reversal of obesity by targeted ablation of adipose tissue. *Nat. Med.* **10,** 625–632.

Laakkonen, P., Porkka, K., Hoffman, J. A., and Ruoslahti, E. (2002). A tumor-homing peptide with a targeting specificity related to lymphatic vessels. *Nat. Med.* **8,** 751–755.

Lou, Q., and Pastan, I. (1999). A Lewis(y) epitope mimicking peptide induces anti-Lewis(y) immune responses in rabbits and mice. *J. Pept. Res.* **53,** 252–260.

Martens, C. L., Cwirla, S. E., Lee, R. Y., Whitehorn, E., Chen, E. Y., Bakker, A., Martin, E. L., Wagstrom, C., Gopalan, P., Smith, C. W., Tate, E., Koller, K. J., Schatz, P. J., Dower, W. J., and Barrett, R. W. (1995). Peptides which bind to E-selectin and block neutrophil adhesion. *J. Biol. Chem.* **270,** 21129–21136.

Matsubara, T., Ishikawa, D., Taki, T., Okahata, Y., and Sato, T. (1999). Selection of ganglioside GM1-binding peptides by using a phage library. *FEBS Lett.* **456,** 253–256.

O, I., Kieber-Emmons, T., Otvos, L., and Blaszczyk-Thurin, M. (2000). Peptide mimicking sialyl-Lewis(a) with anti-inflammatory activity. *Biochem. Biophys. Res. Commun.* **268,** 106–111.

Oku, N., Asai, T., Watanabe, K., Kuromi, K., Nagatsuka, M., Kurohane, K., Kikkawa, H., Ogino, K., Tanaka, M., Ishikawa, D., Tsukada, H., Momose, M., Nakayama, J., and Taki, T. (2002). Anti-neovascular therapy using novel peptides homing to angiogenic vessels. *Oncogene* **21,** 2662–2669.

Oldenburg, K. R., Loganathan, D., Goldstein, I. J., Schultz, P. G., and Gallop, M. A. (1992). Peptide ligands for a sugar-binding protein isolated from a random peptide library. *Proc. Natl. Acad. Sci. USA* **89,** 5393–5397.

Pasqualini, R., and Ruoslahti, E. (1996). Organ targeting *in vivo* using phage display peptide libraries. *Nature* **380,** 364–366.

Scott, J. K., Loganathan, D., Easley, R. B., Gong, X., and Goldstein, I. J. (1992). A family of concanavalin A-binding peptides from a hexapeptide epitope library. *Proc. Natl. Acad. Sci. USA* **89,** 5398–5402.

Smith, G. P. (1985). Filamentous fusion phage: Novel expression vectors that display cloned antigens on the virion surface. *Science* **228,** 1315–1317.

Taki, T., Ishikawa, D., Hamasaki, H., and Handa, S. (1997). Preparation of peptides which mimic glycosphingolipids by using phage peptide library and their modulation on beta-galactosidase activity. *FEBS Lett.* **418,** 219–223.

Takikawa, M., Kikkawa, H., Asai, T., Yamaguchi, N., Ishikawa, D., Tanaka, M., Ogino, K., Taki, T., and Oku, N. (2000). Suppression of GD1alpha ganglioside-mediated tumor metastasis by liposomalized WHW-peptide. *FEBS Lett.* **466,** 381–384.

Yamamoto, M., Kominato, Y., and Yamamoto, F. (1999). Phage display cDNA cloning of protein with carbohydrate affinity. *Biochem. Biophys. Res. Commun.* **255,** 194–199.

Zhang, J., Nakayama, J., Ohyama, C., Suzuki, M., Suzuki, A., Fukuda, M., and Fukuda, M. N. (2002). Sialyl Lewis X-dependent lung colonization of B16 melanoma cells through a selectin-like endothelial receptor distinct from E- or P-selectin. *Cancer Res.* **62,** 4194–4198.

## [5]   Phage Display-Derived Human Antibodies Against Specific Glycosaminoglycan Epitopes

*By* Nicole C. Smits, Joost F. M. Lensen,
Tessa J. M. Wijnhoven, Gerdy B. ten Dam,
Guido J. Jenniskens, and Toin H. van Kuppevelt

### Abstract

Glycosaminoglycans (GAGs) are long unbranched polysaccharides, most of which are linked to a core protein to form proteoglycans. Depending on the nature of their backbone, one can discern galactosaminoglycans (chondroitin sulfate [CS] and dermatan sulfate [DS]) and glucosaminoglycans (heparan sulfate [HS], heparin, hyaluronic acid, and keratan sulfate). Modification of the backbone by sulfation, deacetylation, and epimerization results in unique sequences within GAG molecules, which are instrumental in the binding of a large number of proteins. Investigating the exact roles of GAGs has long been hampered by the lack of appropriate tools, but we have successfully implemented phage display technology to generate a large panel of antibodies against CS, DS, HS, and heparin epitopes. These antibodies provide unique and highly versatile tools to study the topography, structure, and function of specific GAG domains. In this chapter, we describe the selection, characterization, and application of antibodies against specific GAG epitopes.

### Overview

Antibodies are widely used in the research on proteoglycans. They have been applied to evaluate proteoglycan expression patterns in a variety of tissues in health and disease, and have been used as (immuno)precipitating agents. Most of the available antibodies are directed against the core protein of proteoglycans. Only a few antibodies that are reactive with the GAG moiety have been described (Bao *et al.*, 2005; Caterson *et al.*, 1983; David *et al.*, 1992; Sorrell *et al.*, 1990; van den Born *et al.*, 1995). However, these were all generated using proteoglycans rather than isolated GAGs.

As stated earlier, investigating the biological role of GAGs has long been hampered by the lack of appropriate tools. Using phage display technology, we have generated a large panel of epitope-specific antibodies against HS, heparin, CS, and DS (Dennissen *et al.*, 2002; Jenniskens *et al.*, 2000; Smetsers *et al.*, 2003; van De Westerlo *et al.*, 2002; van Kuppevelt *et al.*, 1998) (Table I).

METHODS IN ENZYMOLOGY, VOL. 416                                  0076-6879/06 $35.00
                                                     DOI: 10.1016/S0076-6879(06)16005-X

Characterization of the GAGs bound by the antibodies revealed that specific modification patterns are recognized (Dennissen *et al.*, 2002; ten Dam *et al.*, 2004). For instance, one specific antibody against HS requires 3-*O* sulfates and recognizes an epitope that resembles the antithrombin-III–binding pentasaccharide sequence (ten Dam *et al.*, 2005). A variety of anti-GAG antibodies was used to study the distribution of GAG epitopes in spleen (ten Dam *et al.*, 2003), lung (Smits *et al.*, 2004), kidney (Lensen *et al.*, 2005), and skeletal muscle (Jenniskens *et al.*, 2002). GAG-specific antibodies were also used to study changes in expression pattern in diseased versus healthy tissue, especially melanoma and psoriasis (Smetsers *et al.*, 2003, 2004). Next to immunohistochemical evaluation, antibodies were applied to analyze the biological function of GAGs (e.g., growth factor binding). The endogenous expression of anti-HS antibodies resulted in the functional knockout of specific HS epitopes, which was shown to interfere with ion housekeeping in skeletal muscle cells (Jenniskens *et al.*, 2003). Our panel of antibodies thus provides a unique and highly versatile tool to study the topography, structure, and function of GAGs. Here, we describe the selection and evaluation of single-chain variable fragment (scFv) antibodies against specific GAG epitopes.

Phage Display

One of the most successful applications of phage display has been the selection of specific antibodies from phage display libraries (Harrison *et al.*, 1996; Hoogenboom and Chames, 2000; Winter *et al.*, 1994). DNA's encoding antibodies are cloned into a suitable vector as a fusion to the gene encoding one of the phage coat proteins (pIII, pVI, or pVIII). Upon assembly in bacteria, the antibody-coat protein fusion will be part of the surface of the phage, thus "displaying" the antibody. A phage display library is a large collection of phages, each displaying an individual antibody. Phage display allows the generation of antibodies against "self-antigens," which is favorable because GAGs are largely nonimmunogenic.

The phage display system is illustrated in Fig. 1. First, the phage display library is incubated with the "antigen" (GAGs) immobilized onto a tube. Phages that display antibodies that are specific for the antigen will bind, and nonadherent phages will be washed away. Phages expressing a specific anti-GAG antibody can then be recovered from the tube (e.g., by altering the pH) and multiplied by infection into bacteria. This biopanning procedure is repeated several times. Selected phages are then analyzed for anti-GAG antibodies by enzyme-linked immunosorbent assay (ELISA) and immunohistochemistry (IHC).

TABLE I
CHARACTERISTICS OF DIFFERENT scFv ANTIBODIES[a]

| scFv | $V_H$ | DP gene | $V_H$CDR3 | Class of GAG | References |
|---|---|---|---|---|---|
| RB4EA12 | 3 | 32 | RRYALDY | HS | Jenniskens et al., 2000 |
| HS4C3 | 3 | 38 | GRRLKD | HS | van Kuppevelt et al., 1998 |
| EV3C3 | 3 | 42 | GYRPRF | HS | Smits et al., 2004 |
| HS4E4 | 3 | 38 | HAPLRNTRTNT | HS | Dennissen et al., 2002 |
| AO4B08 | 3 | 47 | SLRMNGWRAHQ | HS | Jenniskens et al., 2000 |
| IO3H10 | 1 | 7 | AKRLDW | CS | Smetsers et al., 2004 |
| LKN1 | 1 | 25 | GIKL | DS | Lensen et al., 2005 |
| MPB49 | 3 | 38 | WRNDRQ | — (Control) | — |

[a] CS, chondroitin sulfate; DS, dermatan sulfate. GAG, glycosaminoglycan; HS, heparan sulfate; all scFv antibodies recognize specific epitopes as based on different staining patterns and different reactivities toward various HS/CS/DS preparations.

Note: Given are the scFv antibody code, $V_H$ germline gene family, DP gene segment number, amino acid sequence of the $V_H$ complementarity determining region 3 (CDR3), and the class of GAG with which the antibody reacts. Selection of the antibodies is described in the indicated reference.

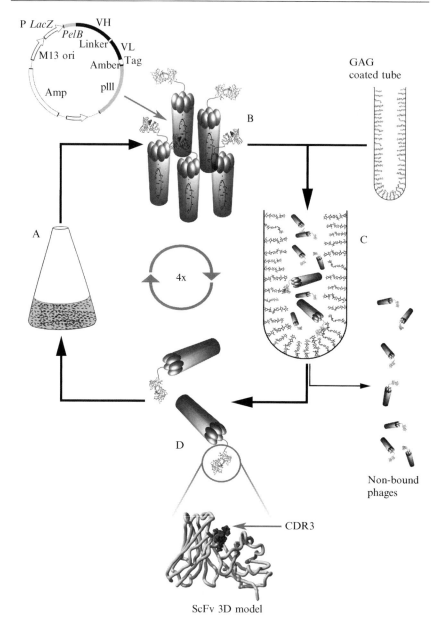

ScFv 3D model

FIG. 1. Schematic outline of the phage display technique. A library of phage-infected bacteria is grown and induced to produce phages expressing ("displaying") scFv antibody fragments on their surface (A). Phages are isolated (B) and used for selection ("panning"; C) against glycosaminoglycans (GAGs), which are immobilized onto the surface of a selection tube.

In principle, any phage display library can be used. In the protocol described here, a human semisynthetic library (Library no. 1 [Nissim *et al.*, 1994]) was used, consisting of more than $10^8$ different clones, each expressing a unique antibody. The semisynthetic Library no. 1 contains 50 different $V_H$ genes combined with a single light-chain gene (DPL 16). In the library, variable parts of the heavy and light chains are joined to each other by a linker sequence to form so-called *scFv*. All antibodies contain a c-*Myc* tag for identification. Because of the use of an amber codon combined with a suppressor *Escherichia coli* strain, one can easily produce soluble scFv antibodies from selected phage clones.

To rescue and amplify phages following selection, a helper phage is required to allow for phage assembly. VCS-M13 is a commonly used helper phage that bears the kanamycin-resistance gene, which, along with the ampicillin-resistance gene carried by compatible phagemids, enables the selection of cells that contain both the helper phage and the phagemid. Helper phages usually have a defective origin of replication (Vieira and Messing, 1987), which allows the preferential packaging of the phagemid DNA over the helper phage DNA and results in a greater output of phagemid phage over helper phage. This is especially important during selection, because only the phagemid contains the DNA encoding the antibody of interest.

### Preparation of Helper Phage VCS-M13

*Materials*

> *Bacterial strain: E. coli* TG1 suppressor strain (K12, *supE, hsdΔ5, thi,* Δ[*lac-pro*AB], F′[*tra*D36, *pro*AB+, *lac*Iq, *lac*-ZΔM15])[1]

---

[1] *E. coli* TG1 is a T-phage–resistant strain that harbors a mutated tRNA gene. The mutated tRNA will suppress the UAG amber (stop codon). To allow expression of scFv-pIII fusion protein on the phage tip, the amber codon will be substituted by a glutamine.

---

Nonbinding phages are washed away, whereas phages displaying antibodies that are reactive with GAGs stay bound to the tube. Thus, the library becomes enriched for phages displaying GAG-reactive antibodies. Selected phages are harvested (D), allowed to infect a fresh *Escherichia coli* culture, and used for another round of panning (A–D). Individual clones can be isolated after four rounds of panning. Insert top left: Schematic representation of the phagemid encoding antibody-expressing phages. At the bottom a three-dimensional model of an scFv antibody showing the light and heavy chain (ribbon) and the CDR3 region is depicted (space filling). CDR3, complementarity determining region 3.

*Solutions*

> Ampicillin (100 mg/ml)
>
> H-top agar: 10 g/liter peptone, 8 g/liter NaCl, 8 g/liter Bacto-agar in $H_2O$; dissolve, bring the volume to 1 liter with $H_2O$. Autoclave and store at 4°. Melt in a microwave before use.
>
> H-bottom plates: 10 g/liter peptone, 8 g/liter NaCl, 15 g/liter Bacto agar in $H_2O$; dissolve, bring the volume to 1 liter with $H_2O$ and autoclave. Mix gently and pour into Petri dishes. (Plates are stored at 4°.)
>
> Kanamycin (25 mg/ml)
>
> Minimal media agar plates: 15 g/liter Bacto agar, 10.5 g/liter $K_2HPO_4$, 4.5 g/liter $KH_2PO_4$, 1 g/liter $(NH_4)_2SO_4$, 0.5 g/liter sodium citrate in 985 ml $H_2O$; dissolve and autoclave. Cool the solution to 60° and add 500 $\mu$l of 20% $MgSO_4$, 0.5 ml of 1% vitamin B1 (thiamine HCl, filter-sterilized 0.2 $\mu$m), and 5 ml 40% glucose (filter sterilized 0.2 $\mu$m). Mix gently and pour into Petri dishes. (Plates are stored at 4°.)
>
> 0.1 $M$ NaOH/1% (w/v) sodium dodecylsulfate (SDS)
>
> Phosphate-buffered saline (PBS)
>
> 20% (w/v) polyethylene glycol 6000 (PEG)/2.5 $M$ NaCl
>
> Tris/EDTA (TE) buffer: 10 m$M$ Tris–HCl, pH 8.0 and 1 m$M$ EDTA, pH 8.0
>
> 1 $M$ Tris–HCl, pH 7.4
>
> 2× TY medium
>
> 2× TY medium containing 100 $\mu$g/ml ampicillin and 1% glucose
>
> TYE plates: 10 g/liter peptone, 5 g/liter Bacto-yeast extract, 8 g/liter NaCl, 15 g/liter Bacto agar in $H_2O$; dissolve, bring the volume to 1 liter, and autoclave. Cool the solution to 60° and add 1 ml of 100 mg/ml ampicillin and 25 ml of 40% glucose to achieve final concentrations of 100 $\mu$g/ml ampicillin and 1% glucose. Mix gently and pour into Petri dishes. (Plates are stored at 4°.)

*Methods*

> *Day 1*
>
> 1. Streak *E. coli* TG1 cells from a glycerol stock on a minimal media agar plate and incubate o/n at 37°. Do not use antibiotics; *E. coli* TG1 has no antibiotic resistance genes.
>
> *Day 2*
>
> 1. Inoculate 5 ml 2× TY medium containing 1% glucose with a single colony of *E. coli* TG1 cells from the minimal media agar plate and incubate o/n at 37°.

*Day 3*

1. Inoculate 50 ml 2× TY medium containing 1% glucose with 500 μl of the o/n culture. Grow the culture, while shaking, at 37° until an absorbance at 600 nm of 0.4–0.5 is reached.
2. Prepare $10^{-6}$, $10^{-7}$, and $10^{-8}$ dilutions of the commercially obtained VCS-M13 preparation (usually in the range of $1 \times 10^{11}$ plaque-forming units [pfu]/ml) in 2× TY medium containing 1% glucose. Add 10 μl of these dilutions to 200 μl *E. coli* TG1 cultures (*Day 3*; *step 1*).
3. Incubate for 30 min at 37°, without shaking, to allow for infection of *E. coli* cells with helper phage.
4. Add 3 ml of liquefied H-top agar (not warmer than 50°) to the infected culture, mix gently, and pour onto H-bottom plates that have been prewarmed to 37°. Incubate o/n at 37°. The aim is to obtain a plate from which single VCS-M13 plaques can be picked easily (i.e., *E. coli* colonies that are growing slower because of infection with VCS-M13).
5. Determine the titer of the VCS-M13 phages from the number of plaques.
6. Separately, inoculate 5 ml 2× TY medium containing 1% glucose with a single colony of *E. coli* TG1 cells from the minimal media agar plate (*Day 1*; *step 1*) and grow o/n at 37°.

*Day 4*

1. Inoculate 50 ml 2× TY medium containing 1% glucose with 500 μl of the o/n culture (*Day 3*; *step 5*). Grow the culture, while shaking, at 37° until an absorbance at 600 nm of 0.4–0.5 is reached (when necessary, the *E. coli* culture can be kept on ice before infection for a moment, but do not exceed 30 min because of the loss of F-pili).
2. Transfer a single VCS-M13 plaque from the H-bottom plate (*Day 3*; *step 4*) to 4 ml of the *E. coli* culture (*step 1*) and incubate, while shaking, for 2 h at 37° (phage infection).
3. Transfer the infected 4 ml culture to a 2-liter Erlenmeyer flask containing 500 ml of prewarmed 2× TY medium and grow the culture, while shaking, for 1 h at 37°.
4. Add 1.2 ml of 25 mg/ml kanamycin to a final concentration of 60 μg/ml and grow, while shaking, o/n at 37°.

*Day 5*

1. Cool the culture (*Day 4*; *step 4*) on ice for 20 min.
2. Collect the bacteria by centrifugation at 5000*g* for 10 min at 4° (after centrifugation phages are present in the supernatant!).

3. Add 100 ml of ice-cold PEG-NaCl to the phage supernatant, mix gently, and incubate on ice for 1 h. In this step, phages are precipitated.
4. Pellet the phages at 10,000g for 30 min at 4°.
5. Resuspend the phage pellet in 40 ml $H_2O$. Transfer the solution to a 50-ml tube and add 8 ml of ice-cold PEG-NaCl. Mix well and incubate on ice for 20 min.
6. Centrifuge the phage solution at 10,000g for 30 min at 4°. Decant the supernatant, centrifuge briefly, and remove residual PEG-NaCl with a capillary.
7. Resuspend the phage pellet in 2.5 ml of TE buffer and centrifuge at 10,000g for 15 min at 4°.
8. Wet a 0.45 disposable filter with TE buffer and filter the phage solution to remove any remaining bacterial debris. Note that it can be difficult to filter the phage solution and that this step may result in the loss of phages that remain in the filter.
9. Titrate the phages to $10^{-12}$ pfu/ml.
10. Store the phage solution at 4° (up to 5 days).

Panning Methodology

Panning methodology (Fig. 1, Table II) involves the physical contact of the phage display library with the antigen (in this case GAGs) and can be regarded as an affinity selection system. In practice, a small volume containing the phage display library is incubated with the antigen immobilized to a tube. Panning consists of several rounds of binding phages to the immobilized GAGs, a defined number of washing steps, and elution of the bound phages by altering the pH. Eluted phages are subjected to another round of panning. During each round of panning, specific binding clones are selected and amplified so they predominate after three to four rounds of panning. The input of each round should be in the range of $10^{12}$ phages.

*Selection of Phages Displaying Specific Antibodies against GAGs Using Panning Methodology*

*Materials*

Bacterial strain: *E. coli* HB2151 nonsuppressor strain (K12, *ara*, *thi*, $\Delta$[*lac-pro*], F'[*pro*AB$^+$, *lac*I$^q$, Z$\Delta$M15])
Bacterial strain: *E. coli* TG1 suppressor strain (K12, *sup*E, *hsd*$\Delta$5, *thi*, $\Delta$[*lac-pro*AB], F'[*tra*D36, *pro*AB+, *lac*Iq, *lac*-Z$\Delta$M15])

TABLE II

TIME SCHEDULE OF EXPERIMENTS DESCRIBED IN THIS CHAPTER

| Day | Activity | Sections | Notes |
|---|---|---|---|
| **Selection of antibodies using panning methodology** | | | |
| 1 | Growth of phage library | A.1.1–1.6 | Inoculate original library or n + 1 generation (from Section D.3.4); grow o/n at 37° |
| | Coating of selection tubes | C.1.1 | |
| | Inoculation of *Escherichia coli* TG1 | A.1.7 | |
| 2 | Isolation of phages | B.1.1–1.8 | Phages are used for selection (see Section C.2.2–2.11) |
| | Preparation selection tube | C.2.1 | To be used for selection (see Section C.2.2–2.9) |
| | Round of panning | C.2.2–2.11 | |
| 3 | Calculate titers | C.3.1 | Store plates; individual clones can be used to inoculate master plate (see Section E.3.1) |
| | Harvest phagemids | D.3.2–3.5 | Phagemids can be stored at –80° or used for a next round of selection (see Section A.1.1–1.6) |
| | Inoculation of master plate | E.3.1 | Stored as back up and used to inoculate induction plate (E.4.1–4.3) |
| **Screening for antibody-expressing clones** | | | |
| 4 | Preparation of induction plate | E.4.1–4.2 | |
| | Storage of master plate | E.4.3 | |
| | Coat ELISA plate | F.4.1 | Coating of ELISA plates with the antigen that was used for selection |
| 5 | ELISA screening | F.5.1–5.10 | Identification of active antibody-expressing clones |
| **Characterization and large-scale production of antibodies** | | | |
| 6 | Streak individual clones | G.6.1 | Streak individual clones from master plate (see Section E.4.3) |
| 7 | Inoculation of o/n cultures | G.7.1 | |
| 8 | Production and isolation of antibodies | G.8.1–8.8 | Identification and characterization of individual antibodies |
| | DNA isolation and sequencing | H.8.1–8.3 | |
| 9 | Aliquot and store antibodies | G.9.1 | Storage at 4° (weeks) or at –20° (months) |
| **Miscellaneous analytical techniques** | | | |
| | Immunofluorescence analysis | I | Characterization of epitope occurrence |
| | Direct ELISA analysis | J.1 | Characterization of epitopes |
| | Competition ELISA analysis | J.2 | Characterization of epitopes |

Glycerol stock of the (semi)-synthetic scFv Library no. 1 (Nissim *et al.*, 1994) (Dr. G. Winter, MRC Centre for Protein Engineering, Cambridge, UK), stored at $-80°$

VCS-M13 helper phages[2] used at a titer of $1 \times 10^{12}$ pfu/ml (Stratagene) (for large amounts, see preparation of helper phage VCS-M13)

*Solutions*

GAGs of interest: 10 mg/ml in $H_2O$
PBS containing 0.1% (v/v) Tween-20
PBS containing 2% (w/v) Marvel (prepare freshly)
PBS containing 4% (w/v) Marvel (prepare freshly)
100 m$M$ triethylamine (prepare freshly)
$2\times$ TY containing 100 $\mu$g/ml ampicillin and 25 $\mu$g/ml kanamycin

*Methods*

Several precautions need to be taken to avoid carryover of phages during the selections. The use of sterile solutions and disposable plastics is highly recommended. Nondisposable plastics should be soaked for 1 h in 2% (v/v) hypochlorite, followed by thorough washing and autoclaving. Glassware should be baked for at least 6 h at 180°. Always use aerosol-resistant pipette tips (Molecular Bio-Products) when working with phages. It is recommended to work in a laminar-flow cabinet. Clean the workplace (bench tops, etc.) with 0.1 $M$ NaOH/1% (w/v) SDS before and after each working day. Clean pipettes and other tools daily by wiping the outside with 0.1 $M$ NaOH/1% (w/v) SDS, followed by a rinse with water.

For a schematic representation of the following steps, see Fig. 1.

*(A) Growth of Antibody Phage Display Library*

1.1. Inoculate 50 ml of $2\times$ TY containing 100 $\mu$g/ml ampicillin and 1% glucose with 50 $\mu$l from a glycerol stock of the (semi)-synthetic scFv Library no. 1 (Nissim *et al.*, 1994).

1.2. Grow the culture, while shaking, at 37° until an optical density at 600 nm of 0.5 is reached.[3] (This takes about 2–3 h.)

---

[2] VCS-M13 phages provide phage coat proteins and enzymes that are crucial for rescue of the phages.

[3] M13 phages infect $F^+$ *E. coli* via sex pili. For optimal infection, *E. coli* needs to be in the log phase (absorbance of 0.4–0.5 at 600 nm) at 37°. When grown to a higher density, sex pili are lost very rapidly. A log phase culture can be held on ice for about 30 min.

1.3. Transfer 10 ml of the culture to a sterile 50-ml tube and add 50 $\mu$l VCS-M13 helper phages.[4] The ratio between bacteria and helper phages must be 1:20 (titer helper phages $10^{12}$ pfu/ml).

1.4. Incubate at 37° for 30 min without shaking (phage infection).

1.5. Centrifuge the infected culture at 4000$g$ for 10 min at room temperature. Decant the supernatant, and resuspend the pellet in 1 ml 2× TY containing 100 $\mu$g/ml ampicillin and 25 $\mu$g/ml kanamycin.

1.6. Add the infected bacterial suspension to 300 ml of prewarmed (30°) 2× TY containing 100 $\mu$g/ml ampicillin and 25 $\mu$g/ml kanamycin (*no glucose*[5]). Incubate o/n, while shaking, at 30°. This culture will be used in section "B. Isolation of Phages" Step 1.1.

Inoculate 10 ml of 2× TY with a single colony of *E. coli* TG1 from a minimal medium plate (do not use antibiotics) and grow o/n, while shaking, at 37°. This culture will be used in section "C. Selection of Phages Binding to GAGs," Day 2, Step 2.8.

## (B) Isolation of Phages

1.1. Cool the culture from section "A. Growth of Antibody Phage Display Library," Step 1.6, on ice for 10 min.

1.2. Centrifuge the culture at 5000$g$ for 15 min at 4°. After centrifugation, the bacterial pellet can be discarded and the supernatant containing the phages transferred into a sterile bucket.

1.3. Add 60 ml of ice-cold PEG-NaCl to the supernatant to precipitate the phages.[6] Mix well by inverting, and leave the bucket on ice for 1 h.

1.4. Pellet the phages at 10,000$g$ for 30 min at 4°. Resuspend the pellet in 40 ml of ice-cold sterile $H_2O$. Transfer the phage suspension to a 50-ml tube and add 8 ml of ice-cold PEG-NaCl. Mix well (as in the preceding Step 1.3) and leave on ice for 20 min.

---

[4] Take the remaining 40 ml of the culture, spin it down, and resuspend the pellet in 1 ml 2× TY. Spread it on TYE plates, and grow overnight at 37°. Harvest the cells in 2 × 5 ml 2× TY medium. Collect the cells in a 50-ml tube and centrifuge for 10 min at 10,000$g$. Resuspend the pellet in 500 $\mu$l of 2× TY, add 500 $\mu$l of 60% glycerol to a final concentration of 30%, mix well, divide this stock in 50-$\mu$l aliquots, and store at −80° as a back up for additional selections (this stock will be the n + 1 generation).

[5] Glucose represses transcription of the scFv-pIII fusion protein through the *Lac* operon in the phagemid.

[6] In this step, phages are precipitated and concentrated. Given that the TG1 suppression of the amber codon is never complete, this step is also necessary for removing any soluble antibodies present in the supernatant.

1.5. Centrifuge the phage mixture at 10,000$g$ for 30 min at 4°. Decant the supernatant, centrifuge briefly, and remove residual PEG-NaCl with a capillary.

1.6. Resuspend the phage pellet in 2.5 ml PBS and centrifuge the phage solution at 10,000$g$ for 5 min.

1.7. Wet a 0.45-$\mu$m disposable filter with PBS and carefully filter the phage solution to remove any remaining bacterial debris. Note that it can be difficult to filter the phage solution.

1.8. This step may result in loss of phages that remain in the filter.

1.9. Store the phage solution at 4° until use. Phages will be used in section "C. Selection of Phages Binding to GAGs," Day 2, Step 2.2.

## (C) Selection of Phages Binding to GAGs

### Day 1

1.1. Coat immunotubes with 3 ml of a 10-$\mu$g/ml solution of the GAG of interest and incubate o/n at room temperature, while rotating on an under-and-over turntable.

### Day 2

2.1. Decant the GAG solution, wash the tube three times with PBS, and block the tube with PBS containing 2% (w/v) Marvel to avoid nonspecific binding of phages to the surface of the tube. Fill the tube to the edge, cover it with Parafilm, and incubate for at least 2 h at room temperature on an under-and-over turntable. This step should be performed early in the day so the tube will be ready when the phages used for biopanning are obtained (section "B. Isolation of Phages," Step 1.8).

2.2. Empty the tube and add 2 ml of PBS containing 4% Marvel and 2 ml of phage supernatant from Step 1.8 in section "B. Isolation of Phages," cover the tube, and incubate for 1 h on an under-and-over turntable at room temperature, followed by standing for 1 h.

2.3. Discard the phage suspension and wash the tube 20 times with PBS  containing 0.1% (v/v) Tween-20. After the last wash, fill the tube with  PBS containing 0.1% (v/v) Tween-20 and rotate on an over-and-under turntable for 10 min at room temperature.

2.4. Empty the tube and wash the tube 20 times with PBS. Fill the tube with PBS and rotate on an over-and-under-table for 10 min at room temperature.

2.5. Empty the tube and add 1 ml of 100 m$M$ triethylamine solution. Cover the tube and rotate for 15 min on an under-and-over turntable at room temperature. In this step, bound phages are eluted.

2.6. Add the eluted phages to a 50-ml tube containing 0.5 ml of 1 $M$ Tris–HCl, pH 7.4 for neutralization. Additionally, add 200 $\mu$l of 1 $M$ Tris–HCl, pH 7.4 to the remaining phages in the immunotube. At this point, phages can be stored at 4° for a short period (up to 2 days) or used directly to infect $E.\ coli$ TG1 cells. The latter is recommended.

2.7. Add the eluted phages (from Step 2.6) to 9 ml of exponentially growing $E.\ coli$ TG1 cells in a 50-ml tube. Add 4 ml TG1 culture to the remaining phages in the immunotube (Step 2.6). Incubate both cultures for 30 min at 37°, without shaking, to allow infection.

2.8. Exponentially growing TG1 culture is prepared as follows: Inoculate 50 ml of 2× TY with 0.5 ml of the o/n culture from section "A. Growth of Antibody Phage Display Library," Step 1.7. Grow the culture, while shaking, at 37° until an absorbance at 600 nm of 0.4–0.5 is reached (this takes ~90 min). It is recommended to inoculate an additional 50 ml of 2× TY with 0.5 ml of the overnight culture 30 and 60 min after the first inoculation and use the culture with an absorbance of 0.4–0.5 for infection.

2.9. Pool both infected $E.\ coli$ TG1 cultures (9 ml and 4 ml from Step 2.8) and take 100 $\mu$l of the pooled cultures to make serial dilutions ($10^2$, $10^4$, $10^6$, $10^8$) in 2× TY to calculate the titer. Plate 100 $\mu$l of the dilutions on TYE plates and grow o/n at 37°.

2.10. Centrifuge the infected TG1 culture for 10 min at 10,000$g$ at room temperature (do not refrigerate!).

2.11. Resuspend the cells in 2 ml 2× TY, spread on TYE plates and incubate o/n at 37°.

*Day 3*

3.1. Count colonies and calculate the titer from the dilutions. An increase in titer is expected after each selection round, indicating enrichment of binding clones. Store plates with separate colonies, because individual clones will be used in section "E. Production of Master Plate and Induction Plate of Phage-Infected Clones," Day 3, Step 3.1.

*(D) Harvesting of Selected Phagemids*

3.2. Add 5 ml 2× TY medium to the TYE plates and scrape the bacterial cells from the plate with a glass spreader. Collect the bacteria in a 50-ml tube and centrifuge at 5000$g$ for 10 min at room temperature.

3.3. Decant the supernatant and resuspend the bacterial pellet in 2 ml 2× TY medium.

3.4. For a next round of selection, take 50 μl of the bacterial suspension and use it to inoculate 50 ml 2× TY containing 100 μg/ml ampicillin and 1% (w/v) glucose as in section "A. Growth of Antibody Phage Display Library," Step 1.1. Repeat this selection procedure for another three selection rounds (all steps of this section).

3.5. Add 1 ml of ice-cold 60% glycerol to the remaining cells, aliquot in 50 μl fractions, snap-freeze in liquid nitrogen, and store at −80°.[7]

## Screening for Clones Expressing Anti-GAG Antibodies

To assess whether the panning experiment was successful, individual clones need to be selected and analyzed. This section describes a procedure to identify antibody-expressing clones after panning. An easy and fast method to test the selected clones for production of antibodies directed against GAGs is ELISA. Antibody-containing supernatants obtained from isopropyl-β-D-thiogalactosidase (IPTG)–induced cultures can be used for this purpose. Clones with desired specificity in ELISA can be further analyzed by DNA sequencing to identify different clones (e.g., with unique CDR3 sequences). Clones of interest can then be used for large-scale production and purification.

### (E) Production of Master Plate and Induction Plate of Phage-Infected Clones

#### Materials

Sterile 96-well flat-bottom assay plates with high-binding surface (ELISA plate, Greiner) 96-wells round-bottom Cellstar plates (master block) (Greiner)

p-Nitrophenyl phosphate disodium salt, hexahydrate (PNPP, MP Biomedicals)

#### Solutions

1 M diethanolamine, pH 9.8. Store in the dark at room temperature.
GAGs of interest 10 mg/ml in H$_2$O
1 mM IPTG in H$_2$O (prepare freshly)
10 mM IPTG in 2× TY

---

[7] Glycerol stocks are used as a backup. When a subsequent selection round fails, use the glycerol stock for a new round of selection.

PBST: 0.05 ml/liter Tween-20 in PBS (PBST)

Secondary antibody solution: Mouse anti-c-*Myc* antibody (9E10, hybridoma culture supernatant[8]) diluted 1:10 with PBST containing 2% (w/v) bovine serum albumin (BSA)

Tertiary antibody solution: Alkaline phosphatase–conjugated rabbit anti-mouse immunoglobulin G (IgG) (DAKO) diluted 1:2000 with PBST containing 2% (w/v) BSA

Substrate solution: Add 1 mg PNPP/ml to 1 *M* diethanolamine solution (prepare freshly)

Washing solution: PBS containing 0.1% (v/v) Tween-20.

*Methods*

DAY 3

3.1. Pick individual bacterial clones with sterile toothpicks from the serial dilution plates of the last two selection rounds (rounds 3 and 4) (section "C. Selection of Phages Binding to GAGs," Day 3, Step 3.1), and inoculate individual wells of a 96-well master block, filled with 500 $\mu$l of 2× TY containing 100 $\mu$g/ml ampicillin and 1% glucose. Repeat this for 94 individual clones. Include a negative control (medium without bacterial clones) and a positive control (a clone that reacts with the antigen). When no positive clone is available, leave the well empty and use a commercially available antibody for further analysis. Secure the lid with tape and grow o/n, while shaking, at 37°.

DAY 4

4.1. Transfer 10 $\mu$l bacterial culture (Day 1, Step 1) to the corresponding wells of a secondary sterile master block containing 1 ml 2× TY containing 100 $\mu$g/ml ampicillin and 0.1% glucose.[9] This plate will be the *induction plate*. Secure the lid with tape, and grow at 37°, while shaking, for 3 h.[10] Do not exceed 200 rpm to prevent contamination.

4.2. When bacterial growth is clearly visible in the induction plate, an absorbance at 600 nm of about 0.9 is reached. For induction, add 100 $\mu$l of 10 m*M* IPTG in 2× TY to each well of the induction plate to reach a final

---

[8] The hybridoma cell line (9E10) (anti-c-*Myc*) is available from the American Type Culture Collection (ATCC). Alternatively, polyclonal rabbit anti-c-*Myc* antibody (A14, Santa Cruz Biotechnology) or purified 9E10 (Sigma) can be used.

[9] Glucose (0.1%) is added to suppress the expression of scFv antibodies until a sufficient number of bacteria is produced for large-scale production of the antibody. The total amount of glucose will be fully metabolized at an absorbance of 0.9 at 600 nm.

[10] Grow the culture in the induction plate for 3 h at 37° while shaking. Bacterial growth should be clearly visible before adding IPTG.

concentration of 1 m$M$ IPTG. Incubate the plate, while gently shaking (do not exceed 200 rpm to prevent contamination) o/n at 30° to induce the production of anti-GAG antibodies.

4.3. Fill a master block with 200 $\mu$l of the o/n culture and add 200 $\mu$l of ice-cold 60% glycerol to all wells to a final concentration of 30% (use a multichannel pipetter). Mix well, and store the plate immediately at −80°. This plate will be the *master plate* and is used as a backup.

DAY 5

5.1. Centrifuge the induction plate (Day 2; Step 2) for 30 min at 5000$g$ to pellet bacteria. The supernatant containing antibodies will be used to screen for positive clones, as described in the following section.

*(F) ELISA Screening for Bacterial Clones Expressing Anti-GAG Antibodies*

*Methods*

DAY 4

4.1. Coat wells from an ELISA plate with 100 $\mu$l of a 10-$\mu$g/ml GAG solution. Incubate o/n at 4°. This step can be performed on Day 2 of the previous section.

DAY 5

5.1. Discard the GAG solution and wash the plate six times with PBST.
5.2. Block the plate with 200 $\mu$l of 2% (w/v) BSA in PBST for 90 min at room temperature.
5.3. Empty the plate and add 100 $\mu$l culture supernatant from the induction plate (section "E. Production of Master Plate and Induction Plate of Phage-Infected Clones," Day 5, Step 5.1) containing soluble antibodies to the corresponding wells of the ELISA plate. Incubate for 2 h at room temperature.[11]
5.4. Discard the culture supernatant and wash the plate six times with PBST.
5.5. Add 100 $\mu$l of the secondary antibody solution to each well and incubate for 1 h at room temperature.
5.6. Discard the secondary antibody solution and wash the plate six times with PBST.

---

[11] Include negative controls (i.e., supernatant without scFv antibodies) and positive controls (if available).

5.7. Add 100 $\mu$l of the tertiary antibody solution to each well and incubate for 1 h at room temperature.

5.8. Discard the tertiary antibody solution and wash the plate six times with PBST. This step is followed by a single wash with 100 $\mu$l of 1 $M$ diethanolamine solution to reduce background signals.

5.9. Add 100 $\mu$l of substrate solution to each well and incubate in the dark at room temperature until color development is clearly visible.

5.10. Read the absorbance on an ELISA reader at 405 nm.

### (G) Large-Scale Production of Periplasmic Fractions Containing Soluble scFv Antibodies

To obtain large amounts of soluble antibodies, periplasmic fractions of bacteria expressing the antibodies are used. Upon induction with IPTG, scFv chains are expressed with the N-terminal *pelB* bacterial leader sequence that targets the chains to the periplasm, where the *pelB* sequence is cleaved off by the enzyme signal peptidase. Within the periplasm, appropriate oxidizing conditions allow the antibody to fold and form functional scFv antibodies. Soluble scFv antibodies are obtained from the bacterial periplasm, through the addition of borate buffer, which selectively releases periplasmic proteins. Antibodies contained in this crude periplasmic fraction can be purified by various ways (e.g., by metal chelate chromatography using a His tag). However, the pHEN-1 vector into which the library is cloned (Fig. 2) yields scFv antibodies with a c-*Myc* tag. To circumvent high background signals in IHC on, for example, tumor tissue (which express c-*Myc*), scFv antibodies are subcloned into pUC119-His-VSV (Fig. 2; Raats, Department of Biochemistry [NWI], Nijmegen Centre for Molecular Life Sciences, Radboud University Nijmegen, Nijmegen, The Netherlands). This vector contains both a His- and a VSV tag for purification and detection. Both the pHEN and the pUC119 vector contain $\beta$-lactamase (ampicillin selection) and a *LacZ* promoter upstream of the antibody gene of interest (IPTG induction of antibody expression).

### Materials

EDTA-free protease inhibitor cocktail tablets (Roche Diagnostics)

Glycerol stock of bacteria producing anti-GAG antibody, stored at $-80°$

Dialysis membrane (Medicell International, molecular weight cutoff $<20$ kDa)

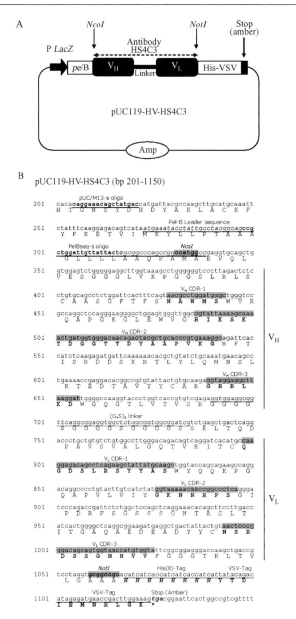

FIG. 2. (A) Schematic representation of plasmid pUC119-HV, harboring the DNA encoding scFv antibody HS4C3 (pUC119-HV-HS4C3). (B) HS4C3-coding sequence cloned into plasmid pUC119-HV (*NcoI-NotI* cloning). Amp, ampicillin resistance–encoding gene; P *LacZ*, *LacZ* promoter; *pel*B, signal peptide of bacterial pectate lyase that targets antibodies to the periplasmic space; (G₄S)₃-linker, a 15 amino acid linker sequence connecting the V<sub>H</sub>

*Solutions*

Ampicillin (100 mg/ml)

0.5 $M$ Boric acid

Borate buffer: Adjust 0.5 $M$ boric acid to pH 8.0 with 0.5 $M$ sodium borate. Store at room temperature. Before use, add 1 EDTA-free protease inhibitor tablet, and 1.6 ml of 5 $M$ NaCl to 16 ml of the borate buffer, and adjust the volume to 40 ml with $H_2O$

LB medium

LB medium containing 1% glucose and 100 $\mu$g/ml ampicillin

0.5 $M$ sodium borate

TYE plates

$2\times$ TY containing 100 $\mu$g/ml ampicillin and 0.1% glucose

$2\times$ TY containing 100 $\mu$g/ml ampicillin and 1% glucose.

*Methods*

DAY 6

6.1. Streak individual clones that express GAG binding antibodies from a master plate on TYE plates and incubate o/n at 37°.

DAY 7

7.1. Inoculate 20 ml $2\times$ TY containing 100 $\mu$g/ml ampicillin and 1% (w/v) glucose with a single colony from a fresh streak of an antibody-expressing clone. Incubate o/n, while shaking, at 37°. Plates can be stored at 4° for up to 1 mo.

DAY 8

8.1. Inoculate 1 liter of $2\times$ TY containing 100 $\mu$g/ml ampicillin and 0.1% (w/v) glucose with 10 ml of the o/n bacterial culture and incubate, while shaking, at 37° until an absorbance at 600 nm of 0.5–0.8 is reached. This takes about 2.5–3.0 h.

8.2. As a backup for additional antibody production, a glycerol stock is prepared. Add 500 $\mu$l of 60% glycerol to 500 $\mu$l of the o/n culture. Mix well, freeze in liquid nitrogen, and store at −80°; 1.5 ml of the remaining o/n culture is used for DNA Miniprep and sequencing as described in the section "Sequence Analysis of Selected Clones."

---

and $V_L$ domains; His-tag, eight histidine residues used for protein purification; VSV-tag, peptide epitope used for immunodetection. Sequences encoding $V_H$ and $V_L$ complementarity determining regions (CDRs) are shaded. Also indicated are the nucleotide sequences of polymerase chain reaction (PCR) primers that can be used for sequencing (pUC/M13-s and PelBseq-s).

8.3. To the 1 liter of induction culture add 1 ml of 1 $M$ IPTG to a final concentration of 1 m$M$ IPTG and incubate the culture, while shaking, for 3 h at 30°.

8.4. Cool the culture on ice for 20 min.

8.5. Collect the bacteria by centrifugation at 5000$g$ for 10 min at 4°. Decant the supernatant, and add 10 ml of ice-cold borate buffer to the bacterial pellet. Borate buffer dissolves the glycocalix. scFv antibodies, which are present in the periplasmic space, become available and can be readily isolated.

8.6. Resuspend the pellet by vortexing and pipetting vigorously. Transfer the mixture to a 50-ml tube and centrifuge at 10,000$g$ for 30 min at 4°. The supernatant is the *periplasmic fraction* containing the soluble scFv antibodies.

8.7. Filter the periplasmic fraction through a 0.45-$\mu$m disposable filter.

8.8. Dialyze the periplasmic fraction o/n against PBS at 4°.

DAY 9

9.1. Divide the periplasmic fraction in aliquots (1 ml). Store at 4° for direct use or freeze and store at −20 to −80°.[12]

*(H) Sequence Analysis of Selected Clones*

For DNA sequence analysis, purified phagemid DNA from an o/n culture of the selected clones is used, which can easily be isolated by a commercially available small-scale plasmid preparation kit. Sequence analysis of the antibody-producing clones is performed with one of the oligonucleotide primers given in the following subsection or any other suitable primer. The sequence can then be analyzed using the alignment program IgBLAST, which can be found on the Internet at http://www.ncbi. nlm.hih.gov/igblast/. $V_H$ CDR3 sequences are especially useful to identify individual clones.

---

[12] The stability of scFv antibodies is highly variable. Whereas some antibodies can be stored at 4° for weeks to months, others stay immunoreactive for only a couple of days. Most antibodies can be stored at −80° for years. Bacterial supernatants containing scFv antibodies can be used for ELISA but are often not suitable for IHC. Periplasmic fractions, containing scFv antibodies, which are more concentrated, are suitable for both. For immunoprecipitation or addition to cell cultures, antibodies may be purified using Protein A (in case of antibodies belonging to the $V_H3$ family) or metal chelating column chromatography.

*Materials*

Plasmid DNA isolation kit (e.g., QIAprep Spin Miniprep Kit, QIAGEN)
Sequencing kit (e.g., ABI Prism)
Sequencing primers (see Fig. 2 for details and location on vector)
Forlinkseq: 5'-GCC ACC TCC GCC TGA ACC-3' (sense)
*PelB*seq: 5'-CCG CTG GAT TGT TAT TAC TC-3' (antisense)

*Solution*

o/n culture of a selected bacterial clone grown at 37°.

*Methods*

8.1. Take 1.5 ml of the o/n culture (large-scale production of periplasmic fractions containing soluble scFv antibodies; Day 8, Step 8.1.) and centrifuge 5 min at 5000$g$.

8.2. Decant the supernatant and use the bacterial pellet for isolation of plasmid DNA according to the manufacturer's protocol.

8.3. Mix suitable amounts of Miniprep DNA and sequencing primer and submit for sequencing according to the manufacturer's protocol.

We have generated a large number of unique scFv antibodies that are specific for unique epitopes on GAG chains. Table I shows a selection of the scFv antibodies obtained. Note that all antibodies are different with respect to their CDR3.

## (I) Evaluation of Specificity of Anti-GAG Antibodies by Immunofluorescence

To investigate GAG specificity of the scFv antibodies, cryosections can be incubated overnight with the glycosidases heparinase-I (0.04 IU/ml in 50 m$M$ NaAc/50 m$M$ Ca[Ac]$_2$, pH 7.0 at 37°) (E.C. 4.2.2.7),[13] heparinase-II (0.04 IU/ml in 50 m$M$ NaPO$_4$, pH 7.1 at 37°), heparinase-III (0.04 IU/ml in 50 m$M$ NaAc/50 mz$M$ Ca[Ac]$_2$, pH 7.0 at 37°) (E.C. 4.2.2.8), or a mix of these enzymes, to digest HS/heparin. CS and DS can be digested from cryosections by o/n incubation with the chondroitin lyases chondroitinase AC, which digests CS (1 IU/ml in 25 m$M$ Tris–HCl, pH 7.5 at 37°), chondroitinase ABC, which digests both CS and DS (1 IU/ml in 25 m$M$ Tris–HCl, pH 8.0 at 37°), or chondroitinase B, which digests DS (1 IU/ml in 25 m$M$ Tris–HCl, pH 7.5 at 37°). As a control, digestion buffer without the enzyme can be used.

---

[13] E.C. is the Enzyme Commission number.

Digested cryosections are incubated with mouse anti-HS stub IgG antibody 3G10 to check for heparinase pretreatment or with mouse anti-CS/DS stub IgG antibody 2B6 to check for chondroitinase pretreatment.

*Materials*

Coverslips
2–5 μm tissue cryosections on glass slides

*Solutions*

Chondroitinase ABC, AC, B (Seikagaku)
Heparinase I, II, III (IBEX)
GAG digestion buffers:
  • Heparinase I, III, digestion buffer: 50 m$M$ NaAc and 50 m$M$, Ca (Ac)$_2$, pH 7.0
  • Heparinase II digestion buffer: 50 m$M$ NaPO$_4$, pH 7.1
  • Chondroitinase AC, B digestion buffer: 25 m$M$ Tris–HCl, pH 7.5
  • Chondroitinase ABC digestion buffer: 25 m$M$ Tris–HCl, pH 8.0
Mouse anti-HS stub IgG antibody 3G10 (Seikagaku)
Mouse anti-CS/DS stub IgG antibody 2B6 (Seikagaku)
Mowiol solution (Calbiochem): (10% [w/v] in 0.1 $M$ Tris–HCl, pH 8.5/ 25% [v/v] glycerol/2.5% [w/v] NaN$_3$)
Primary antibody solution: periplasmic fraction diluted in PBST containing 2% (w/v) BSA, such that the antibody gives maximal specific fluorescence and minimal background staining.
Secondary antibody solution: Mouse anti-c-*Myc* antibody (9E10, hybridoma culture supernatant [see footnote 9] or mouse anti-VSV antibody (P5D4, hybridoma culture supernatant[14]) diluted 1:10 with 2% (w/v) BSA in PBST.
Tertiary antibody solution: Alexa-488–conjugated anti-mouse IgG antibody (Molecular Probes) diluted 1:250 with 2% (w/v) BSA in PBST.

*Methods*

1. Air-dry the cryosections for 30 min before use, to ensure attachment to microslide surface and to preserve the structure.
2. Rehydrate the cryosections with PBS for 5 min.
3. Block free binding sites with 2% (w/v) BSA in PBST for 20 min.
4. Incubate cryosections with the primary antibody solution for 45 min.
5. Remove the primary antibody solution and wash three times for 5 min with PBST.

---

[14] The hybridoma cell line (P5D4 [anti-VSV]) is available from the American Type Culture Collection (ATCC). Alternatively, polyclonal rabbit anti-VSV-G (Sigma) can be used.

6. Incubate cryosections with secondary antibody solution for 30 min.
7. Remove secondary antibody solution and wash three times for 5 min with PBST.
8. Incubate cryosections with tertiary antibody solution for 30 min.
9. Remove tertiary antibody solution and wash three times for 5 min with PBST.
10. Fix cryosections in 96% ethanol for 10 s.
11. Air-dry the sections and use Mowiol solution for embedding; store at −20°.
12. Analyze staining patterns by fluorescence microscopy.[15] Fig. 3 is an example of staining pattern on renal cryosections using scFv antibodies.

To evaluate the specificity of the antibodies, cryosections can be pre-incubated with the glycosidases in the desired buffer for 2 h at the optimal temperature (see earlier discussion).

Cryosections are washed with PBST and blocked with 2% (w/v) BSA in PBST as in Step 4. Proceed with the previous procedure (begin at Step 5). For staining patterns, see Fig. 4

*(J) Characterization of scFv Antibodies by ELISA*

To analyze which chemical groups in HS/heparin or CS/DS are involved in antibody recognition, reactivity with a number of test molecules can be evaluated by additional ELISA experiments: (1) *direct ELISA* in which wells of microtiter plates are coated with the test molecules or (2) *competition ELISA* in which the scFv antibodies and test molecules are simultaneously incubated in wells of a microtiter plate coated with the molecule of interest.

*(J1) Direct ELISA*

*Materials*

See section "F. ELISA Screening of Bacterial Clones Expressing Anti-GAG Antibodies."

*Methods*

DAY 1

1. Coat wells from an ELISA plate with 100 μl of a 10 μg/ml GAG solution. Incubate o/n at 4°.

---

[15] Stained tissue sections can be kept for up to 3 years at 4°. The fluorescent tag (Alexa-488) is very stable, and fading of the signal hardly occurs. Store stained sections at −20°. Background staining can often be eliminated by additional blocking steps with BSA or with 1–5% (v/v) serum from the same species in which the tertiary antibody is raised.

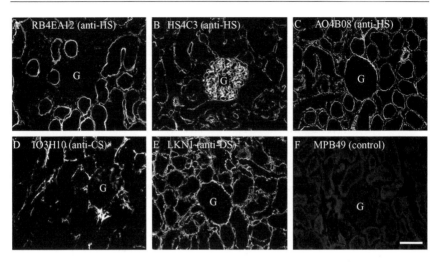

FIG. 3. Immunostaining of normal rat kidney cryosections with six different scFv antibodies. Cryosections were incubated with anti-heparan sulfate (HS) antibodies RB4EA12 (A), HS4C3 (B), AO4B08 (C), anti-chondroitin sulfate (CS) antibody IO3H10 (D), anti-dermatan sulfate (DS) antibody LKN1 (E), and negative control antibody MPB49 (F). G, glomerulus. Note differential staining patterns. Bar = 50 μm.

DAY 2

1. Discard the GAG solution and wash the plate six times with PBST.
2. Block the plate with 200 μl of 2% (w/v) BSA in PBST for 90 min at room temperature.
3. Empty the plate and add 100 μl of primary antibody solution (periplasmic fraction diluted 1:5 with 2% [w/v] BSA in PBST) to the plate. Incubate for 2 h at room temperature.
4. Proceed as in section "F. ELISA Screening for Bacterial Clones Expressing Anti-GAG Antibodies," Day 5, Step 5.4.
5. Analyze the absorbance values to determine apparent affinities/ specificity towards various GAGs.

*(J2) Competition ELISA*

*Materials*

See section "F. ELISA Screening for Bacterial Clones Expressing Anti-GAG Antibodies."

Heparinase III     Chondroitinase B     Chondroitinase ABC

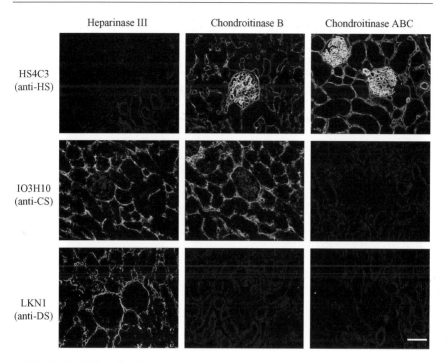

HS4C3 (anti-HS)

IO3H10 (anti-CS)

LKN1 (anti-DS)

FIG. 4. Specificity of anti-glycosaminoglycan antibodies. Normal rat kidney cryosections were treated with heparinase III, chondroitinase B, or chondroitinase ABC. Next, cryosections were stained using anti-heparan sulfate (HS) antibody HS4C3, anti-chondroitin sulfate (CS) antibody IO3H10, or anti-dermatan sulfate (DS) antibody LKN1. Bar = 50 $\mu$m.

*Methods*

DAY 1

1. Coat wells from an ELISA plate with 100 $\mu$l of a 10 $\mu$g/ml GAG solution. Incubate o/n at 4°.

DAY 2

1. Discard the GAG solution and wash the plate six times with PBST.
2. Block the plate with 200 $\mu$l of 2% (w/v) BSA in PBST for 90 min at room temperature.
3. Discard the 2% (w/v) BSA in PBST and add 100 $\mu$l of primary antibody solution (periplasmic fraction diluted 1:5 with 2% [w/v] BSA in PBST) together with the test molecules (serially diluted) to the plate. Incubate for 2 h at room temperature.

4. Proceed as in the section "F. ELISA Screening for Bacterial Clones Expressing Anti-GAG Antibodies," Day 2, Step 4.
5. Analyze the absorbance values to determine apparent affinities/specificity towards various GAGs.

## Acknowledgments

This work was financially supported by the Netherlands Organization for Scientific Research (NWO), grant 902–27–292 (to J. F. M. L. and T. J. M. W.), the Dutch Cancer Society (KWF), grant 2002–2762 (to G. B. t.D), and the International Human Frontier Science Program Organization (HFSP), grant RGP0062/2004-C101 (to G. J. J.). The authors express their gratitude to Dr. G. Winter (Cambridge University, Cambridge, UK) for providing the phage display library. They thank Dr. J. M. H. Raats (Department of Biochemistry, Faculty of Sciences, Nijmegen, The Netherlands) for providing the pUC119-His-VSV vector, and IBEX Technologies (Montreal, PQ, Canada) for providing recombinant heparinase III derived from *Flavobacterium heparinum*.

## References

Bao, X., Pavao, M. S., Dos Santos, J. C., and Sugahara, K. (2005). A functional dermatan sulfate epitope containing iduronate(2-O-sulfate)alpha1–3GalNAc(6-O-sulfate) disaccharide in the mouse brain: Demonstration using a novel monoclonal antibody raised against dermatan sulfate of ascidian. *Ascidia nigra. J. Biol. Chem.* **280,** 23184–23193.

Caterson, B., Christner, J. E., and Baker, J. R. (1983). Identification of a monoclonal antibody that specifically recognizes corneal and skeletal keratan sulfate. Monoclonal antibodies to cartilage proteoglycan. *J. Biol. Chem.* **258,** 8848–8854.

David, G., Bai, X. M., Van der Schueren, B., Cassiman, J. J., and Van den Berghe, H. (1992). Developmental changes in heparan sulfate expression: *In situ* detection with mAbs. *J. Cell Biol.* **119,** 961–975.

Dennissen, M. A., Jenniskens, G. J., Pieffers, M., Versteeg, E. M., Petitou, M., Veerkamp, J. H., and van Kuppevelt, T. H. (2002). Large, tissue-regulated domain diversity of heparan sulfates demonstrated by phage display antibodies. *J. Biol. Chem.* **277,** 10982–10986.

Harrison, J. L., Williams, S. C., Winter, G., and Nissim, A. (1996). Screening of phage antibody libraries. *Methods Enzymol.* **267,** 83–109.

Hoogenboom, H. R., and Chames, P. (2000). Natural and designer binding sites made by phage display technology. *Immunol. Today* **21,** 371–378.

Jenniskens, G. J., Hafmans, T., Veerkamp, J. H., and Van Kuppevelt, T. H. (2002). Spatiotemporal distribution of heparan sulfate epitopes during myogenesis and synaptogenesis: A study in developing mouse intercostal muscle. *Dev. Dyn.* **225,** 70–79.

Jenniskens, G. J., Oosterhof, A., Brandwijk, R., Veerkamp, J. H., and van Kuppevelt, T. H. (2000). Heparan sulfate heterogeneity in skeletal muscle basal lamina: Demonstration by phage display-derived antibodies. *J. Neurosci.* **20,** 4099–4111.

Jenniskens, G. J., Ringvall, M., Koopman, W. J., Ledin, J., Kjellen, L., Willems, P. H., Forsberg, E., Veerkamp, J. H., and Van Kuppevelt, T. H. (2003). Disturbed $Ca^{2+}$ kinetics in N-deacetylase/N-sulfotransferase-1 defective myotubes. *J. Cell. Sci.* **116,** 2187–2193.

Lensen, J. F., Rops, A. L., Wijnhoven, T. J., Hafmans, T., Feitz, W. F., Oosterwijk, E., Banas, B., Bindels, R. J., van den Heuvel, L. P., van der Vlag, J., Berden, J. H., and van Kuppevelt, T. H. (2005). Localization and functional characterization of glycosaminoglycan domains in the normal human kidney as revealed by phage display–derived single chain antibodies. *J. Am. Soc. Nephrol.* **16**, 1279–1288.

Nissim, A., Hoogenboom, H. R., Tomlinson, I. M., Flynn, G., Midgley, C., Lane, D., and Winter, G. (1994). Antibody fragments from a 'single pot' phage display library as immunochemical reagents. *EMBO J.* **13**, 692–698.

Smetsers, T. F., van de Westerlo, E. M., ten Dam, G. B., Clarijs, R., Versteeg, E. M., van Geloof, W. L., Veerkamp, J. H., van Muijen, G. N., and van Kuppevelt, T. H. (2003). Localization and characterization of melanoma-associated glycosaminoglycans: Differential expression of chondroitin and heparan sulfate epitopes in melanoma. *Cancer Res.* **63**, 2965–2970.

Smetsers, T. F., van de Westerlo, E. M., ten Dam, G. B., Overes, I. M., Schalkwijk, J., van Muijen, G. N., and van Kuppevelt, T. H. (2004). Human single-chain antibodies reactive with native chondroitin sulfate detect chondroitin sulfate alterations in melanoma and psoriasis. *J. Invest. Dermatol.* **122**, 707–716.

Smits, N. C., Robbesom, A. A., Versteeg, E. M., Van De Westerlo, E. M., Dekhuijzen, P. N., and Van Kuppevelt, T. H. (2004). Heterogeneity of heparan sulfates in human lung. *Am. J. Respir. Cell. Mol. Biol.* **30**, 166–173.

Sorrell, J. M., Mahmoodian, F., Schafer, I. A., Davis, B., and Caterson, B. (1990). Identification of monoclonal antibodies that recognize novel epitopes in native chondroitin/dermatan sulfate glycosaminoglycan chains: Their use in mapping functionally distinct domains of human skin. *J. Histochem. Cytochem.* **38**, 393–402.

ten Dam, G. B., Hafmans, T., Veerkamp, J. H., and van Kuppevelt, T. H. (2003). Differential expression of heparan sulfate domains in rat spleen. *J. Histochem. Cytochem.* **51**, 727–739.

ten Dam, G. B., Kurup, S., van de Westerlo, E. M., Versteeg, E. M., Lindahl, U., Spillmann, D., and van Kuppevelt, T. H. (2006). 3-O-sulfated oligosaccharide structures are recognized by anti-heparan sulfate antibody HS4C3. *J. Biol. Chem.* **281**, 4654–4662.

ten Dam, G. B., van de Westerlo, E. M., Smetsers, T. F., Willemse, M., van Muijen, G. N., Merry, C. L., Gallagher, J. T., Kim, Y. S., and van Kuppevelt, T. H. (2004). Detection of 2-O-sulfated iduronate and N-acetylglucosamine units in heparan sulfate by an antibody selected against acharan sulfate (IdoA2S-GlcNAc)n. *J. Biol. Chem.* **279**, 38346–38352.

van De Westerlo, E. M., Smetsers, T. F., Dennissen, M. A., Linhardt, R. J., Veerkamp, J. H., van Muijen, G. N., and van Kuppevelt, T. H. (2002). Human single chain antibodies against heparin: Selection, characterization, and effect on coagulation. *Blood* **99**, 2427–2433.

van den Born, J., Gunnarsson, K., Bakker, M. A., Kjellen, L., Kusche-Gullberg, M., Maccarana, M., Berden, J. H., and Lindahl, U. (1995). Presence of N-unsubstituted glucosamine units in native heparan sulfate revealed by a monoclonal antibody. *J. Biol. Chem.* **270**, 31303–31309.

van Kuppevelt, T. H., Dennissen, M. A., van Venrooij, W. J., Hoet, R. M., and Veerkamp, J. H. (1998). Generation and application of type-specific anti-heparan sulfate antibodies using phage display technology. Further evidence for heparan sulfate heterogeneity in the kidney. *J. Biol. Chem.* **273**, 12960–12966.

Vieira, J., and Messing, J. (1987). Production of single-stranded plasmid DNA. *Methods Enzymol.* **153**, 3–11.

Winter, G., Griffiths, A. D., Hawkins, R. E., and Hoogenboom, H. R. (1994). Making antibodies by phage display technology. *Annu. Rev. Immunol.* **12**, 433–455.

# Section III

# Glycosyltransferase Expression Profiling

## [6]   Comprehensive Enzymatic Characterization of Glycosyltransferases with a β3GT or β4GT Motif

*By* Akira Togayachi, Takashi Sato, and
Hisashi Narimatsu

### Abstract

Bioinformatics is a very powerful tool in the field of glycoproteomics, as well as genomics and proteomics. The bioinformatics technique accelerates the comprehensive identification and *in silico* cloning of human glycogenes containing glycosyltransferases, glycolytic enzymes, sugar-nucleotide synthetases, sugar-nucleotide transporters, and so forth. Glycosyltransferase genes play central roles in carbohydrate chain biosynthesis and have been analyzed for their biological functions. At present, over 180 human glycosyltransferases were identified, cloned, and expressed in various expression systems to detect the activity for carbohydrate synthesis. The recombinant proteins for glycosyltransferase were successfully identified for their enzyme activities and substrate specificities. Their substrate specificities were determined using various donor substrates and acceptors. This section reviews the functions, substrate specificities, and enzymatic reactions of glycosyltransferases such as β1,3-glycosyltransferase family and β1,4-glycosyltransferase family.

### In Silico Cloning of Glycosyltransferase Genes

Cloning procedures aided by homology searches in nucleotide sequence databases have greatly accelerated the discovery of novel glycosyltransferase (GT) genes (Narimatsu, 2004). Nevertheless, BLAST-based database searching can be problematic. The rapid increase in the amount of sequence information in the public databases has created a need for more sophisticated bioinformatics software for annotating open reading frames (ORFs). We have developed a novel bioinformatics system for the identification and *in silico* cloning of human glycogenes. EST sequences are automatically assembled with Phrap and the gene region is predicted with GENSCAN software from hits against the genome sequence to obtain the corresponding amino acid sequence of the ORF. From these ORF sequences, candidate GTs are selected based on several parameters. In addition, this system uses profile hidden markov model (HMM) method for clustering GT families. The parameters are described as follows: (1) Motif search is executed to identify candidate sequences that

METHODS IN ENZYMOLOGY, VOL. 416
0076-6879/06 $35.00
DOI: 10.1016/S0076-6879(06)16006-1

possess the motifs of each GT subfamily. These motifs are determined using Multiple EM software for Motif Elicitation. (2) The transmembrane domain, which is localized at the $N$-terminal end, is identified as a region that contains 18–22 hydrophobic amino acid residues. (3) DXD motif, essential for divalent cation binding, is identified. The DXD motif interacts with the phosphate groups of the nucleotide donor through coordination of a divalent cation such as $Mn^{2+}$. (4) The stem region, which is localized between the transmembrane domain and the catalytic domain, is identified. Reliability of candidate genes is scored depending on the number of positive parameters. In addition, the localization of cysteine residues in the sequence is a considerable factor. The results are stored in a database and illustrated graphically. Researchers can browse through these results with their own web browser to easily find new candidate genes. All the present information on human glycogenes is summarized in a database (GlycoGene DataBase: GGDB, http://ggdb.muse.aist.go.jp/).

## Screening Method for Glycosyltransferase Activity

### Enzyme Preparation

For preparation of recombinant enzymes, GT genes are engineered for heterologous expression in mammalian or insect cells as a fusion protein with FLAG Tag at the $N$-terminus. The putative catalytic domain of each GT gene, without the transmembrane domain, is subcloned into pFLAG-CMV1 and/or pFLAG-CMV3 (SIGMA, St. Louis, MO) (Sato et al., 2003) of the mammalian expression vector, or pVL1393-F2, a derivative of pVL1393 (PharMingen, San Diego, CA) (Shiraishi et al., 2001) and pFBIF, which is the destination vector of the Gateway system (Invitrogen, Carlsbad, CA) (Iwai et al., 2002) for the baculovirus expression system (Invitrogen). In the mammalian expression system, each resulting plasmid is transfected into human embryonic kidney (HEK) 293T, Chinese hamster ovary (CHO), COS1, or COS7 cells using Lipofectamine 2000 (Invitrogen) according to the manufacturer's instructions. In both expression systems, each enzyme is purified from culture medium using anti-FLAG M2 agarose affinity gel (Sigma). A 10- to 50-ml volume of culture medium is mixed with anti-FLAG M2 agarose affinity gel and rotated slowly at $4°$ overnight. The gel is then washed two to five times with 50 m$M$ Tris-buffered saline (TBS; 50 m$M$ Tris–HCl, pH 7.4, and 150 m$M$ NaCl) and finally suspended in 100 $\mu$l of TBS. Enzymatic reactions of all GTs are carried out using the suspension as an enzyme source.

### Acceptor Substrates for Glycosyltransferase

The various acceptor substrates, such as monosaccharides, oligosaccharides, glycolipids, and glycoproteins, were purchased from Calbiochem

(La Jolla, CA), Toronto Research Chemicals, Inc. (Ontario, Canada), Seikagaku Kogyo (Tokyo, Japan), TaKaRa (Okaka, Japan), or SIGMA. Oligosaccharides were fluorescently labeled with 2-aminobenzamide (2AB) or pyridylamino-group (PA) and used as acceptors.

## Screening of Substrate Specificities

To determine the substrate specificity of GT, UDP-[$^{14}$C]glucose (Glc), UDP-[$^{14}$C]N-acetylglucosamine (GlcNAc), UDP-[$^{14}$C]galactose (Gal), UDP-[$^{3}$H] N-acetylgalactosamine (GalNAc), UDP-[$^{14}$C]glucuronic acid (GlcA), GDP-[$^{14}$C]mannose (Man), and GDP-[$^{14}$C]fucose (Fuc) (American Radiolabeled Chemicals, Inc., St. Louis, MO) are used as donor substrates. For acceptor substrates, Fucα-para-nitrophenyl (pNp), Glcα-pNp, Glcβ-pNp, GlcNAcα-benzyl (Bz), GlcNAcβ-Bz, Galα-pNp, Galβ-olto- nitrophenyl (oNp), GalNAcα-Bz, GalNAcβ-Bz, GlcAβ-pNp, Manα-pNp, xylose (Xyl)α-pNp, and Xylβ-pNp are used. All acceptor substrates are mixed and dried as preparation of the acceptor mixture for each enzyme reaction. GT reactions are performed in 20 μl of reaction mixture containing 5-μl volume of immobilized enzyme suspension, 10 nmol each of acceptor mixture, various radioactive donor substrates, and 10 mM MnCl$_2$ or MgCl$_2$. Reaction is optimized in various buffer systems (i.e., MES, HEPES, Tris–HCl, and Na-cacodylate buffer at different pH values). After incubation at 37° for various periods, the radioactive reaction products are separated from the free radioactive donor substrates using a Sep-Pak Plus C$_{18}$ Cartridge (Waters, Milford, MA). For conditioning, a Sep-Pak Plus C$_{18}$ Cartridge is washed with 1 ml of methanol and then twice with 1 ml of water. The reaction mixtures are centrifuged and the supernatants are loaded onto the equilibrated cartridge. After washing twice with 1 ml of water, the radioactive reaction products are eluted in 1 ml of methanol. The amount of radioactivity in the eluate is measured by liquid scintillation counter. When the radioactivity is detected in the reaction products, additional enzymatic reactions are performed for individual acceptor substrates to determine the strict substrate specificity. After determination of substrate specificity using monosaccharide, the enzymatic reaction is carried out using larger substrates (i.e., oligosaccharides, glycolipids, glycopeptides, and glycoproteins).

## General Methods for Detection of Reaction Products

### Determination of Products of Glycosyltransferase with Fluorescently Labeled Acceptor Substrates

The GT activity is assayed in various reaction mixtures as described later in this chapter. Various fluorescently labeled oligosaccharides are used as acceptor substrates in the assays. The reaction is terminated by dilution with 80 μl of water and then boiled for 3 min. After centrifugation at 15,000 rpm in a Microfuge tube for 5 min, the supernatant is filtered using an Ultrafree-MC column (Millipore, Bedford, MA). A small aliquot of each supernatant is subjected to high-performance liquid chromatography (HPLC) on a TSK-gel ODS-80T$_S$ QA column (4.6 × 300 mm; Tosoh, Tokyo, Japan). HPLC is also performed on a PALPAK type R column (4.6 × 250 mm; TaKaRa). The reaction products are eluted using suitable buffer conditions. Representative conditions used for the analysis of $\beta$3Gal-T activities on a galacto-LNnT oligosaccharide, are 20 m$M$ ammonium acetate buffer (pH 4.0) containing 7% methanol using a flow rate of 1.0 ml/min, at 50°. Chromatography is monitored with a fluorescence spectrophotometer, JASCO FP-920 (JASCO, Tokyo, Japan). Other suitable conditions involve gradient HPLC using buffer A (5% acetonitrile/0.1% trifluoroacetic acid [TFA]) or buffer B (95% acetonitrile/0.1% TFA). The gradient of 0–100% B in 30–60 min at a flow rate of 1.0 ml/min at 50° is used. The column is monitored with a fluorescence spectrophotometer, JASCO FP-920 (JASCO, Tokyo, Japan).

### Determination of Products of Glycosyltransferase with Radioisotope-Labeled Donor Substrates

We conducted an assay of the incorporation of radioactive sugar into acceptor substrates having hydrophobic functional groups, such as benzyl- or nitrophenyl-labeled acceptor substrates and glycolipid acceptor substrates. The GT activity for these acceptor substrates was assayed in various reaction mixtures containing radioactive donor substrates as described earlier in this chapter. Eluted products from a Sep-Pak Plus C$_{18}$ Cartridge are dried in an evaporator and the residues dissolved in a adequate volume of methanol. The dissolved residues are separated on a HPTLC plate (Silica gel 60 TLC plate, MERCK, Darmstadt, Germany) with a solvent system of chloroform/methanol/0.2% CaCl$_2$ (65:35:8, 65:25:5, or 55:45:10, v/v/v). The radioactivity of each band is quantified with a BAS 2000 or an FLA3000 Imaging Analyzer (Fiji Film, Tokyo, Japan).

### Determination of Products of Glycosyltransferase with Glycoproteins

The transfer of radioactive donor substrate to glycoprotein is performed under various conditions. Glycoprotein acceptors include $\alpha$1-acid glycoprotein (SIGMA), transferrin (SIGMA), ovalbumin (SIGMA), fetuin (SIGMA), ovomucoid (SIGMA), or others. After incubation, the enzyme reaction is terminated by heat treatment at 100° for 3 min and the reaction mixture subjected to sodium dodecylsulfate (SDS)–polyacrylamide gel electrophoresis (PAGE). The gel is then desiccated and the intensity of each radioisotope incorporated glycoprotein band is measured with an FLA3000 Imaging Analyzer (Fuji Film, Tokyo, Japan).

## Determination of Products of Glycosyltransferase by Mass Spectrometry

Reaction mixtures are diluted in 80 ml of $H_2O$ and filtered using Ultrafree-MC column (Millipore). A 50-$\mu$l aliquot of reaction mixture is subjected to HPLC using an ODS-80Ts QA column (4.6 × 250 mm; Tosoh, Tokyo, Japan). The reaction products are eluted with 30 ml of 9% acetonitrile containing 0.1% TFA and $H_2O$ at a flow rate of 1.0 ml/min at 40°. Chromatography was monitored by measuring the absorbance at 210 nm with an SPD-10AVP ultraviolet spectrophotometer (Shimadzu, Kyoto, Japan). Any additional peaks are analyzed by matrix-assisted laser desorption/ionization-time of flight (MALDI-TOF) mass spectrometry (MS) (Reflex IV; Bruker Daltonics, Billerica, MA). Then, 25 pmol of the product is dissolved in 5 $\mu$l of $H_2O$. After adding 45 $\mu$l of 0.1% formic acid and 50 $\mu$l of methanol, the product solution is infused at a rate of 3 ml/min with a capillary voltage of 3 kV. For MALDI-TOF MS analysis, 10 pmol of the product is dried, dissolved in 1 ml of $H_2O$, and then analyzed. However, this detection method is more suitable for qualitative rather than quantitative analysis.

### The Family of $\beta$1,3-Glycosyltransferases

This section mainly reviews a set of GTs transferring sugars via a $\beta$1,3-linkage ($\beta$1,3-GT; $\beta$3GalT, $\beta$1,3-N-acetyl glucosaminyltransferase; $\beta$3GnT, and $\beta$1,3-N-acetylgalactosaminyltransferase; $\beta$3GalNAcT), among the many GTs.

### Cloning of β3Gal-Ts and Their Substrate Specificity

The $\beta$3Gal-T, $\beta$3GalNAc-T, and $\beta$3Gn-T groups have been isolated and reported to date form a gene family with a shared motif ($\beta$3GT motif) (Fig. 1). In 1994, Sasaki et al. first isolated the $\beta$3Gal-T1 gene by expression cloning. Subsequent database expansion led to the report of human and mouse $\beta$3Gal-T2 through $\beta$3Gal-T4 with similar sequences (Amado et al., 1999; Hennet et al., 1998). $\beta$3Gal-T4 was shown to be the human counterpart of rat GD1b/GM1/GA1 synthase, which was cloned by Miyazaki et al. (1997). Using degenerate primers encoding a shared motif, we successfully isolated the $\beta$3Gal-T5 gene by polymerase chain reaction (PCR) cloning (Isshiki et al., 1999). $\beta$3Gal-T5, which is specifically expressed in the digestive organs, synthesizes a type 1 sugar chain, including CA19–9 (sialyl Lewis a). It is now established that $\beta$3Gal-T5 is very important for the expression of these carbohydrate chains in cell lines derived from digestive tract cancers. In addition, $\beta$3Gal-T5 has elongation activity toward core 3 O-glycan structures (Zhou et al., 1999a) and is known as SSEA-3 synthase (Gb5Cer synthase) (Zhou et al., 2000). $\beta$3Gal-T6 has been reported to synthesize the proteoglycan core structure (Gal$\beta$1→3Gal$\beta$1–4Xyl) (Bai et al., 2001).

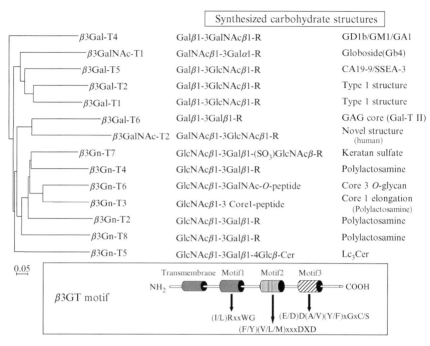

FIG. 1. A phylogenetic tree of human $\beta$1,3-glycosyltransferases. A phylogenetic tree of human $\beta$3-glycosyltransferases ($\beta$3GT) containing $\beta$3Gal-Ts, $\beta$3GalNAc-Ts, and $\beta$3Gn-Ts was constructed by means of the neighbor joining method based on the amino acid sequences. The branch length indicates the evolutionary distance between different sequences. The three $\beta$3GT motifs are *boxed*. The carbohydrate structures, which were synthesized by each enzyme, are shown on the right of the figure.

## Cloning of β3Gn-Ts and Their Substrate Specificity

The first $\beta$3Gn-T, iGn-T, was isolated by the expression cloning method (Sasaki *et al.*, 1997). Three additional $\beta$3Gn-Ts, $\beta$3Gn-T2 to $\beta$3Gn-T4, were subsequently isolated based on structural similarity with the $\beta$3Gal-T family (Shiraishi *et al.*, 2001; Zhou *et al.*, 1999b). The secondary $\beta$3Gn-T, $\beta$3Gn-T1, was isolated based on structural similarity with the $\beta$3Gal-T family (Zhou *et al.*, 1999b). We reported three additional $\beta$3Gn-Ts, $\beta$3Gn-T2 to $\beta$3Gn-T4, which are also structurally related to the $\beta$3Gn-T family (Shiraishi *et al.*, 2001). However, the complementary DNA (cDNA) sequence of $\beta$3Gn-T1 has since been corrected (see "Corrections" in *Proceedings of the National Academy of Science* 2000;97:11673) and has been found to be identical to that of $\beta$3Gn-T2. HUGO Gene Nomenclature Committee (HGNC) designated this polylactosamine synthase as $\beta$3Gn-T2. In addition, the iGn-T is also designated as the first $\beta$3Gn-T, $\beta$3Gn-T1.

*In silico* bioinformatics analysis has been used to identify a set of $\beta$3Gn-T genes having sequences similar to those of $\beta$3Gal-Ts (Isshiki *et al.*, 1999;

Malissard *et al.*, 2002; Shiraishi *et al.*, 2001; Zhou *et al.*, 1999b). β3Gn-T2 was found to have strong activity *in vitro* against oligosaccharide substrates with polylactosamine structures, suggesting that it is the main polylactosamine synthase (Shiraishi *et al.*, 2001; Togayachi *et al.*, 2001). However, β3Gn-T3 and β3Gn-T4 have weak polylactosamine synthase activity. The differential role played by these enzymes in the body is still unclear. Yeh *et al.* (2001) reported that β3Gn-T3 is an enzyme responsible for the further elongation of the core 1 *O*-glycan structure. However, details of the substrate specificity of β3Gn-T4 are still unclear. We have successfully cloned the genes for Lc$_3$Cer synthase (β3Gn-T5) and core 3 *O*-glycan synthase (β3Gn-T6) (Iwai *et al.*, 2002) and the recombinant enzymes have been isolated and characterized. The expression of the HNK-1 and Lewis x antigens on the lacto/neolacto series of glycolipids has been shown to be developmentally and tissue specifically regulated by β3Gn-T5 (synthesizing the lacto/neolacto series of glycolipids) (Togayachi *et al.*, 2001). These antigens appear to be closely involved in important functions such as intercellular recognition for the extension of neurons and the development and differentiation of blood cells. We reported that the newly isolated β3Gn-T6 synthesizes the core 3 *O*-glycan structure (Iwai *et al.*, 2002). We speculate that this enzyme plays an important role in the synthesis and function of mucin *O*-glycan in the digestive organs. In addition, the expression of β3Gn-T6 was markedly down-regulated in gastric and colorectal carcinomas (Iwai *et al.*, 2005). Expression of β3Gn-T7 was reported to be down-regulated upon malignant transformation (Kataoka and Huh, 2002). It is thought that β3Gn-T7 may be responsible for synthesizing the poly-lactosamine structure in the cell (Kataoka and Huh, 2002). Seko and Yamashita (2004) have shown that β3Gn-T7 catalyzes the transfer of GlcNAc to Galβ1→4(SO$_3^-$→6)GlcNAcβ1→3Galβ1→4(SO$_3^-$→6)GlcNAc (L2L2 oligosaccharide) of keratan sulfate (Seko and Yamashita, 2004). β3Gn-T8 transfers GlcNAc to the non-reducing terminus of the Galβ1–4GlcNAc of tetra anntenary *N*-glycan *in vitro* (Ishida *et al.*, 2005). Intriguingly, the level of β3Gn-T8 transcript is significantly higher in colorectal cancer tissues than in normal tissue. Seko and Yamashita (2005) have shown that the complex of β3Gn-T2 and β3Gn-T8 exhibits potent polylactosamine synthesis activity against tetra-antenarial *N*-glycans.

## Cloning of β3GalNAc-Ts and Their Substrate Specificity

β3GalNAc-T1 was initially cloned as β3Gal-T3 by an *in silico* method (Amado *et al.*, 1999; Hennet *et al.*, 1998). Later, however, Okajima *et al.* (2000) reported β3Gal-T3 to be identical to β3GalNAc-T (β3GalNAc-T1), which had been cloned as globoside (Gb4) synthase.

Hiruma *et al.* (2004) cloned the GT capable of synthesizing the Gal-NAcβ1→3GlcNAc structure and designated it β3GalNAc-T2. This structure has not been found in humans and other mammals, although it has been reported in insect glycolipids. Nevertheless, *in vitro* analysis of this enzyme detected the GalNAc$\beta$1→3GlcNAc synthesizing activities on both *O*- and *N*-glycan, suggesting that this carbohydrate structure would be present in humans. Although the $\beta$3GalNAc-T2 gene is expressed in almost all tissues of the body, expression is particularly strong in the testis where it is developmentally regulated, attracting a great deal of interest.

*Assaying for $\beta$3Gal-T Activity*

The basic reaction mixture for assaying $\beta$3Gal-T activity contains 14 m$M$ HEPES buffer (pH 7.4), 75 $\mu M$ UDP-Gal, 11 $\mu M$ MnCl$_2$, 0.01% Triton X-100, and 25 $\mu M$ of various acceptor substrates and the enzyme source. After incubation at 37° for 1–16 h, the enzymatic reactions are terminated by boiling for 3 min. The product is analyzed by various techniques as described earlier in this chapter.

*Assaying of $\beta$3Gn-T Activity*

The $\beta$3Gn-T activity is assayed in a 20-$\mu l$ reaction mixture containing 150 m$M$ sodium cacodylate buffer (pH 7.2), 50 m$M$ UDP-GlcNAc, 10 m$M$ MnCl$_2$, 0.4% Triton CF-54, various acceptor substrates, and the enzyme source. After incubation at 37° for 1–16 h, the reaction is terminated by boiling for 3 min. The product is analyzed by various techniques, as described earlier in this chapter.

The $\beta$3Gn-T activity for glycolipids is assayed in a 25-$\mu l$ reaction mixture containing 150 m$M$ sodium cacodylate buffer (pH 7.2), 480 $\mu M$ UDP-GlcNAc, 175 nCi UDP-[$^{14}$C]GlcNAc (Amersham Biosciences, Piscataway, NJ), 10 m$M$ MnCl$_2$, 0.4% Triton CF-54, 10 nmol glycolipid acceptor substrate, and the enzyme source.

*Assaying of $\beta$3GalNAc-T Activity*

The basic reaction mixture for assaying GalNAc-T activity contains 14 m$M$ HEPES buffer, pH 7.4, an appropriate concentration of UDP-Gal NAc, 10 m$M$ MnCl$_2$, 0.4% Triton CF-54, a suitable amount of acceptor substrate, and the purified enzyme. After incubation at 37° for 1–16 h, the product is analyzed by various techniques, as described earlier in this chapter.

The Family of $\beta$1,4-Glycosyltransferases

The $\beta$1,4-glycosyltransferase gene family members, including seven $\beta$1,4-galactosyltransferases, two $\beta$1,4-*N*-acetylgalactosaminyltransferases ($\beta$4GalNAc-T), and four chondroitin synthases, all share a common WGX EDD/V/W sequence known as a $\beta$4GT motif (Fig. 2). Ramakrishnan and Qasba (2001) performed X-ray crystallographic analysis of $\beta$1,4Gal-T1 and reported that this motif is involved in the binding to a substrate. In this

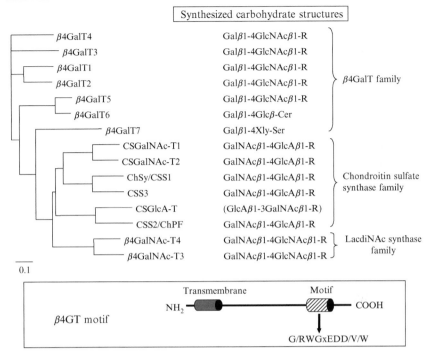

FIG. 2. A phylogenetic tree of human $\beta$1,4-glycosyltransferases. A phylogenetic tree of human $\beta$4-glycosyltransferases ($\beta$4GT) containing $\beta$4Gal-Ts, $\beta$4GalNAc-Ts, and chondroitin sulfate synthase was constructed by means of the neighbor joining method based on the amino acid sequences. The branch length indicates the evolutionary distance between different sequences. The $\beta$4GT motifs are boxed. The carbohydrate structures, which were synthesized by each enzyme, are shown on the right of the figure.

section, we describe methods for studying the enzymatic reactions of $\beta$4GalNAc-Ts.

### $\beta$4GalNAc-T3 and -T4

$\beta$4GalNAc-T3 and -T4, members of the $\beta$4GalNAc-T gene family, transfer GalNAc toward GlcNAc$\beta$- with $\beta$1,4-linkage synthesizing the GlaNAc$\beta$1, 4GlcNAc (LacdiNAc, LDN) structure on $N$-glycans and $O$-glycans *in vivo* (Gotoh *et al.*, 2004; Sato *et al.*, 2003). LDN is known to have a unique terminal structure of $N$-glycans on glycoprotein hormones such as lutropin (Pierce and Parsons, 1981) and thyrotropin (Hiyama *et al.*, 1992). Sulfated LDN is reported to function in the clearance of glycoprotein hormones (Fiete *et al.*, 1991). $\beta$4GalNAc-T3 and -T4 are type II membrane proteins that have a transmembrane region at the $N$-terminus and a predicted catalytic domain at the $C$-terminal region containing a $\beta$1,4-linkage motif and a DLH motif (akin to DXD). The expression patterns of $\beta$4GalNAc-T3 and -T4 differ markedly: $\beta$4GalNAc-T3 is expressed in the stomach and colon, whereas $\beta$4GalNAc-T4 is highly expressed in brain and ovary (Gotoh *et al.*, 2004; Sato *et al.*, 2003). These findings suggest that $\beta$4GalNAc-T3 and -T4 synthesize LDN structures on different carrier proteins in different tissues.

### Preparation of Recombinant Enzymes

Recombinant $\beta$4GalNAc-T3 and -T4 are expressed in HEK293T cells as soluble enzymes with FLAG tag. The predicted catalytic domain of $\beta$4GalNAc-T3 is amplified by PCR from human stomach cDNA as template. The amplified fragment is then inserted into pFLAG-CMV1 to generate the expression vector, pFLAG-CMV1-$\beta$4GalNAc-T3 (Sato *et al.*, 2003). $\beta$4GalNAc-T4 is also amplified from human brain cDNA to construct the expression vector, pFLAG-CMV1-$\beta$4GalNAc-T4 (Gotoh *et al.*, 2004). The resulting plasmids are transfected into HEK293T cells and recombinant enzymes prepared as described earlier.

### Assaying of $\beta$4GalNAc-T Activity

For the reaction in the GalNAc-T assay, 50 m$M$ MES buffer (pH 6.5) containing 0.1% Triton X-100, 1 m$M$ UDP-GalNAc, 10 m$M$ MnCl$_2$, and a suitable concentration of acceptor substrate are used. A 5-$\mu$l volume of enzyme suspension is used in a 20-$\mu$l reaction mixture, which is incubated at 37° for various periods of time. After incubation, the reaction is stopped by heating at 100°. The following monosaccharides and oligosaccharides can be used as acceptor substrate for $\beta$4GalNAc-Ts: GlcNAc$\beta$-Bz, GlcNAc$\beta$1–6 (Gal$\beta$1–3)GalNAc$\alpha$-*p*Np (core 2), GlcNAc$\beta$1–3GalNAc$\alpha$-*p*Np (core 3), GlcNAc$\beta$1–6GalNAc$\alpha$-*p*Np (core 6), bi-antenarial structure of $N$-glycan with GlcNAc at the non-reducing terminus and its derivative (TaKaRa, Shiga, Japan). A glycoprotein, such as asialofetuin or agalactofetuin, can also be used.

### References

Amado, M., Almeida, R., Schwientek, T., and Clausen, H. (1999). Identification and characterization of large galactosyltransferase gene families: Galactosyltransferases for all functions. *Biochim. Biophys. Acta* **1473,** 35–53.

Bai, X., Zhou, D., Brown, J. R., Crawford, B. E., Hennet, T., and Esko, J. D. (2001). Biosynthesis of the linkage region of glycosaminoglycans: Cloning and activity of galactosyltransferase II, the sixth member of the $\beta$1,3-galactosyltransferase family ($\beta$3GalT6). *J. Biol. Chem.* **276,** 48189–48195.

Fiete, D., Srivastava, V., Hindsgaul, O., and Baenziger, J. U. (1991). A hepatic reticuloendothelial cell receptor specific for SO$_4$-4GalNAc $\beta$1,4GlcNAc $\beta$1,2Man $\alpha$ that mediates rapid clearance of lutropin. *Cell* **67,** 1103–1110.

Gotoh, M., Sato, T., Kiyohara, K., Kameyama, A., Kikuchi, N., Kwon, Y. D., Ishizuka, Y., Iwai, T., Nakanishi, H., and Narimatsu, H. (2004). Molecular cloning and characterization of $\beta$1,4-$N$-acetylgalactosaminyltransferases IV synthesizing N,N'-diacetyllactosediamine. *FEBS Lett.* **562**, 134–140.

Hennet, T., Dinter, A., Kuhnert, P., Mattu, T. S., Rudd, P. M., and Berger, E. G. (1998). Genomic cloning and expression of three murine UDP-galactose: $\beta$-$N$-acetylglucosamine $\beta$1,3-galactosyltransferase genes. *J. Biol. Chem.* **273**, 58–65.

Hiruma, T., Togayachi, A., Okamura, K., Sato, T., Kikuchi, N., Kwon, Y. D., Nakamura, A., Fujimura, K., Gotoh, M., Tachibana, K., Ishizuka, Y., Noce, T., Nakanishi, H., and Narimatsu, H. (2004). A novel human $\beta$1,3-$N$-acetylgalactosaminyltransferase that synthesizes a unique carbohydrate structure, GalNAc$\beta$1–3GlcNAc. *J. Biol. Chem.* **279**, 14087–14095.

Hiyama, J., Weisshaar, G., and Renwick, A. G. (1992). The asparagine-linked oligosaccharides at individual glycosylation sites in human thyrotrophin. *Glycobiology* **2**, 401–409.

Ishida, H., Togayachi, A., Sakai, T., Iwai, T., Hiruma, T., Sato, T., Okubo, R., Inaba, N., Kudo, T., Gotoh, M., Shoda, J., Tanaka, N., and Narimatsu, H. (2005). A novel $\beta$1,3-$N$-acetylglucosaminyltransferase ($\beta$3Gn-T8), which synthesizes poly-$N$-acetyllactosamine, is dramatically upregulated in colon cancer. *FEBS Lett.* **579**, 71–78.

Isshiki, S., Togayachi, A., Kudo, T., Nishihara, S., Watanabe, M., Kubota, T., Kitajima, M., Shiraishi, N., Sasaki, K., Andoh, T., and Narimatsu, H. (1999). Cloning, expression, and characterization of a novel UDP-galactose:$\beta$-$N$-acetylglucosamine $\beta$1,3-galactosyltransferase ($\beta$3Gal-T5) responsible for synthesis of type 1 chain in colorectal and pancreatic epithelia and tumor cells derived therefrom. *J. Biol. Chem.* **274**, 12499–12507.

Iwai, T., Inaba, N., Naundorf, A., Zhang, Y., Gotoh, M., Iwasaki, H., Kudo, T., Togayachi, A., Ishizuka, Y., Nakanishi, H., and Narimatsu, H. (2002). Molecular cloning and characterization of a novel UDP-GlcNAc:GalNAc-peptide $\beta$1,3-$N$-acetylglucosaminyltransferase ($\beta$3Gn-T6), an enzyme synthesizing the core 3 structure of $O$-glycans. *J. Biol. Chem.* **277**, 12802–12809.

Iwai, T., Kudo, T., Kawamoto, R., Kubota, T., Togayachi, A., Hiruma, T., Okada, T., Kawamoto, T., Morozumi, K., and Narimatsu, H. (2005). Core 3 synthase is down-regulated in colon carcinoma and profoundly suppresses the metastatic potential of carcinoma cells. *Proc. Natl. Acad. Sci. USA* **102**, 4572–4577.

Kataoka, K., and Huh, N. H. (2002). A novel $\beta$1,3-$N$-acetylglucosaminyltransferase involved in invasion of cancer cells as assayed *in vitro. Biochem. Biophys. Res. Commun.* **294**, 843–848.

Malissard, M., Dinter, A., Berger, E. G., and Hennet, T. (2002). Functional assignment of motifs conserved in $\beta$1,3-glycosyltransferases. *Eur. J. Biochem.* **269**, 233–239.

Miyazaki, H., Fukumoto, S., Okada, M., Hasegawa, T., and Furukawa, K. (1997). Expression cloning of rat cDNA encoding UDP-galactose:GD2 $\beta$1,3-galactosyltransferase that determines the expression of GD1b/GM1/GA1. *J. Biol. Chem.* **272**, 24794–24799.

Narimatsu, H. (2004). Construction of a human glycogene library and comprehensive functional analysis. *Glycoconj. J.* **21**, 17–24.

Okajima, T., Nakamura, Y., Uchikawa, M., Haslam, D. B., Numata, S. I., Furukawa, K., Urano, T., and Furukawa, K. (2000). Expression cloning of human globoside synthase cDNAs. Identification of $\beta$3Gal-T3 as UDP-$N$-acetylgalactosamine:globotriaosylceramide $\beta$1,3-$N$-acetylgalactosaminyltransferase. *J. Biol. Chem.* **275**, 40498–40503.

Pierce, J. G., and Parsons, T. F. (1981). Glycoprotein hormones: Structure and function. *Annu. Rev. Biochem.* **50**, 465–495.

Ramakrishnan, B., and Qasba, P. K. (2001). Crystal structure of lactose synthase reveals a large conformational change in its catalytic component, the $\beta$1,4-galactosyltransferase-I. *J. Mol. Biol.* **310**, 205–218.

Sasaki, K., Kurata-Miura, K., Ujita, M., Angata, K., Nakagawa, S., Sekine, S., Nishi, T., and Fukuda, M. (1997). Expression cloning of cDNA encoding a human $\beta$-1,3-$N$-acetylglucosaminyltransferase that is essential for poly-N-acetyllactosamine synthesis. *Proc. Natl. Acad. Sci. USA* **94**, 14294–14299.

Sato, T., Gotoh, M., Kiyohara, K., Kameyama, A., Kubota, T., Kikuchi, N., Ishizuka, Y., Iwasaki, H., Togayachi, A., Kudo, T., Ohkura, T., Nakanishi, H., and Narimatsu, H. (2003). Molecular cloning and characterization of a novel human $\beta$1,4-$N$-acetylgalactosaminyltransferase, $\beta$4GalNAc-T3, responsible for the synthesis of N,N'-diacetyllactosediamine, GalNAc $\beta$1–4GlcNAc. *J. Biol. Chem.* **278,** 47534–47544.

Seko, A., and Yamashita, K. (2004). $\beta$1,3-$N$-Acetylglucosaminyltransferase-7 ($\beta$3Gn-T7) acts efficiently on keratan sulfate-related glycans. *FEBS Lett.* **556,** 216–220.

Seko, A., and Yamashita, K. (2005). Characterization of a Novel Galactose:{$\beta$}1,3-$N$-Acetylglucosaminyltransferase ({$\beta$}3Gn-T8): The Complex Formation of {$\beta$}3Gn-T2 and {$\beta$}3Gn-T8 Enhances Enzymatic Activity. *Glycobiology* **15,** 943–951.

Shiraishi, N., Natsume, A., Togayachi, A., Endo, T., Akashima, T., Yamada, Y., Imai, N., Nakagawa, S., Koizumi, S., Sekine, S., Narimatsu, H., and Sasaki, K. (2001). Identification and characterization of three novel $\beta$1,3-$N$-acetylglucosaminyltransferases structurally related to the $\beta$1,3-galactosyltransferase family. *J. Biol. Chem.* **276,** 3498–3507.

Togayachi, A., Akashima, T., Ookubo, R., Kudo, T., Nishihara, S., Iwasaki, H., Natsume, A., Mio, H., Inokuchi, J., Irimura, T., sasaki, K., and Narimatsu, H. (2001). Molecular cloning and characterization of UDP-GlcNAc:lactosylceramide $\beta$1,3-$N$-acetylglucosaminyltransferase ($\beta$3Gn-T5), an essential enzyme for the expression of HNK-1 and Lewis X epitopes on glycolipids. *J. Biol. Chem.* **276,** 22032–22040.

Yeh, J. C., Hiraoka, N., Petryniak, B., Nakayama, J., Ellies, L. G., Rabuka, D., Hindsgaul, O., Marth, J. D., Lowe, J. B., and Fukuda, M. (2001). Novel sulfated lymphocyte homing receptors and their control by a Core1 extension $\beta$1,3-$N$-acetylglucosaminyltransferase. *Cell* **105,** 957–969.

Zhou, D., Berger, E. G., and Hennet, T. (1999a). Molecular cloning of a human UDP-galactose:GlcNAc$\beta$1,3GalNAc $\beta$1,3 galactosyltransferase gene encoding an $O$-linked core3-elongation enzyme. *Eur. J. Biochem.* **263,** 571–576.

Zhou, D., Dinter, A., Gutierrez Gallego, R., Kamerling, J. P., Vliegenthart, J. F., Berger, E. G., and Hennet, T. (1999b). A $\beta$-1,3-$N$-acetylglucosaminyltransferase with poly-$N$-acetyllactosamine synthase activity is structurally related to $\beta$-1,3-galactosyltransferases. *Proc. Natl. Acad. Sci. USA* **96,** 406–411.

Zhou, D., Henion, T. R., Jungalwala, F. B., Berger, E. G., and Hennet, T. (2000). The $\beta$1,3-galactosyltransferase $\beta$3GalT-V is a stage-specific embryonic antigen-3 (SSEA-3) synthase. *J. Biol. Chem.* **275,** 22631–22634.

# [7] Autofluorescent Proteins for Monitoring the Intracellular Distribution of Glycosyltransferases

*By* Assou El-Battari

## Abstract

To analyze the Golgi compartmentalization of glycosyltransferases (GTs), we generated versions of several enzymes fused to either the enhanced green fluorescent protein (EGFP) or the red fluorescent protein from *Discosoma* sp. reef coral (DsRed2) and examined their intracellular distribution by confocal fluorescence microscopy in living cells. In a previous work, we have shown that the N-terminal peptides of GTs, encompassing the cytosolic and

METHODS IN ENZYMOLOGY, VOL. 416                    0076-6879/06 $35.00
          DOI: 10.1016/S0076-6879(06)16007-3

the transmembrane domains (CTDs), can serve as Golgi-targeting signals to localize the enzymes to their corresponding compartments within the Golgi apparatus (Zerfaoui *et al.*, 2002). Using sialyl-Lewis x synthesis and selectin binding as functional assays, we show here that by swapping CTDs between GTs, it is possible to mislocalize an enzyme from a Golgi compartment to another, thereby altering the overall cellular glycosylation. On the other hand, we demonstrate that the use of an autofluorescent tag such as EGFP offers numerous advantages including the possibility of (1) facilitating sorting by fluorescence-activated cell sorter (FACS) of stably transfected polyclonal cell population, (2) constantly monitoring the expression of the enzymes in live cells, (3) establishing a direct relationship between the fluorescence intensity and the enzyme activities *in vivo* and *in vitro*, (4) establishing a visual relationship between function and intracellular distribution of a given GT, as well as co-localization with cognate protein acceptors by confocal microscopy, and (5) detecting proteins on blots with highly sensitive commercially available antibodies.

## Overview

The Golgi GTs are type II transmembrane enzymes catalyzing the transfer of monosaccharides to proteins and lipids and sharing some domain features, such as a single transmembrane domain flanked by a short amino-terminal domain (the so-called *cytoplasmic tail*) and a large carboxy-terminal catalytic domain oriented to the lumen of the Golgi apparatus (Colley, 1997; Paulson and Colley, 1989). Extensive studies of the *N*- and *O*-glycosylation pathways in mammalian cells have revealed that glycans are synthesized by an ordered series of sugar transfer reactions, so the final structure of the oligosaccharide is a result of a narrow acceptor specificity. Synthesis of such specific oligosaccharide structures depends not only on the expression level of a given GT but also on how these enzymes segregate into distinct compartments of the Golgi complex. In this regard, GTs are thought to be compartmentalized in the same order in which they act, so enzymes acting early in glycan biosynthesis are localized to *cis* and *medial* compartments, whereas enzymes acting later in the biosynthetic pathway tend to co-localize in the *trans*-Golgi cisternae and the *trans*-Golgi network.

Glycosylation of proteins and lipids plays important roles in many physiological and pathological contexts involving cell–cell and cell–extracellular matrix interactions (Hakomori, 2002). In regard to selectin-mediated adhesion, which is crucial during inflammation and may contribute to metastatic spread of cancers, the synthesis of the major selectin ligand sialyl–Lewis x (sLe$^x$) and its sulfated variants is a result of competition between several GTs of the *O*-glycosylation pathway (Fukuda, 2002). In fact, the synthesis of sLe$^x$ on core-2–branched $O$-glycans of the P-selectin glycoprotein ligand-1 (PSGL-1) is a result of competition between the C2GnT-I, the $\beta$3GlcNAc T-3, and the ST3Gal-I, towards the same oligosaccharide acceptor Gal$\beta$1, 3GalNAc$\alpha$-O-R (Mitoma et al., 2003). This competition can be explained by either substrate specificity, the Golgi compartmentalization of these enzymes, or a combination of these two mechanisms.

We herein report on different ways to simultaneously monitor the localization of a GT of interest and to measure its activity in live cells. Mislocalization studies are presented as a tool for studying the Golgi compartmentalization of GTs.

## Construction of EGFP- and DsRed-Tagged Glycosyltransferases and Their Golgi-Targeting Determinants

The vectors pcDNA3/EGFP and pcDNA3/DsRed2 were constructed by excising the fluorescent protein genes from pEGFP-N1 or pDsRed2-N1 plasmids (Clontech) with *Eco*RI and *Not*I and ligating the excised complementary DNAs (cDNAs) into *Eco*RI/*Not*I–digested pCDNA3.1(+) (Invitrogen). The different GT constructs and their EGFP-tagged variants have been described previously (El-Battari et al., 2003). For mislocalization experiments, the luminal portions of GTs (comprising the stem region and the catalytic domain) were fused with the cytosolic tail and transmembrane domain (denoted *CTD*) of either C2GnT-I or $\beta$1,3GnT-III. For this purpose, two *Age*1 sites encompassing the sequence coding for the luminal portion of GTs were generated by polymerase chain reaction (PCR) and the resulting DNA was inserted in the unique *Age*1 site between the CTD and EGFP sequences. A flowchart of this construction is presented in Fig. 1. Briefly, the construction of fragments encoding the CTD of C2GnT-I (from aa 1 to 32) or $\beta$1,3GnT-III (from aa 1 to 29) were generated by PCR using the sense and antisense primers as follows: for C2GnT-I, sense, 5′-cttaac**ggatcc**gccaccATG CTGAGGACGTTG-3′(1–15); for antisense, 5′-ggtggcg**accggt**GGTAAAAC GGAGAAGGTGAT-3′(79–96); for $\beta$1,3GnT-III, sense, 5′-cttaac**ggatcc**gcc accATGAAGTATCTCCGG-3′(1–15), and antisense, 5′-ggtggcg**accggt**GG CACTAGCAGACTGAAGAG-3′(71–85). (Note that capital letters represent the sequence DNA of GTs [bold GG belong to the multicloning site of the original plasmid pEGFP-N1] and numbers in parentheses correspond to nucleotide positions.) The restriction sites *Bam*H1 (**ggatcc**) and *Age*1 (**accggt**) are represented in bold letters and italicized letters represent the consensus Kozak sequence (Kozak, 1991). The resulting DNAs were gel-purified, restricted by *Bam*H1 and *Age*1, and ligated into *Bam*H1/*Age*1–digested pcDNA3/EGFP. We have previously demonstrated the Golgi-targeting efficiency and compartment specificity of CTDs from various GT by confocal

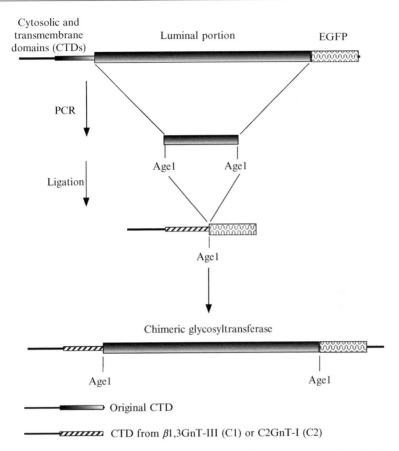

FIG. 1. Schematic representation of the strategy used for the construction of chimeric DNAs coding for the cytosolic and transmembrane domains (CTDs) of either $\beta$1,3GnT-III (C1) or C2GnT-I (C2) fused to the luminal portions of glycosyltransferases. All proteins are C-terminally conjugated to enhanced green fluorescent protein (EGFP).

microscopy, after transiently transfecting CHO cells that express a FLAG-tagged full-length version of the corresponding enzymes. The CTDs used here were found to co-localize with their full-length counterparts (data not shown, see also Zerfaoui *et al.*, 2000). All DNAs produced by PCR were verified by automated DNA sequencing.

*Technical Considerations*

1. In our hands, the best level of expression and stability of GTs were obtained by inserting the cDNAs, not in the original vector pEGFP-N1, but in pCDNA3.1(+)/EGFP, using the *Age*1 site upstream to the coding

sequence of EGFP (or DsRed2, for some GTs). Therefore, all GTs studied here are C-terminally conjugated to the EGFP or DsRed2 through the fusion peptide Pro-Pro-Leu-Ala-Thr, derived from the multicloning site of pEGFP-N1 vector, thus resulting in the fusion protein GTs-$X$-Pro-Pro-Leu-Ala-Thr-**Met**-EGFP(or DsRed2) (where $X$ is the penultimate amino acid of a GT and **Met** is the start codon of EGFP or DsRed2). Using any other combination led to either low expression or clones that do not stably express the transgene.

2. We also found that in most cases, the stem region is not necessary for targeting a reporter protein to the Golgi. However, we found that the Golgi targeting signal (GTsign) can be restricted to the CTDs, although the cytosolic domain alone can serve as a GTsign, provided that a hydrophobic sequence of 18–22 aa is present at its C-terminal end. In general, we found that the Golgi compartment specificity of a given GT is highly dependent on its own CTD (Zerfaoui et al., 2000).

*Stable Transfection of CHO Cells with pcDNA3/GT-EGFP Constructs*

1. The Chinese hamster ovary cells CHO-K1 (ATCC, Rockville, MD) were cultured in Ham F-12 medium (Cambrex, Verviers, Belgium) supplemented with 10% fetal calf serum (FCS) and 100 U/ml penicillin and 100 $\mu$g/ml streptomycin (complete media).

2. Transfer 100 $\mu$l OptiMEM (Invitrogen, Paisley, Scotland) to a 5-ml round-bottom tube polystyrene tube (Falcon) and add 1 $\mu$g DNA and 6 $\mu$l reagent Plus (Lipofectamine transfection kit, Invitrogen) and gently tap the tube five to eight times. Incubate 15 min at room temperature.

3. Dilute 4 $\mu$l Lipofectamine with 100 $\mu$l OptiMEM in another tube and add the mixture dropwise to the tube containing DNA and Reagent Plus. Care is taken not to dispense any Lipofectamine on the walls, because it may hinder efficient transfection. Leave for 15–20 min at room temperature for liposomes/DNA complex to form.

4. Remove the media from culture dishes and wash cell monolayers three times with 1 ml OptiMEM. Add 800 $\mu$l OptiMEM/well and dropwise add the 200 $\mu$l of DNA/Lipofectamine mixture. Incubate between 3 and 24 h at 37°.

5. Replace the transfection medium with complete Ham F-12 medium and incubate cells for an additional 48 h.

Visualization and Isolation of GT-EGFP–Expressing Clones

1. To visualize GT-EGFP in live cells, we used an inverted fluorescence microscope (Zeiss Axiovert 200, Göttingen, Germany). This microscope is equipped with a sensitive polychrome camera (AxioCamMRC, Zeiss) and filter sets for fluorescent probes such as EGFP and its variants (BFP, CFP,

YFP), as well as FITC, rhodamine, DsRed, and DAPI. The MetaMorph software (version 6.0) is used for image acquisition and image analysis. Using this microscope, it is possible to detect a typical Golgi staining as early as 5 h post-transfection (although only a limited number of cells exhibit such a staining). Twenty-four hours later, about 30–50% cells exhibit a strong Golgi staining.

2. The pcDNA3/GT-EGFP plasmids carry the neomycin (G418) resistance gene that allows isolation of drug-resistant clones. Forty-eight hours after transfection, cells are harvested by trypsin/EDTA, resuspended in 1 ml complete Ham F-12 medium (per well), and 200 $\mu$l is transferred to a 150-mm dish containing 20 ml medium supplemented with 1 mg/ml G418. The medium is changed every 2 days during the first 2-wk selection period and then every 3–4 days as soon as individual clones start being formed. Thanks to EGFP fluorescence, it is easy to monitor clone formation throughout the selection period. In our hands, after 4 wk of selection, positive clones represent 5–10% of total drug-resistant clones. At this stage, one can pick up positive clones using cloning cylinders. However, the clones harvested by this technique are very often still heterogenous including clone-to-clone variations, and one needs a second round of cloning using limiting dilution, for example. Again, the EGFP fluorescence can be used to sort/purify a heterogenous polyclonal cell population, by FACS. Furthermore, using this technique, one can not only isolate fluorescent versus nonfluorescent cells but also sort cells with respect to their fluorescence intensity, which is often indicative of the activity level.

3. The G418-resistant transfectants are harvested as described earlier from 150-mm dishes and passaged into 25-cm$^2$ flasks in complete Ham F-12 medium containing 200 $\mu$g/ml G418 and cells are grown until confluency. Cells are then harvested, pelleted at 600 rpm for 5 min, and gently resuspended as $10^6$ cells/ml in 0.2-$\mu$l sterile filtered phosphate buffered saline (PBS)/bovine serum albumin (BSA). Cell suspension is then applied to FACSCalibur analyzer equipped with an argon ion laser, and EGFP-expressing cells were detected by an excitation/emission of 488/507 nm. A cell population with high EGFP fluorescence (superior to $10^2$ log fluorescence intensity) was collected and recovered in complete medium containing 200 $\mu$g/ml G418 and 20% FCS. Fluorescence microscopic examination of these cell lines is presented in Fig. 2. Note some visual differences in the intracellular distribution of GT-EGFP between the clones.

Immunoblotting of GT-EGFP

1. Cells are detached from the flasks with the dissociation solution (Sigma), pelleted, and solubilized in HEPES 25 m$M$, pH 7.4, containing 20% glycerol, 5 m$M$ MgCl$_2$, and 1% n-octylglucoside at a ratio of 1 volume of wet pellet to 4 volumes of the solubilization buffer. The insoluble fractions

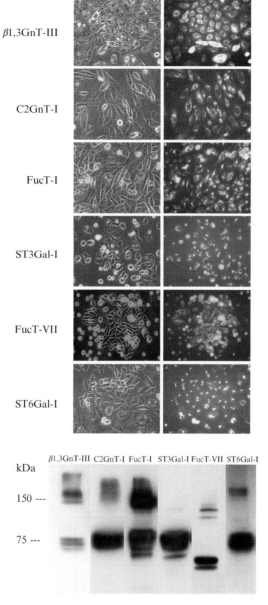

FIG. 2. Visualization of live Chinese hamster ovary (CHO) cells stably expressing different glucosyltransferase enhanced green fluorescent protein (GT-EGFP). Culture
flasks were examined with an inverted fluorescence microscope and a 40× objective. Note
some visual differences in the intracellular distribution of GT-EGFP between cell lines.

are then pelleted at 12,000 rpm. The distribution of EGFP-fused GTs between the soluble and insoluble fraction is monitored by fluorescence spectroscopy using a Perkin Elmer LS3B spectrofluorometer set at an excitation and emission wavelengths of 488 and 507 nm, respectively.

2. Aliquots of 45 $\mu$l are then mixed with 5 $\mu$l of a 10-fold concentrated Laemmli sodium dodecylsulfate (SDS) sample buffer and subjected to 10% polyacrylamide gel electrophoresis (SDS-PAGE) under non-reducing conditions. After electrophoresis, proteins are transferred onto a polyvinylidene diflouride membrane (PVDF, Amersham, Buckinghamshire), stained with Ponceau red, and blocked overnight at 4° in Tris-buffered saline (150 m$M$ NaCl, 20 m$M$ Tris–HCl, pH 7.5) containing 0.05% Tween 20 and 5% nonfat milk. Membranes are then probed with 1/500 anti-EGFP monoclonal antibody (mAb) JL-8 (Clontech) for 1 h followed by goat horseradish peroxidase (HRP)–conjugated anti-mouse antibody (Santa Cruz, Santa Cruz, CA) for an additional hour and visualization of bands is achieved using enhanced chemiluminescence detection kit (Pierce, Rockford, IL).

A blot of GT-EGFP is presented in Fig. 2. Note that under non-reducing conditions, five GTs-EGFP out of the six studied display molecular weights around 75 kDa, yielding a mass of 45–50 kDa for the native enzymes after subtracting the EGFP contribution (25–30 kDa). These values are consistent with amino acid sequences and previously reported data. Among all, FucT-VII has the lowest molecular weight, and ST3Gal-I is the only protein that does not dimerize (see also El-Battari *et al.*, 2003).

*Technical Considerations*

1. It is very convenient to evaluate the solubilization efficiency of EGFP-tagged proteins by fluorescence spectroscopy. This technique allowed us to select the best detergent that yields the maximal extraction from membranes. In this regard, the n-octylglucoside (OG) at a concentration of 1% was found to extract more than 70% of total fluorescence, while CHAPS and Tritons X-100 yielded roughly 50 and 45%, respectively, the remaining fluorescence being in the pellet. We presume that some GTs might partition into Golgi microdomains or "rafts," which are readily extractable by OG but not with Triton X-100 (Baron and Coburn, 2004).

---

For immunoblotting, cells were extracted with n-octylglucoside and subjected to sodium dodecylsulfate–polyacrylamide gel electrophoresis (SDS-PAGE) under non-reducing conditions (in the absence of $\beta$-mercaptoethanol) and proteins were detected using the anti-EGFP monoclonal antibody (mAb) JL-8 (Clontech). Note that under non-reducing conditions, all GTs-EGFP, but FucTVII-EGFP, display molecular weights around 75 kDa, yielding a mass of 45–50 kDa for the native enzymes after substraction of the EGFP contribution (25–30 kDa). The ST3Gal-I is the only protein that does not form dimers.

2. For a given GT, one can establish a proportional relationship between fluorescence and activity. In this regard, before making EGFP-fused GTs, we used to solubilize cells with Triton X-100 to assay for *in vitro* activity of C2GnT-I. Since we found that more than 50% of the C2GnT-I-EGFP is (lost) in the pellet, we shifted to OG, which not only extracts more than 75% protein but the specific activity is generally higher than with Triton (data not shown).

3. We and others had shown that many GTs are processed by proteolysis and released in cell culture media (see El-Battari *et al.*, 2003, as well as references therein). Thanks to EGFP, it is possible to evaluate by spectrofluorometry the amount of released versus cell-associated enzyme. In fact, we have noticed that the cell-to-medium (C/M) ratio of fluorescence is highly dependent on parameters such as cell type, cell density, number of passages, and so on. Therefore, we suggest the C/M ratio to be used as a standardization value to ensure reproducibility of experiments dealing with *in vitro* enzyme assays.

## Confocal Microscopic Examination of Mislocalized FucT-VII

FucT-VII is the latest GT acting on sLe$^x$ precursors by adding an $\alpha(1,3)$-linked fucose to the GlcNAc residue of the sialyl-lactosamine moiety carried by $N$- and $O$-glycoproteins, as well as glycolipids (Lowe, 2003). The PSGL-1, the major P-selectin counter-receptor, interacts with this selectin through a FucT-VII–fucosylated sLe$^x$ determinant borne by a core-2–based oligosaccharide (Moore *et al.*, 1994). Thus, FucT-VII intervenes on PSGL-1 after the $\alpha(2,3)$-sialyl transferases, including the ST3Gal-IV. Accordingly, one would expect FucT-VII to be located after those sialyl transferases in the Golgi, and hence, moving FucT-VII to an earlier compartment such as the *cis/medial* Golgi where C2GnT-I has been located should result in the loss of its function, due to scarcity of sialyl-lactosaminic substrates in this compartment. Based on this assumption and having shown that the CTD of C2GnT-I is specific for the *cis/medial* Golgi compartment, we fused the luminal portion of FucT-VII to the C2GnTI-derived CTD, using the protocol described earlier (see Fig. 3, and the section "Construction of EGFP- and DsRed2-Tagged Glycosyltransferases") and analyzed its intracellular distribution and function. As a control, we fused FucT-I to C2GnT-I CTD, since FucT-I has been reported to localize the *medial*-Golgi (Hartel-Schenk *et al.*, 1991).

1. For this purpose, we used CHO-K1 cells stably expressing C2GnT-I, PSGL-1 (CHO/C2P1), and either FucT-VII, FucT-I, or their corresponding CTD-fused chimeras, all tagged with EGFP. CHO/C2P1 cells were generated by co-transfecting CHO-K1 cells with 5 $\mu$g C2GnT-1/pcDNA1 and 1 $\mu$g pZeoSV/PSGL-1 (Mitoma *et al.*, 2003) and selected in 500 $\mu$g/ml Zeocin for

FucT-VII                          C2-*FucTVII*

FucT-I                            C1-*FucTI*

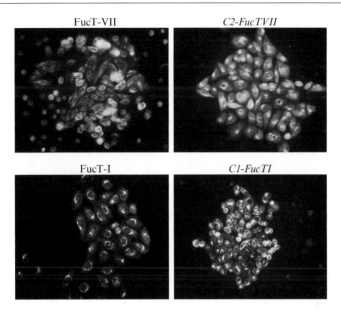

FIG. 3. Altered intracellular distribution of the chimeric fucosyltransferases, made as described in Fig. 1. The luminal portion of FucT-VII was fused to the cytosolic and transmembrane domain (CTD) of C2GnT-I (denoted *C2-FucTVII*) and FucT-I was fused to the CTD of β1,3GnT-III (referred to as *C1-FucTI*). The resulting constructs were stably expressed in CHO cells and examined by fluorescence microscopy as indicated in Fig. 2.

2–3 wk. The Zeocin-resistant cells were then sorted by FACS using the anti-PSGL-1 (PL1) mAb (Immunotech, Marseille, France) and tested for C2GnT-I expression with the anti-C2GnT-I antibody 1719.39 as described (Skrincosky *et al.*, 1997). These cells were then transfected with either FucTVII-EGFP or the chimeric C2-FucTVII-EGFP (CTD of C2GnT-I fused to the luminal portion of FucT-VII, see Fig. 1), together with pSV-Hygro (GenHunter Corp., Nashville, TN) and selected in the presence of 500 μg/ml Hygromycin B and 200 μg/ml Zeocin. The same experiment was performed with FucTI-EGFP and its chimeric form C2-FucT-I-EGFP. The drug-resistant cells were then collected and sorted by FACS on the basis of EGFP as described earlier.

2. Cells were then prepared for confocal fluorescence microscopy. To this end, cell monolayers were fixed in 3.5% paraformaldehyde in PBS for 15 min and permeabilized with 0.5% Triton X-100 in the same buffer containing 1% FCS (PFT medium) during the same period. The rabbit polyclonal antibodies 1719.39, raised against the catalytic domain of C2GnT (Skrincosky *et al.*, 1997), was used to map the intracellular distribution of C2GnT-I. Permeabilized cells were incubated at room temperature for 1 h with the 1719.39 antibody at 1/1000 dilution in PFT medium, rinsed three times in PBS, and

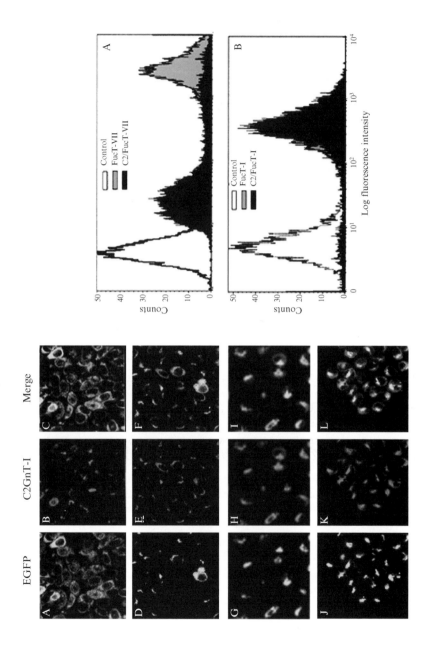

incubated with TRITC-conjugated goat anti-rabbit immunoglobulin G (IgG) (1/200 dilution in PFT medium) for 15 min at room temperature.

3. Intracellular distribution of GT was examined with a Leica confocal microscope (Wetzlar, Germany). Excitation of EGFP was performed using an argon ion laser at 488 nm, and TRITC fluorescence was excited using a green helium neon laser at 543 nm. Images were processed with the Metamorph Imaging system, version 3.5, and transferred to Adobe Photoshop as TIFF files.

Confocal images of co-localization are presented in Fig. 4. FucTVII-EGFP exhibits a broad distribution, which is clearly different from that of C2GnT-I (Fig. 4, compare panels A, B, and C). The CTD of C2GnT-I is able to efficiently mislocalize FucTVII to the same location as C2GnT-I (compare panels D, E, and F). However, and as expected, both FucT-I and its chimeric form C2-FucT-I are localized to the same compartment as C2GnT-I (Fig. 4, from panels G to L). Thus, the C2GnT-derived CTD functions as a *cis/medial*-Golgi determinant because it can target to this compartment a GT (FucT-VII) that is not naturally present at this place.

*Technical Considerations*

The anti-C2GnTI polyclonal antibody was prepared as previously described (Skrinkosky *et al.*, 1997). Briefly, a fusion protein with GT (GT-C2GnTI) was prepared using the bacterial expression vector pGEX-KG, and the recombinant C2GnT-I was eluted from the glutathione-Sepharose column by thrombin cleavage. The eluted protein sample was then resolved by SDS-PAGE and the band corresponding to the C2GnT-I was extracted from polyacrylamide gels by electroelution. The electroeluted fraction was then used for rabbit immunization. The anti-C2GnTI polyclonal antibody is used as a crude rabbit serum preparation. Yet, it has a high titer so it can be used diluted up to 1/5000 dilution, particularly with transfected cells such as CHO or COS series.

---

FIG. 4. Confocal fluorescent micrographs showing the coexpression of C2GnT-I with enhanced green fluorescent protein (EGFP)–tagged proteins, FucT-VII (A, B, C), C2/FucTVII (D, E, F), FucT-I (G, H, I), and C2/FucT-I (J, K, L). Cells were fixed, permeabilized, and stained with polyclonal antibody for C2GnT-I (red) and examined by confocal microscopy. The merged images show overlapping red and green (EGFP) pixels in yellow. Magnification: 500×. Flow cytometric analyses of P-selectin binding and anti-*H* immunoreactivity. CHO/C2P1 cells were transfected with constructs coding for the chimeric C2/FucTI or C2/FucTVII enzymes or their normal counterparts fused to EGFP and assayed for P-selectin binding (A) or for the expression of the blood group *H* determinant (B). Note that no mislocalization or alteration in biological activity can be observed for FucT-I or its chimeric counterpart, although displacing FucT-VII to C2GnT-I compartment dramatically alters the P-selectin–binding activity of cells expressing the chimeric C2/FucTVII. (See color insert.)

## Dual-Color Analysis of Golgi Distribution of EGFP- and DsRed2-Tagged Glycosyltransferases

The mucin $O$-glycosylation pathway involves three different GTs that act on the same precursor Gal$\beta$1,3GalNAc-Ser/Thr (the so-called *core-1 oligosaccharide*). In fact this oligosaccharide can be a substrate for enzymes adding either GlcNAc in a $\beta$1,3 linkage to Gal ($\beta$1,3GnT-III) or in a $\beta$1,6 linkage to GalNAc (C2GnT-I) to initiate the formation of extended or branched structures, respectively. The core-1 oligosaccharide can also be a substrate for the $\alpha$(2,3)-sialyltransferase-I (ST3Gal-I), which adds sialic acid in an $\alpha$2,3 linkage to galactose. These early steps in $O$-glycan synthesis are crucial to the final sequence of a given $O$-glycan. With regard to the selectin ligands, which are mostly O-glycans, the conversion to core-1–extended or core-2–branched chains could end with sequences capped with sLe$^a$ or sLe$^x$, whereas the addition of sialic acid would terminate the chain growth (see Mitoma *et al.*, 2003, and Fig. 1 therein). Mitoma *et al.* (2003) found that when PSGL-1 is expressed with the $\beta$1,3GnT-III, but in the absence of C2GnT-I, the $O$-glycans of the P-selectin ligand become core-1–extended instead of being core-2 branched. On the other hand, preliminary data from our group on PSGL-1, using CHO/C2P1 cells stably expressing PSGL-1 and C2GnT-I, suggest that the PSGL-1 O-glycans remain predominantly core-2 branched, even after stable transfection with either the $\beta$1,3GnT-III or the ST3Gal-I (unpublished results). Together with data from Fukuda's group, these results indicate that when C2GnT-I is present, it takes over the two other GTs, suggesting that the $\beta$1,3GnT-III and ST3Gal-I are placed after C2GnT-I in the Golgi. It is, therefore, highly desirable to study the mechanism underlying this competition and whether this could be due to protein-carrier specificity or to the Golgi compartmentalization of these enzymes.

One way to address this issue is to co-express C2GnT-I with either $\beta$1,3GnT-III or ST3Gal-I, which are differently labeled, and compare their intracellular distribution in live cells by confocal microscopy. Because we already have a green version of our GTs, we had to create red variants of these proteins using the coral red fluorescent protein DsRed2. The DsRed2 excitation/emission wavelengths of 568/583 nm make it ideal for use in conjunction with EGFP. The construction of GTs-DsRed2 was performed by following the same protocol as for GTs-EGFP using our pcDNA3/ DsRed2 vector and the restriction sites *Bam*H1 and *Age*1 (see Fig. 1).

Cells co-expressing EGFP- and DsRed2-tagged proteins were obtained by transiently transfecting cells that stably express C2GnTI-EGFP, with either $\beta$1,3GnTII-DsRed2 or ST3GalI-DsRed2 constructs. Live cells were then observed by an inverted Leica laser scanning microscope setup at 488-nm excitation source and 505- to 550-nm bandpass barrier filter for EGFP and for DsRed2, a 568-nm excitation light from a helium-neon laser, a

575-nm dichroic mirror, and a 580- to 625-nm filter. The cells were illuminated only during image acquisition (3.7 s per frame) and images were collected in z-axis sections (Collazo *et al.*, 2005). Images were processed as described earlier.

Figure 5 presents a series of z-axis confocal slices showing the intracellular distribution of C2GnTI-EGFP versus the $\beta$1,3GnTIII-DsRed (Fig. 5A) and C2GnTI-EGFP versus the ST3GalI-DsRed2 (Fig. 5B). Note that in contrast to C2GnTI-EGFP, which displays a typical Golgi distribution,

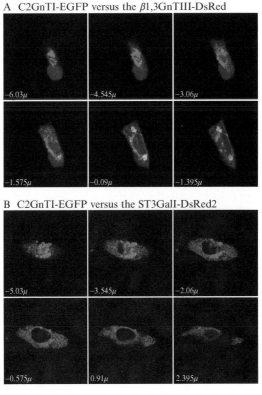

FIG. 5. Dual-color confocal microscopy for simultaneous comparison of intracellular distribution of enhanced green fluorescent protein (EGFP)- and DsRed2-tagged proteins. Cells stably expressing EGFP-conjugated C2GnT-I were transfected with either $\beta$1,3GnT-III or ST3Gal-I, both tagged with DsRed2, and live cells were examined by confocal microscopy. Shown are a series of z-axis slices of the intracellular distribution of C2GnTI-EGFP versus the $\beta$1,3GnTIII-DsRed2 (A) and C2GnTI-EGFP versus the ST3GalI-DsRed2 (B). Note that, in contrast to C2GnTI-EGFP, which displays a typical Golgi distribution, both DsRed2-fused $\beta$1,3GnT-III and ST3Gal-I proteins are concentrated within the two poles of the nucleus. Despite this feature, the ST3Gal-I is localized in the same planes as C2GnT-I, whereas the $\beta$1,3GnTIII distributes to areas relatively far from that of C2GnT-I. (See color insert.)

both DsRed2-fused proteins are concentrated within the two poles of the nucleus. The tendency of DsRed2 to oligomerize in tetramers may be an explanation for the formation of these clusters (Mizuno *et al.*, 2001). Despite this fact, the ST3Gal-I appears to be concentrated in the same planes as C2GnT-I, while the $\beta$1,3GnTIII localizes to a clearly distinct area. Together with the data on the absence of competition with C2GnT-I, we conclude that the $\beta$1,3GnT-III may be located in late compartments. To verify this assumption, we exploited the ability of FucT-I to inhibit sLe$^x$ synthesis (Mathieu *et al.*, 2004) and tentatively suppress this inhibition by mislocalizing FucT-I to the $\beta$1,3GnT-III compartment, through a molecular fusion of FucT-I luminal portion with the CTD of $\beta$1,3GnT-III (see results in the next section).

Note that a new DsRed variant has been created by Clontech, in which the oligomerization property was suppressed. This monomeric red fluorescent protein (DsRed-Monomer) would be a useful tool for use with other autofluorescent tags to explore the fine Golgi compartmentalization of GTs.

## Selectin-Binding Assays for *In Vivo* Activity of Mislocalized Fucosyltransferases

The $\alpha$(1,3)-fucosylation catalyzed by FucT-VII is crucial for selectin-mediated adhesion either in the immune system or in pathologies such as inflammation and cancer. We, therefore, chose selectin binding as a functional assay to evaluate the impact of mislocalizing these GTs on their activities *in vivo*.

1. The human embryonic kidney cells HEK293-T were cultured in Dulbecco's modified Eagle medium (DMEM) supplemented with 10% FCS and 100 U/ml penicillin and 100 $\mu$g/ml streptomycin (complete media).

2. The recombinant P- and E-selectin–immunoglobulin M (IgM) chimeras were produced by transiently transfecting HEK-293T cells with P-selectin–IgM or P-selectin–IgM DNAs in pcDNA1 plasmid (kindly provided by Pr. Fukuda M, The Burnham Institute, La Jolla, CA) as described earlier except that cells were plated in plastic dishes (10-cm diameter) and transfected with a mixture of 5 $\mu$g DNA and 20 $\mu$l Reagent Plus in 800 $\mu$l serum-free DMEM to which 30 $\mu$l Lipofectamine is added after 15 min of incubation at room temperature. This mixture was then added onto cell monolayers in 4 ml serum-free DMEM and dishes were incubated overnight. The transfection medium was then replaced with complete culture medium and collected three to four times every 48 h, centrifuged at 1000 rpm to remove detached cells and cell debris, and tested directly for selectin binding without dilution.

3. To evaluate the effect of mislocalizing the fucosyltransferases on their *in vivo* activities, we used the P-selectin–binding assay for FucT-VII and the

anti-blood group *H* immunoreactivity for FucT-I. Because P-selectin–binding requires the presence of PSGL-1 and the activities of both C2GnT-I and FucT-VII (Moore *et al.*, 1994), we developed a complementation system in which PSGL-1 was co-expressed with C2GnT-I (CHO/C2P1 cells) so that cells do not bind to P-selectin, unless FucT-VII is present. The making of CHO/C2P1 cells is described in the section "Confocal Microscopic Examination of Mislocalized FucT-VII."

4. Cells were detached from the flasks with the dissociation solution, rinsed twice with OptiMEM, and incubated 30 min with the HEK-293T supernatant containing the recombinant P-selectin–Ig*M*, as described earlier. Bound P-selectin-IgM was stained with Cy5-conjugated anti-human IgM (1/200 dilution) in PBS containing 1 mM $Ca^{2+}$ and $Mg^{2+}$ and 1% BSA and analyzed on a FACScan flow cytometer (Becton Dickinson, Mountain View, CA) by measuring the fluorescence of 10,000 cells and displayed on a four-decade log scale.

5. To measure activities of FucT-I and its chimeric variant C2-FucTI, we used a $sLe^x$-expressing cell line made by stably transfecting CHO-K1 cells with 5 $\mu$g FucT-VII DNA and 1 $\mu$g pSV/hygro for selection with Hygromycin B. The bulk Hygromycin-resistant transfectants (referred to as *CHO/F7 cells*) were collected and sorted by FACS using the anti-$sLe^x$ (CSLEX-1) mAb (Pharmingen). Detached cells were incubated with 10 $\mu$g/ml anti-*H* mAb (Institut Mérieux, Lyon, France) for 30 min followed by incubation with Cy5-labeled goat anti-mouse IgM (1/200 dilution) for another 30 min. All incubations were carried out at 4° in PBS containing 1% BSA and fluorescence was analyzed as described earlier.

Figure 4 shows flow cytometric analysis of P-selectin–binding activities of cells stably expressing either FucTVII and C2/FucTVII and the anti-*H* immunoreactivity of cells expressing FucTI or its chimeric variant C2/FucTI. The P-selectin binding to cells expressing C2/FucTVII was dramatically reduced, compared to control (Fig. 4A), whereas no difference in anti-*H* immunoreactivity could be detected between cells expressing the normal and the chimeric C2/FucT-I (Fig. 4B). Because P-selectin binding occurs through $sLe^x$ structures carried by PSGL-1, this result indicates that the chimeric C2/FucTVII, which has been mislocalized to the *cis/medial*-Golgi (i.e., the C2GnT-I compartment), is no longer able to trigger a functional PSGL-1.

*Technical Considerations*

We used Cy5-labeled secondary antibodies because the emission wavelength of Cy5 (670 nm) is far enough from that of EGFP (507 nm) so that overlapping between intracellular (EGFP) and cell surface fluorescence (Cy5) is avoided.

### E-Selectin–Binding Assay to Evaluate the Loss of Function of Mislocalized FucT-I

1. We have shown that introducing the FucT-I gene in cells that interact with E-selectin through sialyl-Lewis x ($sLe^x$) resulted in a complete inhibition of $sLe^x$ synthesis and abolition of E-selectin binding (Mathieu *et al.*, 2004). In that study, we proposed that since FucT-I overlaps very well with C2GnT-I in the *cis/medial*-Golgi, as shown here (see Fig. 4), it may intercept the $sLe^x$ precursors before they enter the late compartments where the $\alpha(2,3)$-sialyltransferases may be located. To further demonstrate that $\beta1,3$GnT-III is indeed localized to late Golgi compartments (see Fig. 5A), we fused the luminal portion of FucT-I with the CTD of the $\beta1,3$GnT-III (denoted *C1/FucTI-EGFP*) so that the $sLe^x$ precursors would be preferentially acted on by the $\alpha(2,3)$-sialyltransferases before FucT-I, and hence, less or no inhibition of the E-selectin binding would be observed.

2. The $sLe^x$-expressing cell line was made by stably transfecting CHO-K1 cells with 5 $\mu$g FucT-VII DNA and 1 $\mu$g pSV/hygro. The bulk Hygromycin-resistant transfectants (referred to as *CHO/F7 cells*) were collected and sorted by FACS using the anti-$sLe^x$ (CSLEX-1) mAb (Pharmingen). CHO/F7 cells were transfected again either with 1 $\mu$g of FucTI-EGFP or the chimeric construct C1/FucTI-EGFP described in Fig. 1. After another round of

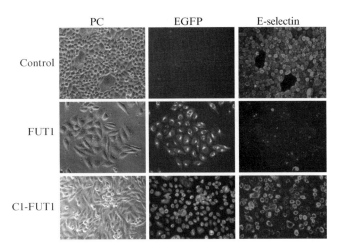

FIG. 6. Altered E-selectin binding to $sLe^x$-expressing cells (CHO/F7, control) following introduction of FucT-I (FUT1) gene and the loss of inhibition in the presence of the chimeric variant fused to the CTD of $\beta1,3$GnT-III (C1-FUT1). Shown are phase contrasts (PCs) of cells and fluorescence of transfectants (enhanced green fluorescent protein [EGFP]), together with their corresponding E-selectin-IgM–binding patterns as revealed by RITC-conjugated secondary antibody (E-selectin). (See color insert.)

selection in 1 mg/ml G418 and 200 $\mu$g/Hygromycin, cells were sorted by FACS on the basis of EGFP fluorescence as previously. Cells were grown and assayed for E-selectin binding.

3. E-selectin binding was performed by incubating cell monolayers with the E-selectin–rich HEK-293T supernatant for 30 min at 4°. Cells were then rinsed three times with PBS containing 1 m$M$ CaCl$_2$ and 1 m$M$ MgCl$_2$ (PBS$^{+/+}$) and incubated at 4° for another 30-min period with RITC-conjugated anti-human IgM antibodies in OptiMEM containing 1 mg/ml BSA (OptiMEM/BSA). Cells were then fixed in 3.5% paraformaldehyde (in PBS$^{+/+}$) for 5 min at room temperature and fluorescence was visualized as described previously.

Shown in Fig. 6 are the CHO/F7 cells that bind E-selectin and lost this property after the introduction of FucT-I gene. However, the C1/FucTI-EGFP construct has no effect on E-selectin binding, suggesting that this variant does not compete with the sLe$^x$-priming $\alpha$(2,3)-sialyltransferases as efficiently as the intact FucT-I. This result is indicative of a late Golgi localization for the $\beta$1,3GnT-III and highlights the efficacy of using the CTD sequences of GTs to enhance or attenuate the competition between these enzymes.

## Acknowledgments

We thank Professor Minoru Fukuda (The Burnham Institute) for providing us with genes of PSGL-1 and all GTs used in this study, as well as the C2GnT-I polyclonal antibody. We are also very grateful to Dr. Fallet Mathieu for confocal microscopy and Mrs. Mathieu Sylvie and Carmona Sylvie for their excellent technical assistance.

## References

Baron, C. B., and Coburn, R. F. (2004). Smooth muscle raft-like membranes. *J. Lipid Res.* **45,** 41–53.

Collazo, A., Bricaud, O., and Desai, K. (2005). Use of confocal microscopy in comparative studies of vertebrate morphology. *Methods Enzymol.* **395,** 521–543.

Colley, K. J. (1997). Golgi localization of glycosyltransferases: More questions than answers. *Glycobiology* **7,** 1–13.

El-Battari, A., Prorok, M., Angata, K., Mathieu, S., Zerfaoui, M., Ong, E., Suzuki, M., Lombardo, D., and Fukuda, M. (2003). Different glycosyltransferases are differentially processed for secretion, dimerization and autoglycosylation. *Glycobiology* **13,** 941–953.

Fukuda, M. (2002). Roles of mucin-type O-glycans in cell adhesion. *Biochim. Biophys. Acta* **1573,** 394–405.

Hakomori, S. (2002). Glycosylation defining cancer malignancy: New wine in an old bottle. *Proc. Natl. Acad. Sci. USA* **99,** 10231–10233.

Hartel-Schenk, S., Minnifield, N., Reutter, W., Hanski, C., Bauer, C., and Morre, D. J. (1991). Distribution of glycosyltransferases among Golgi apparatus subfractions from liver and hepatomas of the rat. *Biochim. Biophys. Acta* **1115,** 108–122.

Kozak, M. (1991). Structural features in eukaryotic mRNAs that modulate the initiation of translation. *J. Biol. Chem.* **266**, 19867–19870.

Lowe, J. B. (2003). Glycan-dependent leukocyte adhesion and recruitment in inflammation. *Curr. Opin. Cell Biol.* **15**, 531–538.

Mathieu, S., Prorok, M., Benoliel, A. M., Uch, R., Langlet, C., Bongrand, P., Gerolami, R., and El-Battari, A. (2004). Transgene expression of alpha(1,2)-fucosyltransferase-I (FUT1) in tumor cells selectively inhibits sialyl-Lewis x expression and binding to E-selectin without affecting synthesis of sialyl-Lewis a or binding to P-selectin. *Am. J. Pathol.* **164**, 371–383.

Mitoma, J., Petryniak, B., Hiraoka, N., Yeh, J. C., Lowe, J. B., and Fukuda, M. (2003). Extended core 1 and core 2 branched O-glycans differentially modulate sialyl Lewis X-type L-selectin ligand activity. *J. Biol. Chem.* **278**, 9953–9961.

Mizuno, H., Sawano, A., Eli, P., Hama, H., and Miyawaki, A. (2001). Red fluorescent protein from *Discosoma* as a fusion tag and a partner for fluorescence resonance energy transfer. *Biochemistry* **40**, 2502–2510.

Moore, K. L., Eaton, S. F., Lyons, D. E., Lichenstein, H. S., Cummings, R. D., and McEver, R. P. (1994). The P-selectin glycoprotein ligand from human neutrophils displays sialylated, fucosylated, O-linked poly-*N*-acetyllactosamine. *J. Biol. Chem.* **269**, 23318–23327.

Paulson, J. C., and Colley, K. J. (1989). Glycosyltransferases. Structure, localization, and control of cell type-specific glycosylation. *J. Biol. Chem.* **264**, 17615–17618.

Skrincosky, D., Kain, R., El-Battari, A., Exner, M., Kerjaschki, D., and Fukuda, M. (1997). Altered Golgi localization of core 2 beta-1,6-N-acetylglucosaminyltransferase leads to decreased synthesis of branched O-glycans. *J. Biol. Chem.* **272**, 22695–22702.

Zerfaoui, M., Fukuda, M., Langlet, C., Mathieu, S., Suzuki, M., Lombardo, D., and El-Battari, A. (2002). The cytosolic and transmembrane domains of the beta 1,6 N-acetylglucosaminyltransferase (C2GnT) function as a cis to medial/Golgi-targeting determinant. *Glycobiology* **12**, 15–24.

Zerfaoui, M., Fukuda, M., Sbarra, V., Lombardo, D., and El-Battari, A. (2000). Alpha (1,2)-fucosylation prevents sialyl Lewis X expression and E-selectin-mediated adhesion of fucosyltransferase VII-transfected cells. *Eur. J. Biochem.* **267**, 53–61.

# [8] Expression Profiling of Glycosyltransferases and Related Enzymes Using *In Situ* Hybridization

*By* Jun Nakayama, Misa Suzuki,
Masami Suzuki, and Minoru Fukuda

## Abstract

*In situ* hybridization (ISH) is a technique used to detect messenger RNAs (mRNAs) expressed in cells in tissue sections with probes specifically hybridizing to an mRNA of interest. Polysialic acid (PSA) is a unique glycan composed of a linear homopolymer of $\alpha 2,8$-linked sialic acid residues and formed by two distinct polysialyltransferases, ST8Sia II and ST8Sia IV. PSA plays an important role in neural development and progression of certain tumors. This chapter describes the use of ISH to detect ST8Sia II and ST8Sia IV mRNAs expressed in human astrocytomas using digoxigenin-labeled RNA probes.

METHODS IN ENZYMOLOGY, VOL. 416                                      0076-6879/06 $35.00
                DOI: 10.1016/S0076-6879(06)16008-5

Overview

ISH is a technique to detect mRNA expressed in cells using tissue sections and a labeled probe specifically hybridizing to an mRNA of interest. Various types of the probes, such as complementary DNAs (cDNAs), oligonucleotides, and RNAs, are applicable to the technique. We preferentially use digoxigenin-labeled RNA probes, because they have several advantages: (1) both antisense and control sense probes can be prepared simultaneously from a single common vector, (2) RNA–RNA hybrids are more stable than DNA–RNA hybrids, and (3) radioisotopes are not necessary to prepare the probe. ISH is useful to investigate the expression patterns of glycosyltransferases and related enzymes crucial for the biosynthesis of carbohydrate chains expressed on the cell surface and/or secreted from cells.

PSA is a developmentally regulated carbohydrate composed of a linear homopolymer of $\alpha$2,8-linked sialic acid residues (Finne, 1982). This unique glycan is mainly attached to the neural cell adhesion molecule (NCAM), where it attenuates homophilic interactions between NCAMs by its large negative charge (Nakayama *et al.*, 1998; Rutishauser and Landmesser, 1996). PSA is highly expressed in embryonic brain, but the degree of polysialylation is significantly decreased in adult brain. Two distinct polysialyltransferases responsible for the biosynthesis of PSA, ST8Sia II (STX) and ST8Sia IV (PST), have been cloned (Eckhardt *et al.*, 1995; Nakayama *et al.*, 1995; Scheidegger *et al.*, 1995; Yoshida *et al.*, 1995). Both catalyze the transfer of multiple $\alpha$2,8-linked sialic acid residues to a glycan containing NeuNAc$\alpha$2→3/6Gal$\beta$1→4GlcNAc→R (Angata *et al.*, 2000), and the expression of ST8Sia II and ST8Sia IV is regulated in a complementary fashion *in vivo*. Mice deficient in ST8Sia II exhibit the improper pathfinding of infrapyramidal mossy fibers and ectopic synapse formation in the hippocampus (Angata *et al.*, 2004), whereas mice deficient in ST8Sia IV exhibit an impairment in long-term potentiation in the hippocampal CA1 region, a phenotype different from that exhibited by ST8Sia II–deficient mice (Eckhardt *et al.*, 2000). However, mice doubly deficient in ST8Sia II and ST8Sia IV show severely impaired brain development, marked by specific wiring defects, progressive hydrocephalus, and postnatal growth retardation, resulting in perinatal lethality (Angata *et al.*, submitted; Weinhold *et al.*, 2005). These results indicate that PSA plays a critical role in brain development.

PSA is also expressed in various tumors such as small and non–small cell lung carcinomas, neuroblastoma, rhabdomyosarcoma, and Wilms' tumor (Fukuda, 1996; Hildebrandt *et al.*, 1998; Roth *et al.*, 1988; Tanaka *et al.*, 2000). In non–small cell lung carcinomas, PSA expression is correlated with poor patient prognosis (Tanaka *et al.*, 2001). Although PSA is abundantly expressed in embryonic brain, it has not been determined whether it is expressed in the most common brain tumor, astrocytoma.

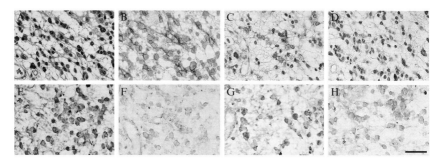

FIG. 1. Expression of polysialic acid (PSA), neural cell adhesion molecule (NCAM), ST8Sia II messenger RNA (mRNA), and ST8Sia IV mRNA in a diffuse astrocytoma. (A) Protoplasmic astrocytoma cells proliferating in a loose microcystic matrix are shown (hematoxylin–eosin staining). (B) NCAM was detected by the 123C3 antibody (Zymed, Carlsbad, CA). (C) PSA was detected using 5A5 antibody (University of Iowa Hybridoma Bank, Iowa City, IA), and (D) 5A5 immunoreactivity was eliminated after pretreatment with endo-N, which cleaves PSA. *In situ* hybridization of ST8Sia IV mRNA using (E) antisense and (F) sense control probes. *In situ* hybridization of ST8Sia II mRNA using (G) antisense and (H) sense control probes. Positive signals were detected in the perinuclear cytoplasm of tumor cells. Bar = 50 $\mu$m. (Reprinted, with permission, from Suzuki *et al.*, 2005.) (See color insert.)

We have demonstrated that NCAM is expressed in 68.2% of 44 patients with astrocytoma, whereas PSA is detected in 30% of patients with NCAM-positive astrocytoma (Suzuki *et al.*, 2005). Clinical surgical records demonstrate that 44.4% of 9 PSA-positive patients were recurred cases, whereas 23.8% of 21 patients expressing NCAM alone were recurred cases, suggesting that PSA expressed on astrocytoma cells was associated with disease recurrence. In one patient with intracranial invasion of the astrocytoma through the corpus callosum, ST8Sia II and ST8Sia IV transcripts were detected in tumor cells (Fig. 1), indicating that PSA expressed by astrocytoma cells is formed by both enzymes (Suzuki *et al.*, 2005). We have used ST8Sia II and ST8Sia IV digoxigenin-labeled RNA probes and ISH to detect corresponding ST8Sia II and ST8Sia IV mRNAs expressed in astrocytoma cells and describe that protocol here.

## Synthesis of RNA Probes by *In Vitro* Transcription

### Materials for RNA Probe Synthesis

Diethylpyrocarbonate (DEPC)-treated water: Add DEPC to distilled water at a 0.1% (v/v) final concentration. Shake solution vigorously, leave at room temperature overnight, and then autoclave to destroy DEPC.

DIG RNA labeling mix (10×) (Roche Diagnostics, Mannheim, Germany).

SP6 RNA polymerase (20 U/$\mu$l) with 10× transcription buffer (Roche Diagnostics).

T7 RNA polymerase (20 U/$\mu$l) with 10× transcription buffer (Roche Diagnostics).

RNase inhibitor, rRNasin (40 U/$\mu$l) (Promega, Madison, WI).

DNase I, RNase-free (10 U/$\mu$l) (Roche Diagnostics).

0.5 *M* ethylenediaminetetraacetic acid (EDTA), pH 8.0 (Amaresco, Solon, OH).

4 *M* LiCl: Dissolve 4.24 g LiCl in distilled water and adjust to 25 ml. Add 25 $\mu$l DEPC to the solution, shake vigorously, and leave at room temperature overnight. Autoclave to destroy DEPC.

Ethanol

*Method for RNA Probe Synthesis*

1. Since glycogenes such as *PST* and *STX* encode structurally related family members (Taniguchi *et al.*, 2002), RNA probes for ISH should be as specific as possible for the gene of interest and should not cross-hybridize with other family members. To avoid such cross-hybridization, it is important to examine the homology of RNA probes by NCBI BLAST searches at *http://www.ncbi.nlm.nih.gov/BLAST/*. In addition, it is recommended that the length and GC content of the DNA template of the RNA probe be approximately 150 base pairs and 55–60%, respectively. We amplified partial cDNA fragments unique to human PST (nucleotides −47 to +113; the first nucleotide of the initiation codon is +1) and human STX (nucleotides +1 to +138) by polymerase chain reaction (PCR) using a PST-specific primer set (5′-GCTCTAGAAGGTGCGGGAGCTGG-3′ and 5′-GGGGTACCGAT-GAGTTGCGTCTCCT-3′) or STX-specific primer set (5′-GCTCTAGA-TGCAGCTGCAGTTCCGGA-3′ and GGGGTACCGTTCACAGCTGA-TCTGATTGT-3′) (Suzuki *et al.*, 2005). (*Xba*I and *Asp*718 sites in these primers are underlined.) The amplicons are subcloned into the *Xba*I and *Asp*718 sites of pGEM-3Zf (+) (Promega) and the resulting vector construct is used as template to synthesize RNA probes.

2. Digoxigenin-labeled antisense RNA probes are obtained by *in vitro* transcription. Specifically, 13 $\mu$l containing 1 $\mu$g *Xba*I-cut linearized vector in DEPC-treated water, 2 $\mu$l of 10x transcription buffer, 2 $\mu$l of 10× DIG RNA labeling mix, 2 $\mu$l of T7 RNA polymerase, and 1 $\mu$l of rRNasin are incubated in RNase-free microcentrifuge tubes for 3 h at 37°. Then, 0.2 $\mu$l of DNAse I is added to the tube and further incubated for 20 min at 37°. Finally, 1 $\mu$l of 0.5 *M* EDTA (pH 8.0), 2.5 $\mu$l of 4 *M* LiCl, and 75 $\mu$l of

chilled 100% ethanol are added and the mixture is stored at $-20°$ overnight. After a 30-min centrifugation (13,000$g$), the supernatant is discarded and 50 $\mu l$ of chilled 70% ethanol is added. RNA is pelleted by centrifugation (13,000$g$) for 5 min and air-dried briefly. The concentration of the resulting RNA probes is adjusted to 20–400 $\mu g/ml$ by adding DEPC-treated water. RNA probes can be stored at $-80°$ until use. Similarly, sense probes are prepared as negative controls by using $Asp718$-cut vector and SP6 RNA polymerase with the DIG RNA labeling mixture.

### In Situ Hybridization with RNA Probes and Immunodetection

*Reagents and Solutions for* In Situ *Hybridization*

The amounts of solutions described here are sufficient for up to 40 slides, including 20 slides for antisense probe and 20 slides for the sense probe. DEPC-treated water, DEPC-treated phosphate-buffered saline (PBS), DEPC-treated 20× SSC, hybridization solution, RNase buffer, and 1 $M$ MgCl$_2$ can be prepared in advance, but others should be made fresh before use.

  Hemo-De or xylene.
  Ethanol.
  DEPC-treated PBS (1×): Dissolve 8.0 g NaCl, 0.2 g KCl, 0.7 g Na$_2$HPO$_4$, and 0.2 g KH$_2$PO$_4$ in distilled water and bring to 1 liter (adjust to pH 7.4 with HCl if necessary). Then, add DEPC (final concentration 0.1% [v/v]) to the solution and shake vigorously. Leave at room temperature overnight and then autoclave to destroy DEPC. Approximately 4.0 liters of DEPC-PBS (1×) is required for one experiment. We usually make the solution in 500-ml bottles.
  DEPC-treated water.
  HCl solution (0.2 N): Add 8.4 ml concentrated HCl to 500 ml DEPC-treated water.
  Proteinase K solution (50 $\mu l/ml$): Add 50 $\mu l$ proteinase K (20 mg/ml) (Amaresco) to 20 ml DEPC-treated PBS (1×).
  Paraformaldehyde solution (4%): Dissolve 20 g paraformaldehyde in 400–450 ml DEPC-treated PBS (1×), and stir at 65° on a hot stirrer (to solubilize paraformaldehyde, add 30 $\mu l$ of 10 N NaOH if necessary). When solution clears, place it on ice. Adjust volume to 500 ml with DEPC-treated PBS (1×).
  0.2% Glycine: Dissolve 2.0 g glycine in 1.0 liter DEPC-treated PBS (1×).
  Triethanolamine (TEA; 0.1 $M$) buffer: Mix 7.5 ml TEA (98% solution; Sigma-Aldrich, St. Louis, MO) and 500 ml DEPC-treated water. Adjust pH to 8 by adding concentrated HCl (~1.75 ml).

Acetic anhydride.

Ethanol.

Chloroform.

SSC (sodium citrate/sodium chloride) (20× stock solution): Dissolve 175.3 g NaCl and 88.2 g trisodium citrate in distilled water. Adjust to pH 7.0 by adding 1 $M$ HCl and bring to 1 liter by adding distilled water. Add DEPC to a final concentration of 0.1% (v/v). After shaking solution vigorously, leave at room temperature overnight and then autoclave to destroy DEPC.

Deionized formamide (Nacali Tesque, Kyoto, Japan).

2× SSC: Mix 50 ml 20× SSC and 450 ml DEPC-treated water.

50% deionized formamide in 2× SSC for prehybridization: Mix 250 ml deionized formamide, 50 ml 20× SSC, and 200 ml DEPC-treated water.

Hybridization solution: Combine the following (Table I) in 1.5-ml RNase-free microcentrifuge tubes. This solution can be kept at −80° for several months.

Formamide.

50% formamide in 2× SSC for washing: Mix 500 ml formamide, 100 ml 20× SSC, and 400 ml autoclaved water.

50% formamide in 1× SSC for washing: Mix 250 ml formamide, 25 ml 20× SSC, and 225 ml autoclaved distilled water.

RNase buffer (10 m$M$ Tris buffer [pH 8.0], 0.5 $M$ NaCl): Dissolve 0.614 g Trizma hydrochloride (Sigma-Aldrich), 0.74 g Trizma base (Sigma-Aldrich), and 29.22 g NaCl in distilled water. Adjust to 1 liter by adding distilled water and autoclave.

RNase A solution: Add 20 $\mu$l RNase A (10 mg/ml) (Amaresco) to 20 ml RNase buffer.

DIG Wash and Block Buffer set (Roche Diagnostics): Contains maleic acid buffer, blocking buffer, washing buffer, and detection buffer for immunodetection of hybridized probes.

TABLE I
COMPOSITION OF HYBRIDIZATION SOLUTION

| Stock solution | | Final concentration |
|---|---|---|
| 10 mg/ml brewer's yeast tRNA (Sigma-Aldrich) | 100 $\mu$l | 1 mg/ml |
| 1.0 $M$ Tris buffer, pH 8.0 (Amaresco) | 20 $\mu$l | 20 m$M$ |
| 0.5 $M$ EDTA, pH 8.0 (Amaresco) | 5 $\mu$l | 2.5 m$M$ |
| 100× Denhardt's (Amaresco) | 10 $\mu$l | 1× |
| 5 $M$ NaCl in DEPC-treated water | 60 $\mu$l | 300 m$M$ |
| Deionized formamide | 500 $\mu$l | 50% |
| DEPC-treated water | 105 $\mu$l | |
| 50% dextran sulfate (Amaresco) | 200 $\mu$l | 10% |
| TOTAL | 1000 $\mu$l | |

Anti-digoxigenin antibody conjugated to alkaline phosphatase (Roche Diagnostics).

Polyvinyl alcohol (PVA) (average molecular weight 70,000–100,000) (Sigma-Aldrich).

1 $M$ MgCl$_2$: Dissolve 2.03 g MgCl$_2 \cdot$ 6H$_2$O in distilled water to a volume of 10 ml.

PVA/detection buffer: Dissolve 4.0 g PVA in distilled water containing 800 $\mu$l 1 $M$ MgCl$_2$ and 4.0 ml detection buffer (10×), and adjust to 40 ml. PVA prevents diffusion of substrate in tissue sections (De Block and Debrouwer, 1993). This buffer is made on a hot-plate stirrer at 90° and cooled to room temperature before use.

Substrate solution: Dissolve 200 $\mu$l 4-nitro blue tetrazolium chloride (NBT) (100 mg/ml) (Roche Diagnostics), 200 $\mu$l 5-bromo-4-chloro-3-indolyl phosphate (BCIP) (50 mg/ml) (Roche Diagnostics), and 40 drops (final concentration 0.2 m$M$) of a ready-to-use alkaline phosphatase–blocking reagent, levamisole (DakoCytomation, Glostrup, Denmark) in 40 ml PVA/detection buffer.

Glysergel (DakoCytomation).

*Method for* In Situ *Hybridization Using RNA Probes*

1. Tissue blocks of primary astrocytomas fixed for 48 h in 20% buffered formalin (pH 7.4) and embedded in paraffin wax are used. Blocks are sectioned at 7-$\mu$m thickness and placed on adhesive-coated slides (MAS-coated Superfrost, Matsunami Glass, Osaka, Japan).

2. Sections are deparaffinized by three 5-min applications of Hemo-De or xylene, followed by three 5-min applications of 100% ethanol, and then incubated three times for 5 min each with DEPC-treated PBS (1×) in a sterilized glass jar.

3. To neutralize basic proteins, tissue sections are treated with 0.2 N HCl for 20 min and then immersed in DEPC-treated PBS (1×) for 5 min.

4. Prepare fresh proteinase K solution (50 $\mu$l/ml) and heat to 37° before use in the next step.

5. On the slide, encircle the area containing tissue specimens using a DAKO Pen (DakoCytomation) and apply 500 $\mu$l of proteinase K solution per slide inside that area. Place slides in a humidified box and incubate 30 min at 37°. Incubation time should be shortened if cryostat sections are used.

6. After careful aspiration of the proteinase K solution, wash slides two times with DEPC-treated PBS (1×) for 5 min each.

7. Post-fix slides with 4% paraformaldehyde in PBS (1×) for 5 min.

8. To quench paraformaldehyde, treat slides twice for 10 min with 0.2% glycine.

9. For the acetylation step, place slides into a jar filled with 0.1 $M$ TEA buffer (pH 8.0) for 5 min. Then, add concentrated acetic anhydride to

the jar to a final concentration of 0.25% (v/v) while shaking vigorously enough to rapidly mix the solution without damaging slides or tissues. Incubate slides for 10 min with occasional shaking.

10. Tissues on slides are dehydrated with a graded ethanol series; that is, 70% ethanol for 1 min, 80% ethanol for 1 min, 95% ethanol for 1 min, and 100% ethanol for 1 min. Then, treat slides with 100% chloroform for 5 min, 100% ethanol for 1 min, and 95% ethanol for 1 min, followed by air-drying for 10–30 min.

11. Incubate slides with 2× SSC in DEPC-treated water for 5 min.

12. For prehybridization, place slides in a sterilized jar filled with 50% deionized formamide in 2× SSC at 45° for 1 h.

13. The digoxigenin-labeled RNA probe (2–4 $\mu$l) is added to 800 $\mu$l of hybridization solution in a 1.5-ml RNase-free microcentrifuge tube. The final concentration of the probe should be approximately 0.1–1.0 $\mu$g/ml. The probes are denatured at 95° for 5 min using a heating block and then immediately transferred to ice.

14. A box humidified with 100% formamide (not water) is prepared.

15. Apply a total volume of 40 $\mu$l of hybridization solution containing the digoxigenin-labeled RNA probe to tissue specimens on each slide. Cover specimens carefully with a 24 × 40-mm coverslip, avoiding bubbles. Place slides in the humidified box and incubate for 48–72 h at 45°. It is recommended that 1.0 liter of 50% formamide in 2× SSC and 500 ml of 50% formamide in 1× SSC used for the following washings be preheated to 45° during this incubation.

16. Remove coverslips carefully from slides by dipping slides into preheated 50% formamide in 2× SSC at 45°. Then place slides in a sterilized jar filled with 50% formamide in 2× SSC for washing at 45° for 1 h.

17. Tissue specimens are incubated twice with RNase buffer for 5 min each at room temperature and then put back in a humidified box.

18. Five hundred microliters of RNase A solution (10 $\mu$g/ml) is applied to each slide in the humidified box, and slides are incubated 30 min at 37°.

19. Remove RNase A solution from slides and place them in a jar filled with 50% formamide in 2× SSC for 1 h at 45° with gentle shaking. Then, wash with 50% formamide in 1× SSC for 1 h at 45°, and follow by immunodetection of the hybridized probes as described next.

*Immunodetection of Hybridized Probes*

1. Wash slides three times with maleic acid buffer for 5 min each and then incubate them with blocking buffer for 20 min.
2. A total volume of 100 $\mu$l of anti-digoxigenin antibody conjugated to alkaline phosphatase diluted in blocking buffer (1:1000) is applied to tissue specimens on each slide in a humidified box and incubated at 4° overnight.

3. Wash slides six times with washing buffer for 10 min each and then incubate with detection buffer for 5 min.
4. Apply 1 ml of freshly prepared substrate solution to each slide and develop the color reaction in a humidified box at 37° from 30 min to overnight.
5. Gently wash slides with running water.
6. Mount specimens with Glysergel prewarmed at 37°.

## Acknowledgment

We thank Dr. Elise Lamar for critical reading of the manuscript.

## References

Angata, K., Suzuki, M., McAuliffe, J., Ding, Y., Hindsgaul, O., and Fukuda, M. (2000). Differential biosynthesis of polysialic acid on neural cell adhesion molecule (NCAM) and oligosaccharide acceptors by three distinct $\alpha 2,8$-sialyltransferases, ST8Sia IV (PST), ST8Sia II (STX), and ST8Sia III. *J. Biol. Chem.* **275**, 18594–18601.

Angata, K., Long, J. M., Bukalo, O., Lee, W., Dityatev, A., Wynshaw-Boris, A., Schachner, M., Fukuda, M., and Marth, J. D. (2004). Sialyltransferase ST8Sia-II assembles a subset of polysialic acid that directs hippocampal axonal targeting and promotes fear behavior. *J. Biol. Chem.* **279**, 32603–32613.

Angata, K., Ranscht, B., Terskikh, A., Marth, J. D., and Fukuda, M. Polysialic acid-directed migration of neural cell is essential for mouse brain development. (submitted).

De Block, M., and Debrouwer, D. (1993). RNA-RNA *in situ* hybridization using digoxigenin-labeled probes: The use of high-molecular-weight polyvinyl alcohol in the alkaline phosphatase indoxyl-nitroblue tetrazolium reaction. *Anal. Biochem.* **215**, 86–89.

Eckhardt, M., Muhlenhoff, M., Bethe, A., Koopman, J., Frosch, M., and Gerardy-Schahn, R. (1995). Molecular characterization of eukaryotic polysialyltransferase-1. *Nature* **373**, 715–718.

Eckhardt, M., Bukalo, O., Chazal, G., Wang, L., Goridis, C., Schachner, M., Gerardy-Schahn, R., Cremer, H., and Dityatev, A. (2000). Mice deficient in the polysialyltransferase ST8SiaIV/PST-1 allow discrimination of the roles of neural cell adhesion molecule protein and polysialic acid in neural development and synaptic plasticity. *J. Neurosci.* **20**, 5234–5244.

Finne, J. (1982). Occurrence of unique polysialosyl carbohydrate units in glycoproteins of developing brain. *J. Biol. Chem.* **257**, 11966–11970.

Fukuda, M. (1996). Possible roles of tumor-associated carbohydrate antigens. *Cancer Res.* **56**, 2237–2244.

Hildebrandt, H., Becker, C., Gluer, S., Rosner, H., Gerardy-Schahn, R., and Rahmann, H. (1998). Polysialic acid on the neural cell adhesion molecule correlates with expression of polysialyltransferases and promotes neuroblastoma cell growth. *Cancer Res.* **58**, 779–784.

Nakayama, J., Fukuda, M. N., Fredette, B., Ranscht, B., and Fukuda, M. (1995). Expression cloning of a human polysialyltransferase that forms the polysialylated neural cell adhesion molecule present in embryonic brain. *Proc. Natl. Acad. Sci. USA* **92**, 7031–7035.

Nakayama, J., Angata, K., Ong, E., Katsuyama, T., and Fukuda, M. (1998). Polysialic acid, a unique glycan that is developmentally regulated by two polysialyltransferases, PST and STX, in the central nervous system: From biosynthesis to function. *Pathol. Int.* **48**, 665–677.

Roth, J., Zuber, C., Wagner, P., Taatjes, D. J., Weisgerber, C., Heitz, P. U., Goridis, C., and Bitter-Suermann, D. (1988). Reexpression of poly(sialic acid) units of the neural cell adhesion molecule in Wilms tumor. *Proc. Natl. Acad. Sci. USA* **85,** 2999–3003.

Rutishauser, U., and Landmesser, L. (1996). Polysialic acid in the vertebrate nervous system: A promoter of plasticity in cell-cell interactions. *Trends Neurosci.* **19,** 422–427.

Scheidegger, E. P., Sternberg, L. R., Roth, J., and Lowe, J. B. (1995). A human STX cDNA confers polysialic acid expression in mammalian cells. *J. Biol. Chem.* **270,** 22685–22688.

Suzuki, M., Suzuki, M., Nakayama, J., Suzuki, A., Angata, K., Chen, S., Sakai, K., Hagihara, K., Yamaguchi, Y., and Fukuda, M. (2005). Polysialic acid facilitates tumor invasion by glioma cells. *Glycobiology* **15,** 887–894.

Tanaka, F., Otake, Y., Nakagawa, T., Kawano, Y., Miyahara, R., Li, M., Yanagihara, K., Nakayama, J., Fujimoto, I., Ikenaka, K., and Wada, H. (2000). Expression of polysialic acid and STX, a human polysialyltransferase, is correlated with tumor progression in non-small cell lung cancer. *Cancer Res.* **60,** 3072–3080.

Tanaka, F., Otake, Y., Nakagawa, T., Kawano, Y., Miyahara, R., Li, M., Yanagihara, K., Inui, K., Oyanagi, H., Yamada, T., Nakayama, J., Fujimoto, I., Ikenaka, K., and Wada, H. (2001). Prognostic significance of polysialic acid expression in resected non-small cell lung cancer. *Cancer Res.* **61,** 1666–1670.

Taniguchi, N., Honke, K., and Fukuda, M. (eds.) (2002). "Handbook of Glycosyltransferases and Related Genes." Springer-Verlag, Tokyo.

Yoshida, Y., Kojima, N., and Tsuji, S. (1995). Molecular cloning and characterization of a third type of N-glycan α2,8-sialyltransferase from mouse lung. *J. Biochem. (Tokyo)* **118,** 658–664.

Weinhold, B., Seidenfaden, R., Rockle, I., Muhlenhoff, M., Schertzinger, F., Conzelmann, S., Marth, J. D., Gerardy-Schahn, R., and Hildebrandt, H. (2005). Genetic ablation of polysialic acid causes severe neurodevelopmental defects rescued by deletion of the neural cell adhesion molecule. *J. Biol. Chem.* **280,** 42971–42977.

# [9]  Expression Profiling of Glycosyltransferases Using RT-PCR

*By* Motohiro Kobayashi

## Abstract

The quantitative reverse-transcription polymerase chain reaction (RT-PCR) is a method used to quantify messenger RNA (mRNA) levels. It is particularly applicable to quantification of mRNAs that are transcribed at very low levels, such as glycosyltransferases (GTs). In this chapter, I describe preliminary experiments for obtaining conditions for a quantitative RT-PCR method to quantify transcript levels of GTs and related genes potentially involved in L-selectin–mediated lymphocyte homing in the mouse gastric mucosa infected by *Helicobacter felis*. This method was developed by modifying conventional RT-PCR protocols and does not require fluorescence-detecting thermal cyclers. The method described here is

METHODS IN ENZYMOLOGY, VOL. 416
Copyright 2006, Elsevier Inc. All rights reserved.
0076-6879/06 $35.00
DOI: 10.1016/S0076-6879(06)16009-7

particularly useful for assaying large numbers of samples that require very accurate analysis.

Overview

The quantitative RT-PCR is a method used to quantify mRNA levels. This technique is particularly useful to quantitate expression levels of transcripts that are expressed at very low levels, such as GTs. Real-time RT-PCR has gained popularity not only in quantifying gene expression but to confirming differential expression of genes detected by microarray analysis. Real-time RT-PCR uses commercially available fluorescence-detecting thermal cyclers (e.g., the ABI Prism 7700 Sequence Detection System from Applied Biosystems) to simultaneously amplify specific DNA sequences and measure their concentration (Bustin, 2000). This method is optimal for measuring DNA concentration over a wide dynamic range, its sensitivity is high, and many samples can be processed simultaneously. However, the thermal cyclers are expensive and maintenance and running costs are considerable.

Alternatively, conventional RT-PCR has been used to estimate the concentration of a particular target DNA relative to a standard one. However, accurate quantification is difficult because of the kinetics of PCR. The amount of PCR product increases exponentially in early cycles of amplification and then reaches a plateau. When that occurs, the ampli-con does not quantitatively reflect the amount of initial DNA template (Yokoi et al., 1993). Thus, to quantify a product, levels must be evaluated as the reaction occurs exponentially (Nakayama et al., 1992). To avoid the plateau effect, several methods have been described, such as the use of an internal standard competing with the target sequence in amplification (Gilliland et al., 1990; Siebert and Larrick, 1992). However, the design of a specific DNA competitor is technically complicated, and validation of quantitation using this technique is labor intensive (Bustin, 2000).

In 1992, Nakayama et al. developed a simple and widely applicable method for quantifying mRNA by nonradioactive RT-PCR without using a specific competitor, thereby avoiding the plateau effect. They compared quantitative RT-PCR analysis with Northern blot analysis in estimating the expression level of macrophage colony-stimulating factor (M-CSF) in mouse cell lines. Their results demonstrate that quantitative RT-PCR is applicable to the accurate quantification of specific mRNA; however, the method is labor intensive when used to analyze numerous samples.

Helicobacter pylori is a Gram-negative microaerophilic bacterium that infects more than 50% of the world's population (Marshall and Warren, 1984) and causes gastric diseases such as chronic gastritis, peptic ulcer, and gastric cancer (Bayerdorffer *et al.*, 1995; Peek and Blaser, 2002; Sipponen and Hyvarinen, 1993). The host responds to *H. pylori* infection primarily by mounting a strong neutrophilic response, and this response is followed by mononuclear cell (chronic inflammatory cell) infiltration. In this pathologic state, as in other chronic inflammatory diseases, L-selectin and its ligands are implicated in lymphocyte recruitment (Rosen, 2004). We previously showed that high endothelial venule (HEV)–like vessels are induced in *H. pylori*–associated chronic gastritis and that peripheral lymph node addressin (PNAd) is expressed on these vessels (Kobayashi *et al.*, 2004). We also showed that the progression of chronic inflammation is highly correlated with the occurrence of PNAd-expressing HEV-like vessels and that eradication of *H. pylori* results in the disappearance of PNAd. These results indicate that at inflammatory sites, lymphocyte recruitment is partly regulated by PNAd.

PNAd is composed of 6-sulfo sialyl Lewis X on extended core 1 *O*-glycans, sialic acid$\alpha2\rightarrow3$Gal$\beta1\rightarrow4$[Fuc$\alpha1\rightarrow3(SO_3\rightarrow6)$]GlcNAc$\beta1\rightarrow3$Gal$\beta1\rightarrow3$Gal NAc$\alpha1\rightarrow$Ser/Thr, and/or that on core 2 branched *O*-glycans, sialic acid$\alpha2\rightarrow3$Gal$\beta1\rightarrow4$[Fuc$\alpha1\rightarrow3(SO_3\rightarrow6)$]GlcNAc$\beta1\rightarrow6$(Gal$\beta1\rightarrow3$)GalNAc$\alpha1\rightarrow$Ser/Thr (Fig. 1). This structure is synthesized by a battery of GTs and sulfotransferases, namely, Core1-$\beta$3GlcNAcT, Core2GlcNAcTs (Yeh *et al.*, 1999), FucT-VII, and GlcNAc6ST-2 (LSST) (Kobayashi *et al.*, 2004; Yeh *et al.*, 2001). Our goal has been to determine to what extent these enzymes participate in PNAd formation on HEV-like vessels formed in *Helicobacter*-induced chronic gastritis.

In this chapter, I describe an economic and accurate method to quantitate mRNA of GTs, taking Core2GlcNAcT-1 for example, using conventional RT-PCR. The method described here forms a basis to quantitate mRNA for GTs and related genes potentially involved in L-selectin-mediated lymphocyte homing, which will be published elsewhere (Kobayashi *et al.*, submitted).

Design of Oligonucleotide Primers

I design oligonucleotide primers based on the following conditions. First, the length of an oligonucleotide primer is 24 nucleotides long. Second, the GC content of a primer should be 50%. An oligonucleotide primer fulfilling these two criteria has a melting temperature (Tm) of 59°, as calculated by Eq. (1), where $n$ is the length of the oligonucleotide.

$$Tm = 2(A + T) + 4(G + C) + 35 - 2n \qquad (1)$$

Third, the length of the expected PCR product should be approximately 200 bp. This condition not only avoids incomplete extension but also reduces amplification time. In this case, a 2% agarose gel should be used

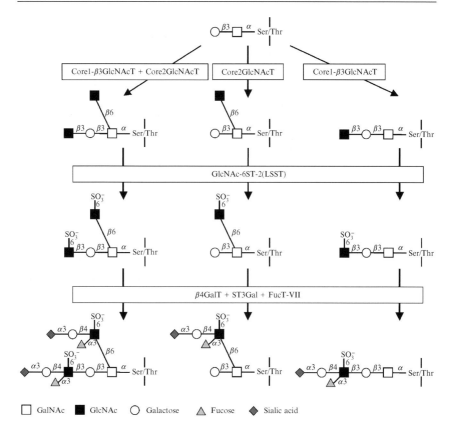

□ GalNAc    ■ GlcNAc    ○ Galactose    △ Fucose    ◆ Sialic acid

Fig. 1. Structure and biosynthesis of 6-sulfo sialyl Lewis X. Core 1 O-glycans can be extended by Core1-β3GlcNAcT and then sulfated by GlcNAc-6ST-2 (LSST). The resultant oligosaccharide can be galactosylated, sialylated, and fucosylated to form 6-sulfo sialyl Lewis X on extended core 1 O-glycans. Alternatively, core 1 O-glycans can be branched by Core2GlcNAcT, and then sulfated by LSST, galactosylated, sialylated, and fucosylated to form 6-sulfo sialyl Lewis X on core 2 branched O-glycans. Core1-β3GlcNAcT can act on core 2 branched O-glycans, leading to bi-antennary O-glycans containing both core 2 branch and core 1 extension. All three O-glycans capped by 6-sulfo sialyl Lewis X function as L-selectin ligands (Based on Yeh et al., 2001). (See color insert.)

for electrophoresis to obtain sharp bands. Fourth, when possible, oligonucleotides binding to sequences located in different exons of the target gene should be used as sense and antisense primers for amplification. In this way, amplification products derived from cDNA rather than contaminating genomic DNA can be easily distinguished. However, cDNA amplified from transcripts of intronless genes cannot be differentiated unambiguously from contaminating genomic sequence. This is unfortunate because some GT genes functioning in 6-sulfo sialyl Lewis X biosynthesis are intronless (Taniguchi *et al.*, 2002). In these circumstances, treating the RNA preparation with RNase-free DNase I is critical. Fifth, and finally, each member of a primer pair should specifically recognize its target sequence in template DNA. Before synthesizing an oligonucleotide primer, it is prudent to scan a DNA database using a BLAST search (http://www.ncbi.nih.gov/BLAST/) to determine whether the proposed sequence occurs only in the desired gene and not in vectors, other genes, or repetitive elements.

Primers we have used in our studies include the following: Core1-$\beta$3Glc NAcT; 5′-ATGTGTTCGCACACACGGACAACA-3′ (5′-primer) and 5′-GAACACGTCGTCGATTGGGAACAT-3′ (3′-primer), Core2GlcNAc T-1; 5′-AGGGTGACCCAGAAGAAATCCAGA-3′ (5′-primer) and 5′-CTCAGGAGCCTGTCAAGCATTTCA-3′ (3′-primer), Core2GlcNAcT-2; 5′-ATTGCGTACTCCATGGTGGTGCAT-3′ (5′-primer) and 5′-ACC AGGAAGCATAGACCACTGACA-3′ (3′-primer), Core2GlcNAcT-3; 5′-TGGTTTCAAGGGAGGAAGAGGACT-3′ (5′-primer) and 5′-GAGG AGGTCTGATAAGCAGTTCCA-3′ (3′-primer), FucT-VII; 5′-GCTGG AGGAGCAACATTCATGGTA-3′ (5′-primer) and 5′-ATCTGAATCA CGCCGATAGCTCAG-3′ (3′-primer), GlcNAc6ST-2 (LSST); 5′-AGCCC TTTGATATGGTGGAGAAGG-3′ (5′-primer) and 5′-TGGTCTTCCTCC TTGATCGTTTCC-3′ (3′-primer), glyceraldehyde-3-phosphate dehydrogenase (G3PDH); 5′-AAGCTTGTCATCAACGGGAAGCC-3′ (5′-primer) and 5′-CTCCACCCTTCAAGTGGGCCC-3′ (3′-primer).

## Subcloning of Target Sequence for Quantitative RT-PCR

In the method described, serially diluted plasmid harboring the target cDNA must be subjected to PCR in parallel with samples to obtain a standard curve. If you do not have an appropriate plasmid DNA, you need to subclone the target sequence into an appropriate vector. To do so, using the same primer pair used for quantitative RT-PCR, one would carry out conventional RT-PCR, run the product on an agarose gel, cut and extract the product from the gel, and subclone it into the appropriate vector using the TA cloning method.

## RNA Extraction from Tissues

Special care should be taken in RNA extraction to maximize RNA integrity because of its chemical instability and the ubiquitous presence of RNase. In our laboratory, we routinely use TRIzol Reagent (Invitrogen) for RNA extraction. The procedure for RNA extraction from tissues is described in the following subsection.

*Materials for RNA Extraction from Tissues*

TRIzol Reagent (Invitrogen)
Chloroform
Isopropyl alcohol
70% ethanol (diluted with diethylpyrocarbonate [DEPC]-treated water)
DEPC-treated water

*RNA Extraction from Tissues*

1. Tissue homogenization: Homogenize tissue samples in at least 1 ml of TRIzol Reagent per 100 mg of tissue using a glass-Teflon or power homogenizer (Polytron, Tekmar's Tissuemizer, or equivalent). An insufficient volume of TRIzol Reagent may result in contamination of genomic DNA.

2. Phase separation: Incubate homogenized samples for 5 min at room temperature. Add 200 $\mu$l chloroform/1 ml TRIzol Reagent. Cap sample tubes securely, shake tubes vigorously by hand for 15 s, and incubate at room temperature for 2–3 min. Centrifuge the samples at 12,000$g$ for 15 min at 2–8°.

3. RNA precipitation: Transfer the colorless upper aqueous phase to a fresh tube. Precipitate RNA in the aqueous phase by mixing with 500 $\mu$l isopropyl alcohol/1 ml TRIzol Reagent used for the initial homogenization. Incubate samples at room temperature for 10 min. Centrifuge at 12,000$g$ for 10 min at 2–8°.

4. RNA wash: Remove the supernatant. Wash the RNA pellet once with 1 ml 70% ethanol/1 ml TRIzol Reagent used for the initial homogenization. Mix the sample by vortexing and centrifuge at 7500$g$ for 5 min at 2–8°.

5. Solubilizing RNA: Air-dry the RNA pellet for 5–10 min. Re-dissolve the pellet in 100 $\mu$l DEPC-treated water.

DNA Digestion

As noted, some GT genes encoding proteins functioning in 6-sulfo sialyl Lewis X biosynthesis are intronless, so we cannot differentiate whether the PCR product originates from reverse-transcribed cDNA or from contaminating genomic DNA. In these circumstances, treat the RNA preparation with RNase-free DNase I.

*Materials for DNA Digestion*

> RNase-free DNase I (Roche)
> 10× universal buffer *M*
> Phenol:chloroform:isoamylalcohol = 25:24:1 (PCI)
> 70% ethanol (diluted with DEPC-treated water)
> DEPC-treated water

*Method of DNA Digestion*

1. Adjust the concentration of each RNA sample to 1 $\mu$g/$\mu$l by adding the appropriate volume of DEPC-treated water. If the concentration of RNA samples is less than 1 $\mu$g/$\mu$l, reduce sample volumes using an evaporator and readjust the concentration.
2. Treat each RNA sample with RNase-free DNase I. Add the following to a nuclease-free microcentrifuge tube:
   > 10× universal buffer *M* 4 $\mu$l
   > RNA sample 10 $\mu$l
   > RNase-free DNase I 4 $\mu$l
   > DEPC-treated water 22 $\mu$l
3. Incubate samples at 37° for 15 min.
4. To remove RNase-free DNase I, perform PCI extraction followed by ethanol precipitation. Dissolve RNA pellet in 22 $\mu$l of DEPC-treated water.

First-Strand cDNA Synthesis

mRNA can exhibit significant secondary structure that affects the ability of the reverse transcriptase to generate first-strand cDNA. This can affect RT-PCR quantification and, therefore, should be minimized. Two types of reverse transcriptases are commonly used: Avian myeloblastosis virus reverse transcriptase (AMV-RT) and Moloney murine leukemia virus reverse transcriptase (MMLV-RT). AMV-RT is robust and retains substantial polymerization activity up to 55°, so it is suitable to eliminate problems associated with RNA secondary structure. In contrast, MMLV-RT and its derivatives have reduced RNase H activity and are, therefore, suitable for amplification of full-length cDNA (Bustin, 2000).

Although the reverse-transcription step can be primed by either oligo (dT) primers, random hexamers, or specific primers, we recommend oligo (dT) or random hexamers. Often one RNA sample is used for several RT-PCRs of different genes of interest, so the ability to use the sample in multiple assays is important. In our laboratory, we routinely use Super-Script II RNase H$^-$ Reverse Transcriptase (Invitrogen) in combination with oligo(dT) primers.

*Materials for First-Strand cDNA Synthesis*

SuperScript II RNase H$^-$ Reverse Transcriptase (Invitrogen)
5× First-Strand Buffer (Invitrogen)
0.1 $M$ DTT (Invitrogen)
Oligo(dT)$_{12-18}$ (50 m$M$)
10 m$M$ dNTP Mix (Invitrogen)
RNase Inhibitor (Invitrogen)
*Escherichia coli* RNase H (Invitrogen)
DEPC-treated water

*Method for First-Strand cDNA Synthesis*

1. Divide the above RNase-free DNase I–treated RNA samples (11 $\mu$l each) into two nuclease-free PCR tubes and add 1 $\mu$l of Oligo(dT)$_{12-18}$ (50 m$M$) to each tube.
2. Heat the mixture to 65° for 5 min and quick chill on ice. Collect the contents of the tube by brief centrifugation and add:
   5× First-Strand Buffer 4 $\mu$l
   0.1 $M$ DTT 2 $\mu$l
   10 m$M$ dNTP Mix 1 $\mu$l
   RNase Inhibitor 1 $\mu$l
3. Mix contents of the tube gently and incubate at 42° for 2 min.
4. Into one tube add 1 $\mu$l (200 U) of SuperScript II RNase H$^-$ Reverse Transcriptase and mix by pipetting gently up and down. Into the other tube add DEPC-treated water instead of SuperScript II RNase H$^-$ Reverse Transcriptase for the negative control.
5. Incubate at 42° for 50 min.
6. Inactivate the reaction by heating to 70° for 15 min.
7. To remove RNA complementary to the cDNA, add 1 $\mu$l of *Escherichia coli* RNase H and incubate at 37° for 20 min.

*PCR*

Because RT-PCR uses single-stranded cDNA as a template, nonspecific annealing of primers tends to occur during the first denaturation step. To avoid this reaction, a hot-start PCR protocol is useful to eliminate nonspecific amplification. The hot-start PCR can be achieved by using neutralizing monoclonal antibody to inhibit polymerase activity during the assembly and warmup phase of the reaction. When the temperature increases, the antibody dissociates from the enzyme and is inactivated during the denaturation step of the first cycle of PCR.

Several experimental conditions must be evaluated in preliminary experiments before undertaking the PCR:

Preliminary Experiment 1

Determine the optimal PCR conditions (primer concentration, $Mg^{2+}$ concentration, annealing temperature, etc.) in advance.

Preliminary Experiment 2

Perform PCR of all samples with serially diluted plasmid harboring the target cDNA. Estimate the range of cDNA concentration that all the samples fall into by comparing the fluorescent intensity of each sample with that of serially diluted plasmid.

Preliminary Experiment 3

Using the plasmid at the highest concentration of the aforementioned range, perform PCR to estimate the cycle number at which amplification reaches a plateau by assaying samples at different cycles of the reaction (e.g., 10, 15, 20, 25, 30 cycles). Then narrow down the cycles (e.g., 20, 21, 22, 23, 24 cycles) at which amplification occurs exponentially, but at which maximum amplification is obtained.

*Materials for PCR*

JumpStart REDTaq DNA Polymerase and accessory $10 \times$ PCR Buffer (Sigma)

$2$ m$M$ dNTP Mix (Invitrogen)

$10$ $\mu M$ sense and antisense primers

Distilled water

*PCR*

1. Add the following to a PCR tube:

   $10 \times$ PCR Buffer 1 $\mu$l

   $2$ m$M$ dNTP Mix 1 ml

   $10$ $\mu M$ sense and antisense primers 0.2 $\mu$l each

   JumpStart REDTaq DNA Polymerase 0.25 $\mu$l

   Reverse-transcribed reaction 0.2 $\mu$l

   Distilled water 7.15 $\mu$l (to 10 $\mu$l)

2. Heat reaction to 94° for 2 min to denature.

3. Perform PCR under appropriate conditions. In our case, denaturation at 94° for 30 s, annealing at 60° for 30 s, and extension at 72° for 1 min. Cycle numbers vary depending on transcript analyzed.

Densitometry Analysis

As noted, it is crucial that transcripts are analyzed during the exponential phase of amplification. Densitometry analysis is useful only when there is a linear correlation between the log of the amount of plasmid and the fluorescent intensity of PCR product.

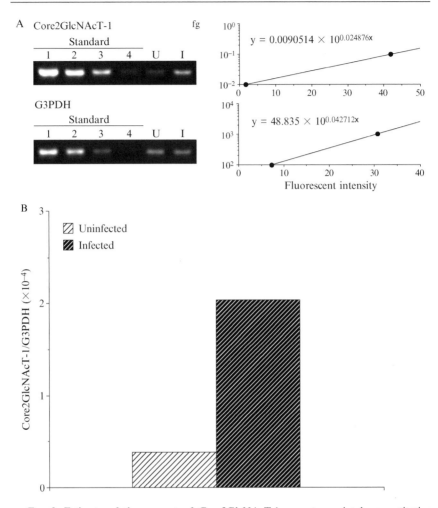

Fig. 2. Estimate of the amount of Core2GlcNAcT-1 gene transcript by quantitative reverse transcriptase polymerase chain reaction (RT-PCR). (A) A standard curve (right panel) was obtained using different amounts of plasmid: 10 fg to 10 pg for G3PDH; 10 ag to 10 fg for Core2GlcNAcT-1 (Standard 1 to 4 in left panel). Amounts varied to accommodate different levels of transcripts. (B) Estimated level of Core2GlcNAcT-1 transcript determined by RT-PCR using standard curves obtained in panel A. The ratio of the expression level of transcript of Core2GlcNAcT-1 relative to G3PDH is shown for uninfected and infected stomachs.

In our preliminary experiments, while transcript of Core2GlcNAcT-1 was increased in a *H. felis*-infected mouse and compared to an uninfected mouse (Fig. 2), key GTs and a sulfotransferase that form the L-selectin ligand did not show obvious up-regulation, probably because relatively few HEV-like vessels were formed in this mouse model, and sample RNA was isolated not from HEV-like vessels but from the whole stomach.

*Materials for Densitometry Analysis*

2% (w/v) agarose gel containing 0.1–1.0 $\mu$g/ml of ethidium bromide
National Institutes of Health (NIH) image software (http://rsb.info. nih.gov/nih-image/)

*Method for Densitometry Analysis*

1. Samples are subjected to 2% agarose gel electrophoresis.
2. Under ultraviolet (UV) light, take a picture using a CCD camera and save the image as a TIFF file.
3. Perform densitometry of each lane by using NIH image software or equivalent densitometry software.
4. Perform regression analysis on the bands originating from serially diluted plasmid harboring cDNA and obtain a standard curve with Eq. (2), where $x$ is fluorescent intensity, $y$ is the amount of DNA, and $a$ and $b$ are given constants.

$$y = a \times 10^{bx} \tag{2}$$

The concentration of the samples is determined by simple interpolation using the standard curve.
5. Quantification is achieved by comparing the amount of each gene with an endogenous standard (G3PDH) and expressing the result as a simple ratio.

## Acknowledgments

The author thanks Professors Minoru Fukuda and Jun Nakayama for encouragement, Dr. Junya Mitoma for useful discussion, and Dr. Elise Lamar for critical reading of the manuscript.

## References

Bayerdorffer, E., Neubauer, A., Rudolph, B., Thiede, C., Lehn, N., Eidt, S., and Stole, M. (1995). Regression of primary gastric lymphoma of mucosa-associated lymphoid tissue type after cure of *Helicobacter pylori* infection. MALT Lymphoma Study Group. *Lancet* **345**, 1591–1594.

Bustin, S. A. (2000). Absolute quantification of mRNA using real-time reverse transcription polymerase chain reaction assays. *J. Mol. Endocrinol.* **25,** 169–193.

Gilliland, G., Perrin, S., Blanchard, K., and Bunn, H. F. (1990). Analysis of cytokine mRNA and DNA: Detection and quantitation by competitive polymerase chain reaction. *Proc. Natl. Acad. Sci. USA* **87,** 2725–2729.

Kobayashi, M., Lee, H., Schaffer, L., Gilmartin, T. J., Head, S. R., Takaishi, S., Wang, T. C., Nakayama, J., and Fukuda, M. (2006). A distinctive set of genes is up-regulated during the inflammation-carcinoma sequence in mouse stomach infected by *Helicobacter felis.* submitted.

Kobayashi, M., Mitoma, J., Nakamura, N., Katsuyama, T., Nakayama, J., and Fukuda, M. (2004). Induction of peripheral lymph node addressin in human gastric mucosa infected by *Helicobacter pylori. Proc. Natl. Acad. Sci. USA* **101,** 17807–17812.

Marshall, B. J., and Warren, J. R. (1984). Unidentified curved bacilli in the stomach of patients with gastritis and peptic ulceration. *Lancet* **i,** 1311–1315.

Nakayama, H., Yokoi, H., and Fujita, J. (1992). Quantification of mRNA by non-radioactive RT-PCR and CCD imaging system. *Nucleic Acid. Res.* **20,** 4939.

Peek, R. M., Jr., and Blaser, M. J. (2002). *Helicobacter pylori* and gastrointestinal tract adenocarcinomas. *Nat. Rev. Cancer* **2,** 28–37.

Rosen, S. D. (2004). Ligands for L-selectin: Homing, inflammation, and beyond. *Annu. Rev. Immunol.* **22,** 129–156.

Siebert, P. D., and Larrick, J. W. (1992). Competitive reverse transcription-polymerase chain reaction (RT-PCR) can be used to obtain quantitative information of mRNA levels comparable to traditional RNA blot techniques, with the added advantages of PCR. *Nature* **359,** 557–558.

Sipponen, P., and Hyvarinen, H. (1993). Role of *Helicobacter pylori* in the pathogenesis of gastritis, peptic ulcer and gastric cancer. *Scand. J. Gastroenterol. Suppl.* **196,** 3–6.

Taniguchi, N., Honke, K., and Fukuda, M. (2002). "Handbook of Glycosyltransferases and Related Genes." Springer-Verlag, Tokyo, Japan.

Yeh, J. C., Hiraoka, N., Petryniak, B., Nakayama, J., Ellies, L. G., Rabuka, D., Hindsgaul, O., Marth, J. D., Lowe, J. B., and Fukuda, M. (2001). Novel sulfated lymphocyte homing receptors and their control by a Core1 extension $\beta$1,3-*N*-acetylglucosaminyltransferase. *Cell* **105,** 957–969.

Yeh, J. C., Ong, E., and Fukuda, M. (1999). Molecular cloning and expression of a novel $\beta$-1,6-*N*-acetylglucosaminyltransferase that forms core 2, core 4, and I branches. *J. Biol. Chem.* **274,** 3215–3221.

Yokoi, H., Natsuyama, S., Iwai, M., Noda, Y., Mori, T., Mori, K. J., Fujita, K., Nakayama, H., and Fujita, J. (1993). Non-radioisotopic quantitative RT-PCR to detect changes in mRNA levels during early mouse embryo development. *Biochem. Biopys. Res. Commun.* **195,** 769–775.

[10]   Expression Profiling of Glycosyltransferases and
Related Enzymes Using Gene Microarrays

*By* Jennifer A. Hammond and Steven R. Head

## Abstract

Gene expression analysis has become a standard tool for studies of biological models of disease processes, growth, and development and has been used increasingly in the field of clinical research. The completion of the human genome sequence is now providing the glycomics research community with the tools to develop a synergism with related fields such as MS glycan profiling and proteomics. The use of microarray technology to measure changes in gene expression is a useful approach for the study of the expression profiles of glycosyltransferases, carbohydrate binding proteins, and related enzymes. The availability of standardized microarray platforms, high-quality reagents for sample preparation and hybridization, and image and statistical analysis tools has contributed to recent advances in the application of this technology to glycobiology. Most microarray experiments today are done within specialized core facilities equipped with instrumentation, software, and personnel with expertise in their application. The focus of this chapter is on aspects of microarray analysis common to all platforms and within the realm of the glycobiologist: consideration of experimental design issues and methods for the preparation of RNA samples from a variety of tissue sources suitable for microarray analysis.

## Overview

It is beyond the scope of this chapter to describe how to analyze RNA samples using all of the microarray platforms available. It is also not appropriate to select one platform to focus on since no single platform may be available to all users or agreed on as the "best" platform to use. In addition, methods and approaches to data analysis and interpretation are complex and rapidly evolving, so the reader is advised to consult with microarray core laboratory personnel for assistance in interpreting the typically large data sets generated by microarray experiments. The focus of this chapter is on aspects of microarray analysis common to all platforms and within the realm of the glycobiologist: consideration of experimental design issues and the preparation of RNA samples suitable for microarray analysis. *A key point to remember is that data can be analyzed and reanalyzed as often as needed, but*

METHODS IN ENZYMOLOGY, VOL. 416
0076-6879/06 $35.00
DOI: 10.1016/S0076-6879(06)16010-3

*the quality of the microarray data can never be better than the quality of the RNA samples and the experimental design used to generate the data ("garbage in = garbage out").*

Gene expression microarrays were initially developed on two technological platforms: (1) Affymetrix GeneChips composed of arrays of short oligonucleotides (25 mers) synthesized *in situ* using photolithographic techniques (Lockhart *et al.*, 1996) and (2) arrays of polymerase chain reaction (PCR) products derived from complementary DNA (cDNA) libraries (typically 200–1000 base pairs in length) spotted onto glass slides (Perou *et al.*, 1999; Schena *et al.*, 1995; Sudarsanam *et al.*, 2000). Recently, spotted arrays of oligonucleotides (30–70 bases) have been used as an alternative to cDNA-based microarrays (Chambers *et al.*, 1999; Hughes *et al.*, 2001; Stingley *et al.*, 2000) and a variety of filter-based arrays have also been developed and commercialized.

Affymetrix GeneChip arrays used oligonucleotide "probe sets" consisting of multiple "probe pairs" (generally 11 pairs per target gene) to interrogate each targeted messenger RNA (mRNA) sequence. Improvements in array manufacturing technology now allow an entire transcriptome to be analyzed on a single array. Each probe pair consists of one perfect match (PM) and one mismatch (MM) 25-base oligonucleotide. The "perfect match" oligonucleotide is complementary to a given portion of the targeted gene, typically in the $3'$ portion of the transcript, while the "mismatch" oligonucleotide is identical in sequence to the perfect match probe except for a single mismatched base at the 13th position in the center of the probe. The hybridization intensities from the PM and MM probes for a given gene may be used to calculate a gene expression "signal." The signal is a quantitative metric calculated for each probe set, representing the relative abundance of an mRNA transcript. Several software programs are commonly used to calculate hybridization signals including GeneChip Operating Software (GCOS v1.0; Affymetrix), dChip (Li, 2001a,b), RMA, and GCRMA (Irizarry and Wong, 2003). These signal generating algorithms continue to evolve and the reader is advised to consult with microarray core laboratory personnel for assistance in determining the most appropriate methods for their microarray data.

In addition to (and independent of) the hybridization signal, a metric termed "detection call" is calculated by GCOS for each probe set to indicate whether or not the transcript can be detected as "Present" or "Absent" with a predefined level of statistical confidence. These Present/Absent calls are determined by an algorithm based on the difference in signals between the PM and MM probes. It is the combined use of PM and MM probes that make the Present/Absent calls possible. The key point is that alternative microarray platforms generally do not include mismatched control sequences for each gene. They depend on analysis of the distribution of signal intensities from negative controls to estimate the likelihood that a specific transcript is actually being detected.

The use of the PM/MM strategy to increase specificity in combination with interrogating each transcript with multiple probe pairs increases the sensitivity of the Affymetrix GeneChip technology to detect low-level transcripts. GeneChip arrays have been reported to detect as little as 0.075 pM (one transcript per 7 cells) of target mRNA (Lipshutz *et al.*, 1999). A more recent report suggests 0.5 pM (1–3 transcripts/cell) may be a more realistic assessment of this technology's practical detection limit (Chudin *et al.*, 2002). Our own analysis of spike-in data on the Hu-133A array provided by Affymetrix indicates that transcripts spiked in at 1 pM are called as "Present" only 50% of the time. This concentration represents an equivalent of 2–6 copies/cell. Transcripts concentrations $\geq 4$ pM are called *Present* nearly 100% of the time. Transcripts that are expressed at low concentrations in cells may show signal changes that correlate with expression changes (as reported by Lipshutz *et al.* [1999] and Chudin *et al.* [2002]) despite being called "Absent" by the Affymetrix detection call algorithm. Despite the inability of the Affymetrix detection call algorithm to confidently detect transcripts present at very low concentrations, our analysis indicates that the algorithm compares favorably to alternative methods such as calculating a z-score to estimate confidence of whether a transcript is present or not (data not shown).

In the spotted cDNA and oligonucleotide microarray formats, either PCR products or synthetic oligonucleotides (usually 30–70 bases in length) are printed onto glass slides coated with either poly-l-lysine or a silane derivative to link the DNA to the surface. Technology for manufacturing these arrays is well developed (http://cmgm.stanford.edu/pbrown/mguide/index.html) and capable of printing 20,000 or more spots on a standard 25- × 75-mm glass slide. Usually one PCR product or oligonucleotide sequence is printed per target gene (often with several replicate spots/array).

A different hybridization strategy is frequently (but not always) used with spotted arrays. Typically, labeled cDNA from two samples (one labeled with Cy3 and the other labeled with Cy5) are co-hybridized to a single array. After hybridization, each spot on the array is scanned at two wavelengths to collect fluorescence signals from both dye labels. After signal normalization, relative Cy3 and Cy5 signal intensities are compared for each spot. The data that are generated describe the relative abundance of each of the transcripts in the two samples (i.e., differential gene expression) as a function of the relative amounts of fluorescent signal.

Spotted array technology with dual labeling has been tested for the ability to discriminate between genes that have a high level of sequence homology. Reports demonstrate discrimination of genes on cDNA arrays ranging from 70–94% sequence identity (Miller *et al.*, 2002; Schena *et al.*, 1996). The ability of spotted array technologies to discriminate related gene sequences probably varies with the sequence, as other reports show a failure to distinguish genes with 88% identity (i.e., *Saccharomyces cerevisiae* ADH1 and ADH2) (DeRisi *et al.*, 1997). On the other hand, we did a retrospective analysis of yeast GeneChip data generated in our Microarray Core Facility at Scripps. The results were that the Affymetrix GeneChips successfully documented reciprocal regulation of ADH1 and ADH2. Therefore, it is important to keep in mind that there are differences in all these technologies for gene expression detection and that validation of experimental results for any technology remains a critical step.

Several publications (Chambers *et al.*, 1999; Hughes *et al.*, 2001; Stingley *et al.*, 2000) describe the use of spotted oligonucleotide arrays for gene expression studies. In addition, several companies (Sigma, Qiagen, MWG Biotech, GE Healthcare, Agilent) have commercialized oligonucleotide arrays or oligonucleotide sets specifically for printing as gene probes. Printed oligonucleotide arrays have potential advantages over cDNA arrays including the elimination of problems associated with clone sequence verification, PCR amplification artifacts (i.e., multiple bands, missing bands), and PCR contamination. The use of synthetic oligonucleotides provides more flexibility in probe design to facilitate targeting of specific sequences. This design feature also permits selection for more uniform hybridization temperatures, as well as the elimination of DNA sequences that are repetitive, low complexity, or contain significant secondary structure.

New technologies for array synthesis using digital light processing (developed by Texas Instruments using electronically controlled arrays of mirrors) (Nuwaysir, 2002) and piezoelectric printing (Hughes *et al.*, 2001) have enabled lower costs for custom array designs by eliminating the need to manufacture photolithographic masks for *in situ* array synthesis. Affymetrix leads the industry with respect to the highest density arrays on the market (5 $\mu M$ features). It is likely that improvements to technology will continue to increase the amount of information that can be obtained from a single chip hybridization and reduce relative costs of microarray experiments.

## Experimental Design

### Introduction

A comprehensive discussion of microarray experimental design issues is beyond the scope of this chapter; however, some basic issues to be considered when planning a microarray experiment are discussed in this section.

Microarray experiments are expensive to perform and time consuming to analyze and interpret. A single hybridization can generate many thousands of data points, making the experimental design and analysis method chosen of utmost importance. The reader is advised to consult with microarray core laboratory personnel before performing a microarray experiment to discuss the specifics of the proposed experiment and obtain advice on experimental design issues. For a given experiment, the "right" design will depend not only on the specific hypothesis being tested, but also on resources and long-range plans. Key factors to be considered in the design of a microarray experiment include choice of an appropriate experimental and control samples for analysis and the number and types of replicates.

*Replicates*

There are two main categories of experimental replicates: biological and technical. The function of a biological replicate is to compare states using independent samples that differ in an observed biological phenomenon. Technical replicates function to measure the precision of the microarray platform used in the analysis. Because the goal of most microarray experiments is to measure changes in gene expression, not to evaluate the precision of a particular microarray platform, it is generally more useful to focus on biological replicates with the understanding that the variance in gene expression measurements for independent biological samples includes not only the biological variance contributed by the samples themselves but the technical variance contributed by the sample labeling, hybridization, detection, and chip-to-chip differences.

The question of how many replicates to do depends on how small the differences are that you want to detect, as well as the level of variance in the measurements taken from the replicate samples. In a well-controlled system such as the analysis of whole kidney samples from healthy congenic mice that are age- and sex-matched littermates, biological variance will be relatively low. However, if collecting samples requires considerable technical skill (such as dissecting retinas from E10 mouse embryos) or requires extensive manipulation of the samples (e.g., activation of lymphocytes followed by cell sorting), the variance may be increased considerably, requiring a higher level of replication to achieve confidence in the resulting data. Similarly, analysis of clinical samples may result in high variability because of technical issues of collecting samples (such as an ultrasound-guided needle biopsy of a kidney or liver), heterogeneity in the samples themselves (such as tumors or diseased organs/tissues), as well as differences in the genetic makeup, natural history, and environmental exposure of the patients from whom the samples are collected.

In general, we recommend at least three independent biological replicates as a minimum for any microarray experiment. That is not to say that no useful data can be collected from less than three replicates, but the options for statistical analysis increase substantially at this level. For experiments in which more variance is anticipated, the number of replicates should be increased. Power calculations may be performed for microarray experiments but do require a data set from which to estimate the variance.

*Sample Pooling*

There are two reasons for pooling sample: (1) If the amount of material obtainable from any one sample is insufficient for hybridization, then pooling is probably unavoidable. Independent biological replicates may be obtained by creating multiple pools. (2) In an attempt to increase the statistical power of an experiment using a defined number of replicates through analysis of multiple pooled rather than multiple individual replicates (Kendziorski *et al.*, 2003; Peng *et al.*, 2003). The effectiveness of this approach is controversial however and may not be an appropriate strategy (Shih *et al.*, 2004). Pooling of samples should never be done if one of the goals is to identify new phenotypes based on gene expression data, for example, screening a large number of tumor samples to look for gene expression patterns associated with clinical outcome. Pooling the samples could obscure subgroups of patients with similar phenotypes but distinct gene expression patterns.

Materials for RNA Handling and Storage

Diethylpyrocarbonate (DEPC)-treated water, stored at room temperature (Ambion, Austin, TX)

RNAseZap RNAse Decontamination Solution, stored at room temperature (Ambion)

TE Buffer: 10-m$M$ Tris–HCl, pH 8.0, 1.0 m$M$ EDTA, pH 8.0, stored at room temperature (Teknova, Hollister, CA)

THE RNA Storage Solution, 1 m$M$ sodium citrate, pH 6.4 ± 0.2, stored at room temperature (Ambion)

Methods for RNA Handling and Storage

1. RNA handling. Special care should be taken when working with RNA, due to its chemical instability and the general presence of RNases in the laboratory. Whenever working with RNA, a dedicated bench space should be used that has been treated with commercially available RNase-inactivating

agents (e.g., RNAseZap, Ambion). Gloves should always be worn and changed frequently to prevent contamination by RNases commonly found on the body. Commercially available RNase-free plasticware and reagents (e.g., water) should always be used whenever working with RNA.

It is critical that RNA be preserved and prevented from degrading. When thawing RNA frozen in solution, the sample should be kept on ice during thawing and throughout the entire experiment, unless otherwise stated in the protocol. Prolonged exposure to high temperatures ($>65°$) can affect RNA integrity and should be avoided whenever possible.

2. RNA storage. It is imperative that RNA be suspended and stored in an RNase-free solution. Commonly used solutions include nuclease-free (DEPC-treated) $H_2O$, TE Buffer, or other commercially available RNA storage solutions (e.g., THE RNA Storage Solution, Ambion). Once suspended in an appropriate solution, RNA that is to be stored for up to 1 mo should be frozen at $-20°$. RNA stored for archival purposes for longer than 1 mo should be frozen at $-80°$ to minimize potential long-term degradation. Freeze–thaw cycles should be kept to a minimum whenever possible.

## Methods for RNA Analysis

1. RNA analysis. To assess the quality of RNA, both purity and degradation should be taken into consideration. To assess for purity, a spectrophotometer is generally used to determine the 260/280 ratio. A peak of 260 nm is the wavelength at which DNA and RNA absorb light, whereas 280 nm is the peak wavelength for proteins. A commonly acceptable ratio for pure RNA is 1.8–2.0. Lower ratios are an indication of contamination with protein.

The assessment of RNA degradation is somewhat more subjective. Total RNA is made up of ~3% mRNA and >80% ribosomal RNA, with most of the ribosomal RNA (rRNA) derived from the 28S and 18S ribosomes (in mammalian systems). mRNA quality is generally assessed by electrophoresis of total RNA followed by staining with ethidium bromide. The integrity of the 28S and 18S bands is assessed under the assumption that integrity of the rRNA reflects that of the underlying mRNA. The 28S and 18S rRNAs are approximately 5 and 2 kb in size, and anything other than sharp bands of this size on the gel are indicative of RNA degradation. This type of visual assessment is fairly subjective; therefore, the use of improved analytical tools, if available is recommended. The Agilent 2100 Bioanalyzer uses a combination of microfluidics, capillary electrophoresis, and fluorescence to evaluate both RNA concentration and integrity.

Using as little as 50 ng of total RNA, the Bioanalyzer produces a gel-like image and an electropherogram for each sample. For RNA samples of high quality, the electropherogram should show two strong, distinct bands representing the rRNA and a light smear behind the ribosomal bands representing the mRNA. Degradation of the total RNA by RNases can be seen by a shift in the RNA size distribution towards smaller fragments and a decrease in fluorescence signal of ribosomal peaks.

Materials for RNA Extraction from Tissue

> Buffer RLT (Qiagen, Valencia, CA) with 10 $\mu$l $\beta$-mercaptoethanol (Sigma, St. Louis, MO) added per 1 ml Buffer RLT, stored at room temperature
> Buffer RPE (Qiagen) with appropriate amount of 100% ethanol (Invitrogen, Carlsbad, CA) added, stored at room temperature
> Chloroform, stored at room temperature (Invitrogen)
> DEPC-treated water, stored at room temperature (Ambion)
> DNase I Buffer (10×), stored at $-20°$ (Ambion)
> DNase I Enzyme, stored at $-20°$ (Ambion)
> DNase Inactivation Reagent, stored at $-20°$ (Ambion)
> Ethanol: 100%, stored at $-20°$ (Invitrogen)
> Ethanol: 80%, working solution prepared with DEPC-treated water (Ambion) and 100% EtOH (Invitrogen), stored at $-20°$
> Isopropanol, stored at room temperature (Invitrogen)
> RNA*later* RNA Stabilization Reagent, stored at room temperature (Qiagen)
> RNeasy spin column, stored at room temperature (Qiagen)
> TRIzol Reagent, stored at $4°$ (Invitrogen)

Methods for RNA Extraction from Tissue

1. Tissue harvest. Our laboratory uses two methods to preserve RNA integrity when harvesting tissue depending on the size of the tissue pieces. Whatever the method of storage when harvesting, it is crucial that the tissue be preserved immediately following sacrifice and extraction.

2. Immersion in RNA*later*. Upon extraction from the source, immediately slice the tissue into pieces no wider than 0.5 cm and drop into RNA*later* RNA Stabilization Reagent (Qiagen). The volume of RNA*later* should be at least 10 times the volume of the tissue. Slicing of tissue can be performed while submerged in RNA*later* solution in a small RNase-free

dish. Tissue can be stored submerged in RNA*later* until homogenization according to the following: up to 4 weeks at 2–8°, up to 7 days at 18–25°, or up to 1 day at 37°. For archival storage at −20°, first incubate the sample overnight in the reagent at 2–8° and then transfer the tissue, in the reagent, to −20° for storage. For archival storage at −80°, first incubate the sample overnight in the reagent at 2–8°, then remove the tissue from the RNA*later* and transfer it to −80° for storage.

3. Snap-freezing in liquid nitrogen. For tissue pieces wider than 0.5 cm, tissue should be immersed in liquid nitrogen upon extraction from the source and kept in the nitrogen until the procedure is completed. Upon completion of the harvest procedure, remove the tissue pieces from the liquid nitrogen and transfer to empty falcon tubes kept on dry ice. Keep the tissue frozen on dry ice, at −20° or −80° until the homogenization procedure is ready to be performed. If tissue homogenization will not be performed within 1–2 h, it is recommended that the frozen tissue be stored at −80°.

4. Tissue homogenization. For tissues that have been snap-frozen, the homogenization should be done by mortar and pestle that have been cooled to freezing temperature in a liquid nitrogen bath. Prepare falcon tubes containing at least 1 ml TRIzol reagent (Invitrogen) per 100 mg tissue to be homogenized. Transfer the frozen tissue to the mortar and grind with the pestle until reduced to a layer of very fine dust. Using a spatula that is RNase-free, transfer the dust to the falcon tubes containing the TRIzol reagent. Vortex mixture thoroughly.

For tissues that have been stored in RNA*later*, a handheld tissue grinder is recommended. The tissue should be removed from the RNA*later* solution and homogenized in the presence of 1 ml TRIzol/100 mg tissue until the tissue is completely dissolved in solution. Once homogenized, aliquot solution to 1.5 ml Eppendorf tubes (1 ml solution per tube) and leave in TRIzol at room temperature for 5 min.

5. Phase separation. Add 200 $\mu$l chloroform for every 1 ml TRIzol solution, vortex for 15 s, and leave at room temperature for 2–3 min. Centrifuge samples at 12,000$g$ for 15 min at 2–8°.

6. RNA precipitation. Following centrifugation, there will be three phases visible within the tube. Transfer the topmost aqueous phase to a fresh 1.5 ml Eppendorf tube, being careful not to contaminate the solution with the other phases. Contamination will be obvious by the visual presence of any flakes or opaque liquid. Add 500 $\mu$l isopropanol per 1 ml TRIzol solution to the new Eppendorf tube and incubate at room temperature for 10 min. Centrifuge samples at 12,000$g$ for 10 min at 2–8°.

7. RNA wash and resuspension. Following centrifugation, remove the supernatant and discard. Wash RNA pellet with 500 $\mu$l 80% EtOH

per 1 ml TRIzol solution and vortex. Centrifuge samples at 7500g for 5 min at 2–8°. Remove supernatant and discard. Allow any remaining EtOH to evaporate at room temperature for 2–3 min. Transfer tubes to a 70° heat block and let sit for 2–3 min. Redissolve the RNA pellet in 81 $\mu$l of DEPC-treated $H_2O$.

8. DNAse treatment (using DNAse treatment kit from Ambion). Add 8 $\mu$l of 10× DNase I Buffer to each 81 $\mu$l solution. Add 2 $\mu$l of DNase I Enzyme to each solution. Vortex, quick-spin, and incubate at 42° for 25 min. Add 9 $\mu$l of DNase Inactivation Reagent and incubate at room temperature for 2–3 min.

9. RNeasy Column Purification (using RNeasy Column Protocol from Qiagen). Add 350 $\mu$l Buffer RLT with $\beta$-mercaptoethanol added (10 $\mu$l per 1 ml Buffer RLT). Add 250 $\mu$l 100% EtOH. Apply entire volume to RNeasy column and spin at full speed for 1 min. Reapply entire volume to RNeasy column and spin at full speed for 1 min. Transfer column to a new 2-ml collection tube. Add 750 $\mu$l Buffer RPE (making sure EtOH has been added) and spin at full speed for 1 min. Discard flow-through and add 750 $\mu$l Buffer RPE (with EtOH added) and spin at full speed for 1 min. Discard flow-through and spin at full speed for 1 min. Transfer column to a new labeled 1.5-ml Eppendorf tube. Add 56 $\mu$l DEPC-treated $H_2O$ and let sit for 2 min at room temperature. Spin at full speed for 2 min. Discard column and transfer tube to ice.

10. Quantification and quality control. Quantify each sample using NanoDrop or spectrophotometer. Run 5 $\mu$l of each sample on an agarose gel to assess RNA quality.

### Materials for RNA Extraction from Cell Pellets

In addition to reagents listed in the section "Materials for RNA Extraction from Tissue," earlier in this chapter, phosphate-buffered saline (PBS), stored at room temperature (BioWhittaker, Walkersville, MD)

### Methods for RNA Extraction from Cell Pellets

1. Cell harvest. For cultures of cells, first pellet cells out of growth media. Wash cell pellet three times with sufficient amount of PBS and resuspend in RNA*later* (Qiagen). Store the cells suspended in RNA*later* until homogenization according to the following: up to 4 wk at 2–8°, up to 7 days at 18–25°, or up to 1 day at 37°. For archival storage at −20°, first incubate the sample overnight in the reagent at 2–8°, then transfer the tissue, in the reagent, to −20° for storage. For archival storage at −80°, first incubate the sample overnight in the reagent at 2–8°, then remove the tissue from the RNA *later* and transfer it to −80° for storage.

2. Homogenization. Quantify the suspended cells using a NanoDrop or spectrophotometer. Pellet the cells and resuspend in TRIzol reagent (Invitrogen) at a volume of $5 \times 10^6$ cells per 1 ml TRIzol. Once homogenized, aliquot solution to 1.5-ml Eppendorf tubes (1 ml solution per tube) and leave in TRIzol at room temperature for 5 min. Follow steps 5 through 9 as outlined in the protocol for RNA extraction from tissue, earlier in this chapter.

### Materials for RNA Extraction from Blood Using PAXgene System

Buffer BR1 (Resuspension Buffer), stored at room temperature (Qiagen)

Buffer BR2 (Binding Buffer), stored at room temperature (Qiagen)

Buffer BR3 (Wash Buffer), stored at room temperature (Qiagen)

Buffer BR4 (Wash Buffer) (Qiagen) with 4 volumes 100% ethanol (Invitrogen) added, stored at room temperature

Buffer BR5 (Elution Buffer), stored at room temperature (Qiagen)

DEPC-treated water, stored at room temperature (Ambion)

DNase I stock solution (Qiagen) with 70 $\mu$l Buffer RDD (Qiagen) added, dispensed into 10-$\mu$l aliquots and stored at $-20°$

Ethanol: 100%, stored at $-20°$ (Invitrogen)

PAXgene Blood RNA Collection Tube, stored at room temperature (Qiagen)

PAXgene spin column, stored at room temperature (Qiagen)

Proteinase K, stored at room temperature (Qiagen)

### Methods for RNA Extraction from Blood Using PAXgene System

1. Preparation. Draw 2.5 ml of blood directly into PAXgene tube and invert the tube 10 times immediately. Do not shake. Before starting RNA purification procedure, ensure that the tubes have been incubated at room temperature for at least 2 h to ensure complete cell lysis. This incubation can be carried out either before or after storage at $-20°$ or below. If tubes were immediately frozen after blood collection, then after removal from storage, first equilibrate to room temperature for at least 2 h and then incubate at room temperature for an additional 2 h. After thawing, invert the tubes 10 times. Set shaker-incubator to 55°. Set BioOven to 65°. Thaw DNase I stock solution for on-column DNase digestion.

2. Isolation of total RNA. Centrifuge the PAXgene Blood RNA tube for 10 min at 3345$g$ at room temperature, brake on, using a swing-out rotor with adapters for round-bottom tubes. Remove the supernatant by decanting into a 15-ml conical tube and discard. Dry the rim of the tube with a Kim wipe. Add 5 ml RNase-free water to the pellet, and close the tube using a fresh secondary Hemogard closure. Thoroughly resuspend the pellet by pulse vortexing with vortexer set at ~2200. Centrifuge for 10 min at 3345$g$, room temperature, brake on.

Remove the entire supernatant by decanting and discard. Add 360 $\mu$l Buffer BR1 and vortex until the pellet is visibly dissolved. Pipette the sample into a 1.5-ml microcentrifuge tube. Add 300 $\mu$l Buffer BR2 and 40 $\mu$l Proteinase K. Mix by vortexing, and incubate for 10 min at 55° using a shaker-incubator at 1400 rpm (setting 4.0) with tubes taped to surface. Centrifuge the sample for 3 min at maximum speed in a microcentrifuge, and transfer the supernatant to a fresh 1.5-ml microcentrifuge tube. Add 350 $\mu$l 100% EtOH. Mix by vortexing, and quick-spin for 1–2 s to remove drops from inside of the tube lid.

Add 700 $\mu$l sample to a PAXgene spin column, place in a 2-ml processing tube, and centrifuge for 1 min at maximum speed. Place the PAXgene spin column in a new 2-ml processing tube and discard the old processing tube containing the flow-through. Add remaining sample to the PAXgene spin column, and centrifuge for 1 min at maximum speed. Place the PAXgene spin column in a new 2-ml processing tube and discard the old processing tube containing the flow-through. Add 350 $\mu$l Buffer BR3 to the PAXgene spin column. Centrifuge for 1 min at maximum speed.

Transfer the PAXgene spin column to a new processing tube and discard the old processing tube containing the flow-through. Add 10 $\mu$l DNase I stock solution to 70 $\mu$l Buffer RDD. Mix by gently flicking the tube, and centrifuge briefly to collect residual liquid from the sides of the tube. Add the DNase I incubation mix (80 $\mu$l) directly onto the PAXgene spin column membrane and incubate on the benchtop for 15 min. Add 350 $\mu$l Buffer BR3 to the PAXgene spin column. Centrifuge for 1 min at maximum speed. Place the PAXgene spin column in a new 2-ml processing tube and discard the old processing tube containing the flow-through. Add 500 $\mu$l Buffer BR4 to the PAXgene spin column and centrifuge for 1 min at maximum speed. Place the PAXgene spin column in a new 2-ml processing tube and discard the old processing tube containing the flow-through. Add another 500 $\mu$l Buffer BR4 to the PAXgene spin column. Centrifuge for 3 min at maximum speed to dry the PAXgene spin column membrane.

Place the PAXgene spin column in a new 2-ml processing tube and discard the old processing tube containing the flow-through. Centrifuge for 1 min at maximum speed. Place the PAXgene spin column in a 1.5-ml elution tube and discard the old processing tube containing the flow-through. Add 40 μl Buffer BR5 directly onto the PAXgene spin column membrane. Centrifuge for 1 min at maximum speed to elute the RNA. Do not discard the eluate. Add another 40 μl Buffer BR5 directly onto the PAXgene spin column membrane. Centrifuge for 1 min at maximum speed to elute further RNA. Do not discard the eluate. Incubate the eluate for 5 min at 65° in BioOven. After incubation, chill immediately on ice. Quantify RNA using NanoDrop or spectrophotometer.

Materials for RNA Extraction from Blood Using CPT System

    23G needle attached to 5-ml syringe (BD, Franklin Lakes, NJ)
    Buffer RLT (Qiagen) with 10 μl β-mercaptoethanol (Sigma) added
        per 1 ml Buffer RLT, stored at room temperature
    Buffer RPE (Qiagen) with appropriate amount of 100% ethanol
        (Invitrogen) added, stored at room temperature
    Chloroform, stored at room temperature (Invitrogen)
    DEPC-treated water, stored at room temperature (Ambion)
    Ethanol: 100%, stored at −20° (Invitrogen)
    Ethanol: 70%, working solution prepared with DEPC-treated water
        (Ambion) and 100% EtOH (Invitrogen), stored at −20°
    Isopropanol, stored at room temperature (Invitrogen)
    PBS, stored at room temperature (BioWhittaker)
    RNA*later* RNA Stabilization Reagent, stored at room temperature
        (Qiagen)
    RNeasy spin column, stored at room temperature (Qiagen)
    TRIzol Reagent, stored at 4° (Invitrogen)
    Vacutainer CPT Cell Preparation Tube (with sodium citrate), stored
        at room temperature (BD)

Methods for RNA Extraction from Blood Using CPT System

    1. Preparation. Draw 8 ml of blood into the CPT tube, invert the tube 10 times immediately after draw (do not shake), keep at room temperature, and centrifuge within 2 h. Bring TRIzol (Invitrogen) and isopropanol to room temperature. Cool Eppendorf centrifuges. Set BioOven to 55°.
    2. Isolation of cells. Centrifuge within 2 h from the time of blood draw in a centrifuge with a swing-out bucket rotor for 20 min, room temperature, at 1700g. Using a 3-ml transfer pipette, aliquot approximately 3 ml of clear plasma from the uppermost layer to a 4-ml tube, avoid disturbing the whitish cell layer, and freeze at −20°. Recap the tube with the stopper and invert 10 times. Pour off the cell/plasma mixture into a 15-ml tube. Add PBS to

bring the volume to 15 ml, cap tube, and mix cells by inverting tube five times. Centrifuge for 15 min, room temperature, at 300$g$. Aspirate as much supernatant as possible without disturbing the cell pellet, leaving a few microliters of supernatant with the cell pellet. Resuspend the pellet by gently vortexing or tapping tube with finger. Add PBS to bring volume up to 10 ml, cap tube, and mix cells by inverting five times. Centrifuge for 10 min, room temperature, at 300$g$. Aspirate as much supernatant as possible without disturbing the cell pellet, leaving a few microliters of supernatant with the cell pellet. Resuspend the pellet by gently vortexing or tapping tube with finger. Add 1 ml of RNA*later* solution (Qiagen) to the cell pellet, resuspend, and then transfer into a 2-ml tube for storage at $-20°$ until total RNA isolation procedure is to be performed.

3. Isolation of total RNA. Centrifuge PBMC/RNA*later* mixture at 14,000$g$, room temperature, for 2 min to pellet the cells. Remove supernatant using pipette. In fume hood, add 1 ml TRIzol reagent. Using a 23G needle attached to a 5-ml syringe, homogenize the cells by carefully aspirating and dispensing the TRIzol through the needle. Check visually to ensure complete homogenization. Wear eye protection during this step, and do not recap the needle. In fume hood, add 200 $\mu$l chloroform, cap securely, and vortex lightly for 15–20 s. Spin at 12,000$g$ at 4° for 10 min. Carefully remove the upper aqueous layer down to the interphase using a P200 pipette tip and transfer into a new 1.5-ml tube ($\sim$500 $\mu$l volume). Save the tubes containing the TRIzol/chloroform mixture for subsequent DNA and protein extraction (if needed). Add 500 $\mu$l isopropanol (at room temperature) and mix by inversion (add more isopropanol to tubes that have total volumes that are <1 ml). Incubate at room temperature for 10 min. Spin at 12,000$g$ at 4° for 10 min. Carefully decant the supernatant by pouring off. Add 500 $\mu$l 70% EtOH. Do not resuspend pellet. Spin at 7000$g$ at 4° for 10 min. Carefully decant the supernatant by pouring off and turn tubes upside down on Kim wipes. Using a P10 pipette, aspirate any remaining EtOH and then immediately add 100 $\mu$l DEPC-treated $H_2O$. Do not allow the RNA pellet to dry completely, do not speed-vac, and do not resuspend the pellet. Incubate the tubes at 55° in BioOven for 10 min. Pellet usually dissolves by itself, tap the tube a couple of times, and then quick spin. Store sample at 4° or $-20°$ until ready to perform the RNeasy cleanup.

4. Cleanup of total RNA (using RNeasy Column Protocol from Qiagen). Add 350 $\mu$l Buffer RLT (with $\beta$-mercaptoethanol added; 10 $\mu$l BME:1 ml Buffer RLT) to each tube containing sample (total RNA in 100 $\mu$l $H_2O$). Add 250 $\mu$l cold 100% EtOH, and mix. Add mixture to RNeasy column. Spin at 10,000$g$ at room temperature for 1 min. Reapply flow-through to column and spin at 10,000$g$ at room temperature for 1 min. Place column in a new 2-ml collection tube. Add 500 $\mu$l Buffer RPE (make sure EtOH has been

added), spin at 10,000g at room temperature for 1 min. Discard flow-through. Add 500 $\mu$l Buffer RPE, and spin at 10,000g at room temperature for 1 min. Discard flow-through. Spin column at 10,000g at room temperature for 2 min. Heat an aliquot of $H_2O$ to 70°. Place column in a new 1.5-ml microcentrifuge tube. Add 50 $\mu$l $H_2O$ heated at 70° to the column membrane, and incubate at room temperature for 1 min. Spin column at maximum speed at room temperature for 2 min. Quantify RNA using NanoDrop or spectrophotometer.

## Acknowledgment

This work was supported by the National Institute of General Medical Sciences Grant GM62116 to the Consortium for Functional Glycomics.

## References

Chambers, J., Angulo, A., Amaratunga, D., Guo, H., Jiang, Y., Wan, J. S., Bittner, A., Frueh, K., Jackson, M. R., Peterson, P. A., Erlander, M. G., and Ghazal, P. (1999). DNA microarrays of the complex human cytomegalovirus genome: Profiling kinetic class with drug sensitivity of viral gene expression. *J. Virol.* **73**(7), 5757–5766.

Chudin, E., Walker, R, Kosaka, A., Wu, S. X., Rabert, D., Chang, T. K., and Kreder, D. E. (2002). Assessment of the relationship between signal intensities and transcript concentration for Affymetrix GeneChip arrays. *Genome Biol.* **3**(1), RESEARCH 0005.

DeRisi, J. L., Iyer, V. R., and Brown, P. O. (1997). Exploring the metabolic and genetic control of gene expression on a genomic scale. *Science Oct 24;* **278**(5338), 680–686.

Hughes, T. R., Mao, M., Jones, A. R., Burchard, J., Marton, M. J., Shannon, K. W., Lefkowitz, S. M., Ziman, M., Schelter, J. M., Meyer, M. R., Kobayashi, S., Davis, C., Dai, H., He, Y. D., Stephaniants, S. B., Cavet, G., Walker, W. L., West, A., Coffey, E., Shoemaker, D. D., Stoughton, R., Blanchard, A. P., Friend, S. H., and Linsley, P. S. (2001). Expression profiling using microarrays fabricated by an ink-jet oligonucleotide synthesizer. *Nat. Biotechnol.* **19**(4), 342–347.

Irizarry, R. A., Bolstad, B. M., Collin, F., Cope, L. M., Hobbs, B., and Speed, T. P. (2003). Summaries of Affymetrix GeneChip probe level data. *Nucleic Acids Res. Feb. 15;* **31**(4), e15.

Kendziorski, C. M., Zhang, Y., Lan, H., and Attie, A. D. (2003). The efficiency of pooling mRNA in microarray experiments. *Biostatistics* **4**(3), 465–477.

Li, C., and Wong, W. H. (2001a). A Model-based analysis of oligonucleotide arrays: Expression index computation and outlier detection. *Proc. Natl. Acad. Sci. USA* **98**, 31–36.

Li, C., and Wong, W. H. (2001b). Model-based analysis of oligonucleotide arrays: Model validation, design issues and standard error application. *Genome Biol.* **2**(8), 0032.1–0032.11.

Lipshutz, R. J., Fodor, S. P. A., Gingeras, T. R., and Lockhart, D. J. (1999). High density synthetic oligonucleotide arrays. *Nat. Genet.* **21**, 20–24.

Lockhart, D. J., Dong, H., Byrne, M. C., Follettie, M. T., Gallo, M. V., Chee, M. S., Mittmann, M., Wang, C., Kobayashi, M., Horton, H., and Brown, E. L. (1996). Expression monitoring by hybridization to high-density oligonucleotide arrays. *Nat. Biotechnol.* **13**, 1675–1680.

Miller, N. A., Gong, Q., Bryan, R., Ruvolo, M., Turner, L. A., and LaBrie, S. T. (2002). Cross-hybridization of closely related genes on high-density macroarrays. *Biotechniques* **32**(3), 620–625.

Nuwaysir, E. F., Huang, W., Albert, T. J., Singh, J., Nuwaysir, K., Pitas, A., Richmond, T., Gorski, T., Berg, J. P., Ballin, J., McCormick, M., Norton, J., Pollock, T., Sumwalt, T., Butcher, L., Porter, D., Molla, M., Hall, C., Blattner, F., Sussman, M. R., Wallace, R. L., Cerrina, F., and Green, R. D. (2002). Gene expression analysis using oligonucleotide arrays produced by maskless photolithography. *Genome Res.* **12**(11), 1749–1755.

Peng, X., Wood, C. L., Blalock, E. M., Chen, K. C., Landfield, P. W., and Stromberg, A. J. (2003). Statistical implications of pooling RNA samples for microarray experiments. *BMC Bioinformatics* **4**, 26.

Perou, C. M., Jeffrey, S. S., van de Rijn, M., Rees, C. A., Eisen, M. B., Ross, D. T., Pergamenschikov, A., Williams, C. F., Zhu, S. X., Lee, J. C., Lashkari, D., Shalon, D., Brown, P. O., and Botstein, D. (1999). Distinctive gene expression patterns in human mammary epithelial cells and breast cancers. *Proc. Natl. Acad. Sci. USA* **96**(16), 9212–9217.

Schena, M., Shalon, D., Davis, R. W., and Brown, P. O. (1995). Quantitative monitoring of gene expression patterns with a complementary DNA microarray. *Science* **270**(5235), 467–470.

Schena, M., Shalon, D., Heller, R., Chai, A., Brown, P. O., and Davis, R. W. (1996). Parallel human genome analysis: Microarray-based expression monitoring of 1000 genes. *Proc. Natl. Acad. Sci. USA* **93**(20), 10614–10619.

Shih, J. H., Michalowska, A. M., Dobbin, K., Ye, Y., Qiu, T. H., and Green, J. E. (2004). Effects of pooling mRNA in microarray class comparisons. *Bioinformatics* **20**(18), 3318–3325.

Stingley, S. W., Ramirez, J. J., Aguilar, S. A., Simmen, K., Sandri-Goldin, R. M., Ghazal, P., and Wagner, E. K. (2000). Global analysis of herpes simplex virus type 1 transcription using an oligonucleotide-based DNA microarray. *J. Virol.* **74**(21), 9916–9927.

Sudarsanam, P., Iyer, V. R., Brown, P. O., and Winston, F. (2000). Whole-genome expression analysis of snf/swi mutants of *Saccharomyces cerevisiae*. *Proc. Natl. Acad. Sci. USA* **97**(7), 3364–3369.

# Section IV

# CHO Mutants in Glycan Biosynthesis

## [11] Lectin-Resistant CHO Glycosylation Mutants

By Santosh Kumar Patnaik and Pamela Stanley

Abstract

Chinese hamster ovary (CHO) mutant cells with a wide variety of alterations in the glycosylation of proteins and lipids have been isolated by selection for resistance to the cytotoxicity of plant lectins. These CHO mutants have been used to characterize glycosylation pathways, to identify genes that code for glycosylation activities, to elucidate functional roles of glycans that mediate biological processes, and for glycosylation engineering. In this chapter, we briefly describe the available panel of lectin-resistant CHO mutants and summarize their glycan alterations and the biochemical and genetic bases of mutation.

Introduction

CHO cells have been very successfully used as somatic cell genetic tools. A significant proportion of the CHO genome is functionally hemizygous (Siminovitch, 1976), which in combination with a high frequency of segregation-like events (Worton, 1978) makes it easy to obtain loss-of-function (recessive) mutant cells. A large number of CHO mutants affected in a wide range of cellular processes have been selected from CHO cell populations (Gottesman, 1985; Hanada and Nishijima, 2000; Nishimoto, 1997). This chapter describes what is known about CHO mutants that were mainly selected for resistance to toxic plant lectins (Stanley, 1983, 1984). Their resistance primarily arises from reduced cell surface binding of the lectin used in selection due to an alteration in the glycans to which that lectin binds. Most of these mutants are affected in $N$- and $O$-glycan synthesis pathways. Some are also affected in the synthesis of glycolipids, glycosaminoglycans (GAGs) or glycosylphosphatidylinositol (GPI) membrane anchors. Chapters 12 and 13 in this volume discuss mutants affected in the GPI and GAG pathways, respectively. Some CHO glycosylation mutants have a defect in one of the components of the conserved oligomeric complex (COG) associated with the Golgi compartment (Oka and Krieger, 2005). However, the affects on glycosylation due to COG defects are indirect and these mutants are not described here.

METHODS IN ENZYMOLOGY, VOL. 416
Copyright 2006, Elsevier Inc. All rights reserved.

0076-6879/06 $35.00
DOI: 10.1016/S0076-6879(06)16011-5

Glycosylation in CHO Cells

CHO cells have a range of complex and oligomannosyl $N$-glycans with few hybrid structures (Lee *et al.*, 2001). Polylactosamine chains containing up to six units (Kawar *et al.*, 2005) and polysialic acid (Hong and Stanley, 2003; Muhlenhoff *et al.*, 1996) have been observed as minor species. $O$-glycans of CHO cells include mucins containing up to four sugars but not the core 2 structure (Bierhuizen and Fukuda, 1992; Sasaki *et al.*, 1987), and $O$-fucosylated (Moloney *et al.*, 2000), $O$-glucosylated (Moloney *et al.*, 2000), or $O$-mannosylated (Patnaik and Stanley, 2005) structures. The major glyco-lipid of CHO cells is $GM_3$, $\alpha2,3$-sialylated lactosylceramide (Stanley, 1980; Warnock *et al.*, 1993). CHO cells also contain heparan sulfate and chondroitin sulfate GAGs (Esko *et al.*, 1985). CHO cells do not express $\alpha1,2$-, $\alpha1,3$-, or $\alpha1,4$-fucosyltransferases (Howard *et al.*, 1987), ST6Gal $\alpha2,6$-sialyltransferases that transfer sialic acid (SA) to galactose (Gal) (Sasaki *et al.*, 1987), or GlcNAc-TIII that transfers the bisecting $N$-acetylglucosamine (GlcNAc) (Campbell and Stanley, 1984). CHO cells express the six known $\beta1,4$-galactosyltransferases; however, *B4galt6* is not expressed in the Pro$^-$5 parent CHO line (Lee *et al.*, 2001). CHO cells lack sulfotransferases required to gener-ate sulfated glycolipids or sulfated $N$- or $O$-glycans (Brockhausen *et al.*, 2001).

Isolation and Characterization of Lectin-Resistant Mutants

CHO parent populations Pro$^-$5 CHO, Gat$^-$2 CHO, or CHO-K1 were subjected to selection with and without prior mutagenesis using agents such as 5-azacytidine (5-azaC) and $N$-methyl-$N'$-nitro-$N$-nitrosoguanidine (MNNG). The selective agents were used in single-step or multistep nega-tive selection strategies employing cytotoxic plant lectins (Stanley, 1984) or bacterial toxins in the case of CHO2.38 (Ashida *et al.*, 2006). Most mutants were isolated in a single step for resistance to a single lectin, but others harboring mutations in up to four genes were isolated following selection with multiple lectins in combination or sequentially (Stanley, 1989). The change in lectin sensitivity of a mutant to a panel of lectins reflects the nature of the underlying alteration in glycosylation. Thus, a particular glycosylation gene mutation generates a characteristic lectin-resistance (Lec$^R$) phenotype. Table I presents the Lec$^R$ phenotype for many mutant lines. Such phenotypic profiles allow rapid classification of new isolates (Stanley, 1989), have provided initial clues to the nature of the biochemical activity affected in the different mutants (e.g., Campbell and Stanley, 1984), have aided in the functional cloning of glycosylation genes (e.g., Kumar *et al.*, 1990), and are of value in designing selections for the isolation and characterization of new glycosylation mutants.

TABLE I
LECTIN-RESISTANCE PHENOTYPE OF CHO GLYCOSYLATION MUTANTS

| Cell line | L-PHA (μg/ml) | WGA (μg/ml) | ConA (μg/ml) | Ricin (ng/ml) | LCA (μg/ml) | PSA (μg/ml) | E-PHA (μg/ml) | MOD (pg/ml) | Abrin (ng/ml) |
|---|---|---|---|---|---|---|---|---|---|
| CHO | 5 | 2 | 18 | 5 | 18 | 50 | 35 | 2.5 | 2.5 |
| Lec1 | >1000R | 30R | 6S | 100R | >200R | 9R | >10R | 4R | 300R |
| Lec1A | >300R | 9R | 5S | 10R | 35R | 5R | >10R | 2R | 3R |
| Lec2 | S | 11R | — | 100S | 2S | 2S | — | 5S | >10S |
| Lec2B | S | 25R | S | 3S | S | ? | ? | ? | ? |
| Lec3 | S | 5R | — | 10S | 2S | 2S | — | 25S | ? |
| Lec4 | >1000R | R | S | S | S | 2S | R | — | ? |
| Lec5 | 7R | R | R | 3R | 3R | S | 2R | <2R | <6R |
| Lec8 | 10R | 100R | S | R | 10R | 2S | >10R | 2R | R |
| Lec9 | R | R | — | 10R | R | — | — | <5R | 15R |
| LEC10 | 2S | S | — | 20R | — | — | 10S | — | 20R |
| LEC11 | 4R | 8R | — | 25S | 3R | — | R | 2R | 5S |
| LEC12 | 3R | 50R | — | 4S | 2R | — | ? | 4R | S |
| Lec13 | — | — | — | — | 27R | 48R | ? | — | S |
| Lec13A | R | — | — | — | 3R | 9R | ? | S | — |
| LEC14 | — | — | — | 4R | 2R | 10R | ? | — | — |
| Lec15 | R | — | — | 3R | — | ? | 2S | — | 4R |
| LEC16 | R | S | — | 35R | R | ? | — | — | 3R |
| LEC17 | — | — | — | S | — | — | ? | — | 4R |
| LEC18 | — | — | — | 10R | 16R | 39R | ? | — | — |
| Lec19 | 3S | 2S | S | 10R | S | ? | 2S | 2R | 10R |
| Lec20 | 2R | S | S | 2R | S | ? | 7R | 3R | 2R |
| Lec21 | — | — | — | 2R | ? | ? | — | 300R | — |
| Lec22 | S | — | — | 5R | — | ? | — | — | 4R |

(continued)

TABLE I (continued)

| Cell line | L-PHA (μg/ml) | WGA (μg/ml) | ConA (μg/ml) | Ricin (ng/ml) | LCA (μg/ml) | PSA (μg/ml) | E-PHA (μg/ml) | MOD (pg/ml) | Abrin (ng/ml) |
|---|---|---|---|---|---|---|---|---|---|
| Lec23 | >58R | 10R | 5S | 4R | — | ? | 8R | R | 8R |
| Lec24 | 2R | 2R | R | 12R | 4R | ? | 2R | R | 12R |
| Lec25 | R | — | — | 4R | — | ? | R | — | 3R |
| Lec26 | — | — | — | 10R | S | ? | — | — | 16R |
| Lec27 | — | — | — | 12R | — | ? | — | R | 4R |
| Lec28 | R | S | S | 4R | S | ? | S | R | 12R |
| LEC29 | R | R | R | 10S | R | ? | ? | ? | ? |
| LEC30 | 10R | 50R | R | S | 4R | ? | ? | ? | ? |
| LEC31 | ? | ? | ? | ? | ? | R | 3-7R | 4R | ? |
| Lec32 | ? | 17R | ? | 500R | ? | ? | ? | ? | ? |

—, same as wild-type CHO; ?, unknown; L-PHA, leukophytohemagglutinin from *Phaseolus vulgaris*; WGA, wheat germ agglutinin; ConA, concanavalin A; Ricin, *Ricinus communis* lectin II; LCA, *Lens culinaris* lectin; PSA, *Pisum sativum* lectin; E-PHA, erythrophytohemagglutinin from *Phaseolus vulgaris*; MOD, modeccin.

*Note:* Fold resistance (R) or sensitivity (S) compared to wild-type parental CHO $D_{10}$ value (lectin concentration at 10% relative plating efficiency); S or R given alone are less than twofold different from wild-type CHO.

Most of the mutants have been named numerically with the prefix Lec or LEC (Table I). This nomenclature was first defined in 1983 by Stanley. Lec mutants are so called because they have lost an activity (loss of function), whereas LEC mutants have acquired an activity not detected in parent cells and are gain of function. All Lec mutants are recessive in that their defect is complemented in somatic cell hybrids formed with parent cells. Some loss-of-function mutants with distinct but related Lec$^R$ phenotypes belong to the same complementation group. Thus, complementation analysis identifies mutants that, in spite of having apparently different phenotypes, are mutated in the same gene. For example, the Lec1 and Lec1A mutants belong to the same complementation group (number 1) and they are both mutated in the *Mgat1* gene (Chaney and Stanley, 1986; Chen and Stanley, 2003; Chen *et al.*, 2001b). Mutants with multiple glycosylation defects belong to multiple complementation groups (e.g., Lec3.2.8.1). All LEC mutants are dominant in that somatic cell hybrids formed with parent CHO exhibit the mutant phenotype. However, one dominant mutant was found to arise from a loss-of-function mutation in a negative regulatory factor (Zhang *et al.*, 1999).

The mutants in Table I have been characterized extensively at the structural, biochemical, and/or genetic level. Figure 1 summarizes the sites of many of the glycosylation defects in terms of the sugar added in gain-of-function mutants (+) or the sugar that is not transferred (−) in the case of loss-of-function mutants to *N*-glycan, *O*-fucose, *O*-mannose, and mucin *O*-glycan structures. Tables II and III summarize what is known about the glycan structural changes in each mutant, the biochemical basis of these changes, and the molecular basis of mutation where known. The precise genetic change leading to the activation of a glycosylation gene in the gain-of-function mutants is largely unknown. On the other hand, the mutation in the glycosylation gene affected in many of the loss-of-function mutants has been identified (Table IV). Whereas revertants generally do not arise at high frequency, some null mutant phenotypes are "leaky," perhaps because of redundancy at a biochemical level. Some mutants, however, revert at a frequency of $10^{-4}$ (Chen *et al.*, 2005). Interestingly, all the different genes sequenced from clones and independent CHO mutants show remarkably little variation in sequence, thus allowing point mutations to be easily identified in glycosylation mutants. However, this is likely to be because these isolates have not been in continuous culture for more than about 2–4 mo.

*Availability and Culture of the Mutants*

The Pro$^-$5 and CHO-K1 parental CHO cells and the Lec1, Lec2, and Lec8 mutants are available from the American Type Culture Collection (ATCC;

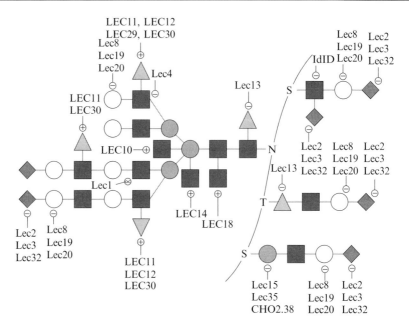

FIG. 1. This diagram illustrates the altered *N*-glycans, mucin *O*-glycans, *O*-fucose, and *O*-mannose glycans in many of the CHO glycosylation mutants described here. The major glycolipid in CHO cells is $GM_3$, which is affected in mutants defective for transfer or synthesis of CMP-SA or UDP-Gal. Proteoglycans are affected in mutants defective for the synthesis or transfer of UDP-Gal. A loss or reduction of a sugar residue at a particular position is indicated with the minus (−) sign, whereas the gain of a sugar residue is indicated with a plus (+) sign. Asparagine, threonine, or serine residues bearing the different glycans are indicated by their amino acid code. Sugar symbols: Gray triangle, fucose; gray circle, mannose; white circle, galactose; black square, *N*-acetylglucosamine; white square, *N*-acetylgalactosamine; gray trapezoid, sialic acid.

Manassas, VA). Other mutants are available from investigator laboratories. The MI8–5, Lec9, Lec15, and Lec35 mutants are temperature sensitive and should be grown at 34°. These four mutants and the ldlD mutant, as well as cells derived from CHO-K1, should be grown in monolayer, whereas the other mutants can be cultured in suspension or monolayer. For certain purposes, dialyzed serum should be used when culturing Lec13 cells as fucose in serum can be used by the cells to generate GDP-fucose, partially bypassing the defect in Lec13 (Chen *et al.*, 2001a). Similarly, the ldlD mutation can be bypassed in the presence of undialyzed serum (Kingsley *et al.*, 1986). Lec3 cells are known to salvage SA from serum proteins and lipids for conversion to CMP-SA (Hong and Stanley, 2003).

TABLE II
GLYCOSYLATION DEFECTS IN LECTIN-RESISTANT CHO MUTANTS

| CHO line | Biochemical change | Genetic change | Predicted N-glycans[a] | Predicted O-glycans[a] | Most recent reference |
|---|---|---|---|---|---|
| Gat⁻2 (parent) | — | — | | | Lee et al., 2001 |
| Pro⁻5 (parent) | ↓ Gal on N-glycans | No expression of B4galt6 | | | Lee et al., 2001 |
| Lec1 | ↓ GlcNAc-TI | Insertion/deletion in Mgat1 ORF | Oligomannosyl | | Chen and Stanley, 2003 |
| Lec1A | Kₘ mutant of GlcNAc-TI | Point mutation in Mgat1 ORF | ↓ Complex and hybrid ↑ Oligomannosyl | | Chen et al., 2001b |

(continued)

TABLE II (continued)

| CHO line | Biochemical change | Genetic change | Predicted N-glycans[a] | Predicted O-glycans[a] | Most recent reference |
|---|---|---|---|---|---|
| Lec2 | ↓ CMP-sialic acid Golgi transporter | Mutation in Slc35a1 ORF | | | Eckhardt et al., 1998 |
| Lec3 | ↓ UDP-GlcNAc 2-epimerase | Mutation in Gne ORF (epimerase domain) | | | Hong and Stanley, 2003 |
| Lec4 | ↓ GlcNAc-TV | Deletion in Mgat5 ORF | | | Weinstein et al., 1996 |
| Lec4A | Mislocalized GlcNAc-TV | Point mutation in Mgat5 ORF | | | Weinstein et al., 1996 |

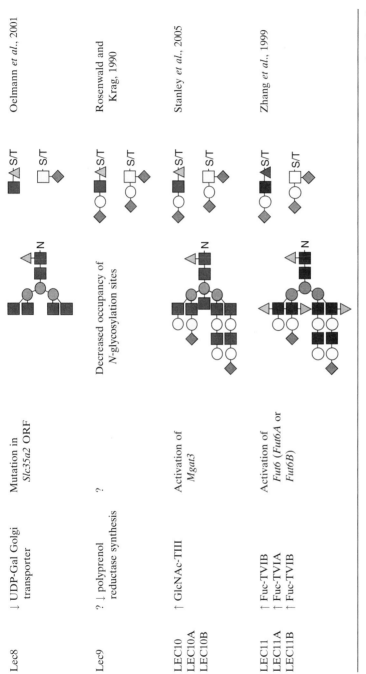

| | | | | |
|---|---|---|---|---|
| Lec8 | ↓ UDP-Gal Golgi transporter | Mutation in *Slc35a2* ORF | | Oelmann *et al.*, 2001 |
| Lec9 | ? ↓ polyprenol reductase synthesis | ? | Decreased occupancy of *N*-glycosylation sites | Rosenwald and Krag, 1990 |
| LEC10 LEC10A LEC10B | ↑ GlcNAc-TIII | Activation of *Mgat3* | | Stanley *et al.*, 2005 |
| LEC11 LEC11A LEC11B | ↑ Fuc-TVIB ↑ Fuc-TVIA ↑ Fuc-TVIB | Activation of *Fut6* (*Fut6A* or *Fut6B*) | | Zhang *et al.*, 1999 |

(continued)

TABLE II (continued)

| CHO line | Biochemical change | Genetic change | Predicted N-glycans[a] | Predicted O-glycans[a] | Most recent reference |
|---|---|---|---|---|---|
| LEC12 | ↑ Fuc-TIX | Activation of Fut9 | | | Patnaik et al., 2000 |
| Lec13 | ↓ GDP-Man-4,6-dehydratase | ↓ Expression of Gmds | | | Ohyama et al., 1998; Sullivan et al., 1998 |
| LEC14 | ↑ GlcNAc-TVII | ? | | | Raju and Stanley, 1998 |
| Lec15 | ↓ Dol-P-Man synthase | Allelic loss and Dpm2 inactivation | Decreased occupancy of N-glycosylation sites | | Maeda et al., 1998 |

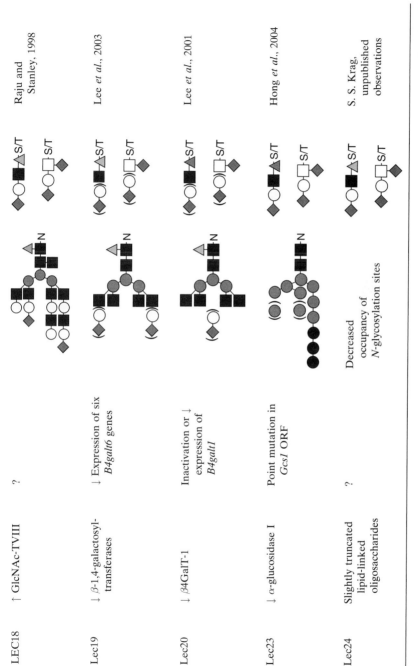

(continued)

TABLE II (continued)

| CHO line | Biochemical change | Genetic change | Predicted N-glycans[a] | Predicted O-glycans[a] | Most recent reference |
|---|---|---|---|---|---|
| LEC29 | ↑ Fuc-TIX | Activation of Fut9 | | | Patnaik et al., 2000 |
| LEC30 | ↑ Fuc-TIV and Fuc-TIX | Activation of Fut4 and Fut9 | | | Patnaik et al., 2000 |
| LEC31 | ↑ α-1,3-fucosyltransferase | Activation of Fut9 | | | S. K. Patnaik and P. Stanley, unpublished observations |

| Mutant | Enzyme defect | Gene | Other effects | Reference |
|---|---|---|---|---|
| Lec32 | ↓ CMP-sialic acid synthetase | ↓ expression of *Cmas* | | Potvin *et al.*, 1995 |
| Lec35 | Accumulation of Man$_5$GlcNAc$_2$-P-P-dolichol | Inactivation of *Mpdu1* | Decreased occupancy of N-glycosylation sites | Anand *et al.*, 2001 |
| CHO2.38 | ↓ Dol-P-Man synthase | Mutation in *Dpm3* | Decreased occupancy of N-glycosylation sites | Ashida *et al.*, 2006 |
| ldlD | ↓ UDP-Gal/GlcNAc-4-epimerase | ? | | Kingsley *et al.*, 1986 |
| MI8-5 | ↓ glucosylated Man$_9$GlcNAc$_2$-P-P-dolichol | ? | Decreased occupancy of N-glycosylation sites | Quellhorst *et al.*, 1999 |

[a] Complex N-glycans of CHO parent cells may have polylactosamines (Lee *et al.*, 2001); other N-glycans include oligomannosyl and hybrid types; O-mannose glycans on α-dystroglycan will also be affected, as will the synthesis of GM$_3$ in mutants that do not transfer Gal or SA. Sugar symbols: gray triangle, fucose; gray circle, mannose; white circle, mannose; black square, N-acetylglucosamine; white square, N-acetylgalactosamine; gray trapezoid, sialic acid; black circle, glucose.

TABLE III
CHO MUTANTS WITH MULTIPLE GLYCOSYLATION DEFECTS

| CHO line | Biochemical change | Genetic change | Predicted N-glycans[a] | Predicted O-glycans[a] | Most recent reference |
|---|---|---|---|---|---|
| Lec3.2 | ↓ UDP-GlcNAc 2-epimerase, ↓ CMP-sialic acid Golgi transporter | Mutations in *Gne* and *Slc35a1*[b] | | | Hong and Stanley, 2003 |
| Lec3.2.1 | ↓ UDP-GlcNAc 2-epimerase, ↓ CMP-sialic acid Golgi transporter, ↓ GlcNAc-TI | Mutations in *Gne, Mgat1* and *Slc35a1*[b] | | | Hong and Stanley, 2003 |
| Lec3.2.8 | ↓ CMP-sialic acid Golgi transporter, ↓ UDP-Gal Golgi transporter | Mutations in *Gne, Slc35a2, Slc35a1*[b] | | | Hong and Stanley, 2003 |
| Lec3.2.8.1 | ↓ UDP-GlcNAc 2-epimerase, | Mutations in *GNE, Mgat1, Slc35a2, Slc35a1*[b] | | | Hong and Stanley, 2003 |

| Mutant | Enzyme/transporter changes | Mutation | Reference |
|---|---|---|---|
| | ↓ CMP-sialic acid Golgi transporter, ↓ UDP-Gal Golgi transporter, ↓ GlcNAc-TI | | |
| Lec4.8 | ↓ GlcNAc-TV, ↓ UDP-Gal Golgi transporter | Mutations in *Mgat5* and *Slc35a2* | Oelmann *et al.*, 2001 |
| Lec4A.8 | Mislocalized GlcNAc-TV, ↓ UDP-Gal Golgi transporter | Mutations in *Mgat5* and *Slc35a2* | Oelmann *et al.*, 2001 |
| LEC10.8 | ↑ GlcNAc-TIII, ↓ UDP-Gal Golgi transporter | Mutations in *Mgat3* and *Slc35a2* | Stanley *et al.*, 1991 |
| ldlD.Lec1 | ↓ UDP-Glc-4 epimerase, ↓ GlcNAc-TI | Mutation in *Mgat1* and ? | Chen and Stanley, 2003 |

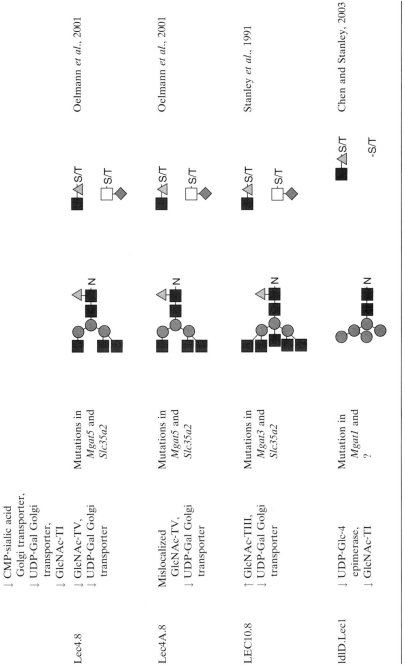

(continued)

TABLE III (*continued*)

| CHO line | Biochemical change | Genetic change | Predicted N-glycans[a] | Predicted O-glycans[a] | Most recent reference |
|---|---|---|---|---|---|
| Lec35.Lec1 | ↓ GlcNAc-TI and no synthesis of Man₆GlcNAc₂-P-P-dolichol | Inactivation of *Mpdul* and mutation in *Mgatl* | ■-N | S/T S/T | Chen and Stanley, 2003 |

[a] Complex N-glycans of CHO parent cells may have polylactosamines (Lee *et al.*, 2001); other N-glycans include oligomannosyl and hybrid types; O-mannose glycans on α-dystroglycan will also be affected, as will the synthesis of GM₃ in mutants that do not transfer Gal or SA.
[b] Based on genetic complementation analysis (Stanley, 1989).
Sugar symbols: gray triangle, fucose; gray circle, mannose; white circle, galactose; black square, N-acetylglucosamine; white square, N-acetylgalactosamine; gray trapezoid, sialic acid; black circle, glucose.

TABLE IV
GENE MUTATIONS IN LOSS-OF-FUNCTION CHO GLYCOSYLATION MUTANTS

| Mutant (clone number) | Parental CHO | Gene | DNA sequence alteration[a] | Protein sequence alteration | References |
|---|---|---|---|---|---|
| Lec1A (2A and 2C) | Pro⁻5 CHO | Mgat1 | 634G>A | 212D>N | Chen et al., 2001b |
| Lec1A (3E) | Gat⁻2 CHO | Mgat1 | 907C>A | 303R>W | Chen et al., 2001b |
| Lec1A (5J) | Pro⁻5 CHO | Mgat1 | 907C>A | 303R>W | Chen et al., 2001b |
| Lec1 (1C) | Pro⁻5 CHO | Mgat1 | 330–362 33 bp insertion | Frameshift and premature stop at codon 114 | Chen and Stanley, 2003 |
| Lec1 (1N) | Gat⁻2 CHO | Mgat1 | 784C>T | 262R>stop | Chen and Stanley, 2003 |
| Lec1 (3C) | Pro⁻5 CHO | Mgat1 | 702–705 4 bp insertion | Frameshift and premature stop at codon 245 | Chen and Stanley, 2003 |
| Lec2 (4C) | Pro⁻5 CHO | Slc35a1 | 575–751 deletion | 192G-251F>V | Eckhardt et al., 1998 |
| Lec2 (1E3) | CHO-K1 | Slc35a1 | 195–574 duplication | Frameshift at codon 193 | Eckhardt et al., 1998 |
| Lec2 (6B2) | CHO-K1 | Slc35a1 | 752–886 deletion | 251F–296T>S | Eckhardt et al., 1998 |
| Lec2 (8G8) | CHO-K1 | Slc35a1 | 194–195 deletion | Frameshift at codon 67 | Eckhardt et al., 1998 |
| Lec2 (9D3) | CHO-K1 | Slc35a1 | 566G>A | 189G>E | Eckhardt et al., 1998 |
| Lec3 (4B) | Pro⁻5 CHO | Gne | 103G>T | 35E>stop | Hong and Stanley, 2003 |
| Lec3 (6F) | Gat⁻2 CHO | Gne | 394G>A | 135G>E | Hong and Stanley, 2003 |

(continued)

TABLE IV (*continued*)

| Mutant (clone number) | Parental CHO | Gene | DNA sequence alteration[a] | Protein sequence alteration | References |
|---|---|---|---|---|---|
| Lec4 | Pro⁻5 CHO | *Mgat5* | 403–1224 822 bp insertion | Frameshift at codon 135 and premature stop at codon 156 | Weinstein et al., 1996 |
| Lec4A | Pro⁻5 CHO | *Mgat5* | 563T>G | 188L>R | Weinstein et al., 1996 |
| Lec8 | Pro⁻5 CHO | *Slc35a2* | 275–374 deletion | 92E>stop | Oelman et al., 2001 |
| Lec8 (5H) | Pro⁻5 CHO | *Slc35a2* | 636–638 deletion | 213S deletion | Oelman et al., 2001 |
| Lec8 (1C) | Gat⁻2 CHO | *Slc35a2* | 844G>A | 281G>D | Oelman et al., 2001 |
| Lec15 (B4-2-1) | Pro⁻5 CHO | *Dpm1* | 29G>A | 10G>E | Pu et al., 2003 |
| Lec20 (6A) | Gat⁻2 CHO | *B4galt1* | 311 bp deletion of exons 4 and 5 in coding region | Truncated protein of 214 aa | Lee et al., 2001 |
| Lec23 (11C) | Pro⁻5 CHO | *Gcs1* | 1320C>T | 440S>F | Hong et al., 2004 |
| Lec35.1 | CHO-K1 | *Mpdu1* | Gross gene disruption | | Anand et al., 2001 |
| Lec3.2.1 | Pro⁻5 CHO | *Mgat1* | ~400 bp deletion in coding region | Not known | Chen and Stanley, 2003 |
| Lec3.2.8.1 | Pro⁻5 CHO | *Mgat1* | 1154G insertion | Frameshift at codon 385 and premature stop at codon 392 | Chen and Stanley, 2003 |
| Lec9.1 | Pro⁻5 CHO | *Mgat1* | 310C insertion | Frameshift at codon 104 and premature stop at codon 116 | Chen and Stanley, 2003 |
| LEC10.8 | Pro⁻5 CHO | *Slc35a2* | 364T>C | 122Y>H | Oelman et al., 2001 |
| CHO2.38 | CHO-K1[b] | *Dpm3* | 108–115 8 bp deletion | Frameshift at codon 36 | Ashida et al., 2006 |

[a] A of the ATG start codon is nucleotide number 1.
[b] Transfected with multiple copies of various genes.

TABLE V

VARIOUS APPLICATIONS OF SOME OF THE CHO GLYCOSYLATION MUTANTS

| Mutant | Biochemical alteration | Observation | References |
|---|---|---|---|
| Lec1 | No GlcNAcT-I | An adhesin of *Candida glabrata* specifically recognizes asialo-lactosyl *N*-glycans on human epithelial cells | Cormack *et al.*, 1999 |
| Lec2 | ↓ CMP-sialic acid Golgi transporter | Identification of a novel type of CDG IIf resulting from mutation in CMP-sialic acid transporter | Martinez-Duncker *et al.*, 2005 |
| Lec4 | Loss of GlcNAc-TV | *Gly-2* of *C. elegans* encodes GlcNAc-TV | Warren *et al.*, 2002 |
| Lec8 | ↓ UDP-Gal Golgi transporter | Molecular cloning of *Arabidopsis* UDP-Gal transporter genes | Bakker *et al.*, 2005 |
| LEC10 | Expression of GlcNAcT-III | Bisecting GlcNAc of complex *N*-glycans of human IgG1 is not important for antibody-dependent cellular cytotoxicity | Shinkawa *et al.*, 2003 |
| LEC11 | Expression of FucT-VI | Neuronal protection in stroke by CR1 possessing sLe^X moieties | Huang *et al.*, 1999 |
| LEC12 | Expression of FucT-IX | VIM-2 glycan may be a ligand for E-selectin | Patnaik *et al.*, 2004 |
| Lec13 | ↓ GDP-Man-4,6-dehydratase | Fucose facilitates Jagged1-induced Notch signaling | Chen *et al.*, 2001a |
| Lec15 | ↓ Dol-P-Man synthase | Protein *C*-mannosylation uses dolichol-phosphate mannose as a precursor | Doucey *et al.*, 1998 |
| Lec20 | ↓ β4GalT-I | Fringe modulation of Jagged1 induced Notch signaling requires the action of β4GalT-I | Chen *et al.*, 2001a |
| Lec23 | ↓ α-glucosidase I | ERp57 interacts with both calnexin and calreticulin in the absence of their glycoprotein substrates | Oliver *et al.*, 1999 |
| Lec3.2.8.1 | ↓ UDP-GlcNAc 2-epimerase, ↓ CMP-sialic acid Golgi translocase, ↓ UDP-Gal Golgi translocase, no GlcNAc-TI | Sialylation has no effect on the overall structure of the stalk-like region of CD8 | Merry *et al.*, 2003 |
| Lec35 | No synthesis of Man$_6$GlcNAc$_2$-P-P-dolichol | Requirement for *O*-mannosylation of α-dystroglycan for binding LCMV virus | Imperiali *et al.*, 2005 |

*Applications of the Mutants*

The CHO mutants have been used to address a large variety of functional questions in glycobiology over the years. Examples of some of these applications are listed in Table V. They include glycosylation engineering of experimental and therapeutic glycoproteins, as well as cloning and characterization of glycosylation genes. Panels of mutants have been used to characterize glycans involved in biological processes such as the binding of an amebic lectin (Ravdin *et al.*, 1989), recognition of sialylated Le$^X$ by selectins (Polley *et al.*, 1991), binding of different galectins to the mixture of cell surface glycans (Patnaik *et al.*, 2005), and the functional glycosylation of $\alpha$-dystroglycan (Patnaik and Stanley, 2005).

Acknowledgments

This work was supported by National Cancer Institute grant RO1 36434 to P.S. We thank Subha Sundaram for assisting with the preparation of this manuscript.

References

Anand, M., Rush, J. S., Ray, S., Doucey, M. A., Weik, J., Ware, F. E., Hofsteenge, J., Waechter, C. J., and Lehrman, M. A. (2001). Requirement of the Lec35 gene for all known classes of monosaccharide-P-dolichol–dependent glycosyltransferase reactions in mammals. *Mol. Biol. Cell* **12,** 487–501.

Ashida, H., Maeda, Y., and Kinoshita, T. (2006). DPM1, the catalytic subunit of dolichol-phosphate mannose synthase, is tethered to and stabilized on the endoplasmic reticulum membrane by Dpm3. *J. Biol. Chem.* **281,** 896–904.

Bakker, H., Routier, F., Oelmann, S., Jordi, W., Lommen, A., Gerardy-Schahn, R., and Bosch, D. (2005). Molecular cloning of two Arabidopsis UDP-galactose transporters by complementation of a deficient Chinese hamster ovary cell line. *Glycobiology* **15,** 193–201.

Bierhuizen, M. F., and Fukuda, M. (1992). Expression cloning of a cDNA encoding UDP-GlcNAc:Gal beta 1-3-GalNAc-R (GlcNAc to GalNAc) beta 1-6GlcNAc transferase by gene transfer into CHO cells expressing polyoma large tumor antigen. *Proc. Natl. Acad. Sci. USA* **89,** 9326–9330.

Brockhausen, I., Vavasseur, F., and Yang, X. (2001). Biosynthesis of mucin type O-glycans: Lack of correlation between glycosyltransferase and sulfotransferase activities and CFTR expression. *Glycoconj. J.* **18,** 685–697.

Campbell, C., and Stanley, P. (1984). A dominant mutation to ricin resistance in Chinese hamster ovary cells induces UDP-GlcNAc:glycopeptide beta-4-N-acetylglucosaminyltransferase III activity. *J. Biol. Chem.* **259,** 13370–13378.

Chaney, W., and Stanley, P. (1986). Lec1A Chinese hamster ovary cell mutants appear to arise from a structural alteration in N-acetylglucosaminyltransferase I. *J. Biol. Chem.* **261,** 10551–10557.

Chen, J., Moloney, D. J., and Stanley, P. (2001a). Fringe modulation of Jagged1-induced Notch signaling requires the action of beta 4galactosyltransferase-1. *Proc. Natl. Acad. Sci. USA* **98,** 13716–13721.

Chen, W., and Stanley, P. (2003). Five Lec1 CHO cell mutants have distinct Mgat1 gene mutations that encode truncated N-acetylglucosaminyltransferase I. *Glycobiology* **13**, 43–50.

Chen, W., Tang, J., and Stanley, P. (2005). Suppressors of alpha(1,3)fucosylation identified by expression cloning in the LEC11B gain-of-function CHO mutant. *Glycobiology* **15**, 259–269.

Chen, W., Unligil, U. M., Rini, J. M., and Stanley, P. (2001b). Independent Lec1A CHO glycosylation mutants arise from point mutations in N-acetylglucosaminyltransferase I that reduce affinity for both substrates. Molecular consequences based on the crystal structure of GlcNAc-TI(,). *Biochemistry* **40**, 8765–8772.

Cormack, B. P., Ghori, N., and Falkow, S. (1999). An adhesin of the yeast pathogen Candida glabrata mediating adherence to human epithelial cells. *Science* **285**, 578–582.

Doucey, M. A., Hess, D., Cacan, R., and Hofsteenge, J. (1998). Protein C-mannosylation is enzyme-catalysed and uses dolichyl-phosphate-mannose as a precursor. *Mol. Biol. Cell* **9**, 291–300.

Eckhardt, M., Gotza, B., and Gerardy-Schahn, R. (1998). Mutants of the CMP-sialic acid transporter causing the Lec2 phenotype. *J. Biol. Chem.* **273**, 20189–20195.

Esko, J. D., Stewart, T. E., and Taylor, W. H. (1985). Animal cell mutants defective in glycosaminoglycan biosynthesis. *Proc. Natl. Acad. Sci. USA* **82**, 3197–3201.

Gottesman, M. M. (1985). "Molecular Cell Genetics." Wiley, New York.

Hanada, K., and Nishijima, M. (2000). Selection of mammalian cell mutants in sphingolipid biosynthesis. *Methods Enzymol.* **312**, 304–317.

Hong, Y., and Stanley, P. (2003). Lec3 CHO mutants lack UDP-GlcNAc 2-epimerase activity due to mutations in the epimerase domain of the Gne gene. *J. Biol. Chem.* **278**, 53045–53054.

Hong, Y., Sundaram, S., Shin, D. J., and Stanley, P. (2004). The Lec23 Chinese hamster ovary mutant is a sensitive host for detecting mutations in alpha-glucosidase I that give rise to congenital disorder of glycosylation IIb (CDG IIb). *J. Biol. Chem.* **279**, 49894–49901.

Howard, D. R., Fukuda, M., Fukuda, M. N., and Stanley, P. (1987). The GDP-fucose: N-acetylglucosaminide 3-alpha-L-fucosyltransferases of LEC11 and LEC12 Chinese hamster ovary mutants exhibit novel specificities for glycolipid substrates. *J. Biol. Chem.* **262**, 16830–16837.

Huang, J., Kim, L. J., Mealey, R., Marsh, H. C., Jr., Zhang, Y., Tenner, A. J., Connolly, E. S., Jr., and Pinsky, D. J. (1999). Neuronal protection in stroke by an sLex-glycosylated complement inhibitory protein. *Science* **285**, 595–599.

Imperiali, M., Thoma, C., Pavoni, E., Brancaccio, A., Callewaert, N., and Oxenius, A. (2005). O Mannosylation of alpha-dystroglycan is essential for lymphocytic choriomeningitis virus receptor function. *J. Virol.* **79**, 14297–14308.

Kawar, Z. S., Haslam, S. M., Morris, H. R., Dell, A., and Cummings, R. D. (2005). Novel poly-GalNAcbeta1–4GlcNAc (LacdiNAc) and fucosylated poly-LacdiNAc N-glycans from mammalian cells expressing beta1,4-N-acetylgalactosaminyltransferase and alpha1,3-fucosyltransferase. *J. Biol. Chem.* **280**, 12810–12819.

Kingsley, D. M., Kozarsky, K. F., Hobbie, L., and Krieger, M. (1986). Reversible defects in O-linked glycosylation and LDL receptor expression in a UDP-Gal/UDP-GalNAc 4-epimerase deficient mutant. *Cell* **44**, 749–759.

Kumar, R., Yang, J., Larsen, R. D., and Stanley, P. (1990). Cloning and expression of N-acetylglucosaminyltransferase I, the medial Golgi transferase that initiates complex N-linked carbohydrate formation. *Proc. Natl. Acad. Sci. USA* **87**, 9948–9952.

Lee, J., Park, S. H., Sundaram, S., Raju, T. S., Shaper, N. L., and Stanley, P. (2003). A mutation causing a reduced level of expression of six beta4-galactosyltransferase genes is the basis of the Lec19 CHO glycosylation mutant. *Biochemistry* **42**, 12349–12357.

Lee, J., Sundaram, S., Shaper, N. L., Raju, T. S., and Stanley, P. (2001). Chinese hamster ovary (CHO) cells may express six $\beta$4-galactosyltransferases ($\beta$4GalTs). Consequences of the loss of functional $\beta$4GalT-1, $\beta$4GalT-6, or both in CHO glycosylation mutants. *J. Biol. Chem.* **276**, 13924–13934.

Maeda, Y., Tomita, S., Watanabe, R., Ohishi, K., and Kinoshita, T. (1998). DPM2 regulates biosynthesis of dolichol phosphate-mannose in mammalian cells: Correct subcellular localization and stabilization of DPM1, and binding of dolichol phosphate. *EMBO J.* **17**, 4920–4929.

Martinez-Duncker, I., Dupre, T., Piller, V., Piller, F., Candelier, J. J., Trichet, C., Tchernia, G., Oriol, R., and Mollicone, R. (2005). Genetic complementation reveals a novel human congenital disorder of glycosylation of type II, due to inactivation of the Golgi CMP-sialic acid transporter. *Blood* **105**, 2671–2676.

Merry, A. H., Gilbert, R. J., Shore, D. A., Royle, L., Miroshnychenko, O., Vuong, M., Wormald, M. R., Harvey, D. J., Dwek, R. A., Classon, B. J., Rudd, P. M., and Davis, S. J. (2003). O-glycan sialylation and the structure of the stalk-like region of the T cell co-receptor CD8. *J. Biol. Chem.* **278**, 27119–27128.

Moloney, D. J., Shair, L. H., Lu, F. M., Xia, J., Locke, R., Matta, K. L., and Haltiwanger, R. S. (2000). Mammalian Notch1 is modified with two unusual forms of O-linked glycosylation found on epidermal growth factor-like modules. *J. Biol. Chem.* **275**, 9604–9611.

Muhlenhoff, M., Eckhardt, M., Bethe, A., Frosch, M., and Gerardy-Schahn, R. (1996). Autocatalytic polysialylation of polysialyltransferase-1. *EMBO J.* **15**, 6943–6950.

Nishimoto, T. (1997). Isolation and characterization of temperature-sensitive mammalian cell cycle mutants. *Methods Enzymol.* **283**, 292–309.

Oelmann, S., Stanley, P., and Gerardy-Schahn, R. (2001). Point mutations identified in Lec8 Chinese hamster ovary glycosylation mutants that inactivate both the UDP-galactose and CMP-sialic acid transporters. *J. Biol. Chem.* **276**, 26291–26300.

Ohyama, C., Smith, P. L., Angata, K., Fukuda, M. N., Lowe, J. B., and Fukuda, M. (1998). Molecular cloning and expression of GDP-D-mannose-4,6-dehydratase, a key enzyme for fucose metabolism defective in Lec13 cells. *J. Biol. Chem.* **273**, 14582–14587.

Oka, T., and Krieger, M. (2005). Multi-component protein complexes and Golgi membrane trafficking. *J. Biochem. (Tokyo)* **137**, 109–114.

Oliver, J. D., Roderick, H. L., Llewellyn, D. H., and High, S. (1999). ERp57 functions as a subunit of specific complexes formed with the ER lectins calreticulin and calnexin. *Mol. Biol. Cell* **10**, 2573–2582.

Patnaik, S. K., and Stanley, P. (2005). Mouse large can modify complex N- and mucin O-glycans on alpha-dystroglycan to induce laminin binding. *J. Biol. Chem.* **280**, 20851–20859.

Patnaik, S. K., Potvin, B., and Stanley, P. (2004). LEC12 and LEC29 gain-of-function Chinese hamster ovary mutants reveal mechanisms for regulating VIM-2 antigen synthesis and E-selectin binding. *J. Biol. Chem.* **279**, 49716–49724.

Patnaik, S. K., Potvin, B., Carlsson, S., Sturm, D., Leffler, H., and Stanley, P. (2005). Complex N-glycans are the Major Ligands for Galectin-1, Galectin-3 and Galectin-8 on Chinese Hamster Ovary Cells. *Glycobiology* (in press).

Patnaik, S. K., Zhang, A., Shi, S., and Stanley, P. (2000). alpha(1,3)fucosyltransferases expressed by the gain-of-function Chinese hamster ovary glycosylation mutants LEC12, LEC29, and LEC30. *Arch. Biochem. Biophys.* **375**, 322–332.

Polley, M. J., Phillips, M. L., Wayner, E., Nudelman, E., Singhal, A. K., Hakomori, S., and Paulson, J. C. (1991). CD62 and endothelial cell-leukocyte adhesion molecule 1 (ELAM-1) recognize the same carbohydrate ligand, sialyl-Lewis x. *Proc. Natl. Acad. Sci. USA* **88**, 6224–6228.

Potvin, B., Raju, T. S., and Stanley, P. (1995). Lec32 is a new mutation in Chinese hamster ovary cells that essentially abrogates CMP-N-acetylneuraminic acid synthetase activity. *J. Biol. Chem.* **270**, 30415–30421.

Pu, L., Scocca, J. R., Walker, B. K., and Krag, S. S. (2003). A single point mutation resulting in an adversely reduced expression of DPM2 in the Lec15.1 cells. *Biochem. Biophys. Res. Commun.* **312**, 555–561.

Quellhorst, G. J., Jr., O'Rear, J. L., Cacan, R., Verbert, A., and Krag, S. S. (1999). Nonglucosylated oligosaccharides are transferred to protein in MI8–5 Chinese hamster ovary cells. *Glycobiology* **9**, 65–72.

Raju, T. S., and Stanley, P. (1998). Gain-of-function Chinese hamster ovary mutants LEC18 and LEC14 each express a novel N-acetylglucosaminyltransferase activity. *J. Biol. Chem.* **273**, 14090–14098.

Ravdin, J. I., Stanley, P., Murphy, C. F., and Petri, W. A., Jr. (1989). Characterization of cell surface carbohydrate receptors for Entamoeba histolytica adherence lectin. *Infect. Immun.* **57**, 2179–2186.

Rosenwald, A. G., and Krag, S. S. (1990). Lec9 CHO glycosylation mutants are defective in the synthesis of dolichol. *J. Lipid Res.* **31**, 523–533.

Sasaki, H., Bothner, B., Dell, A., and Fukuda, M. (1987). Carbohydrate structure of erythropoietin expressed in Chinese hamster ovary cells by a human erythropoietin cDNA. *J. Biol. Chem.* **262**, 12059–12076.

Shinkawa, T., Nakamura, K., Yamane, N., Shoji-Hosaka, E., Kanda, Y., Sakurada, M., Uchida, K., Anazawa, H., Satoh, M., Yamasaki, M., Hanai, N., and Shitara, K. (2003). The absence of fucose but not the presence of galactose or bisecting N-acetylglucosamine of human IgG1 complex-type oligosaccharides shows the critical role of enhancing antibody-dependent cellular cytotoxicity. *J. Biol. Chem.* **278**, 3466–3473.

Siminovitch, L. (1976). On the nature of hereditable variation in cultured somatic cells. *Cell* **7**, 1–11.

Stanley, P. (1980). Altered glycolipids of CHO cells resistant to wheat germ agglutinin. *In* "American Chemical Society Symposium Series, No. 128, Cell Surface Glycolipids" (C. C. Sweeley, ed.), pp. 213–221. American Chemical Society.

Stanley, P. (1983). Selection of lectin-resistant mutants of animal cells. *Methods Enzymol.* **96**, 157–184.

Stanley, P. (1984). Glycosylation mutants of animal cells. *Annu. Rev. Genet.* **18**, 525–552.

Stanley, P. (1989). Chinese hamster ovary cell mutants with multiple glycosylation defects for production of glycoproteins with minimal carbohydrate heterogeneity. *Mol. Cell. Biol.* **9**, 377–383.

Stanley, P., Sundaram, S., and Sallustio, S. (1991). A subclass of cell surface carbohydrates revealed by a CHO mutant with two glycosylation mutations. *Glycobiology* **1**, 307–314.

Stanley, P., Sundaram, S., Tang, J., and Shi, S. (2005). Molecular analysis of three gain-of-function CHO mutants that add the bisecting GlcNAc to N-glycans. *Glycobiology* **15**, 43–53.

Sullivan, F. X., Kumar, R., Kriz, R., Stahl, M., Xu, G. Y., Rouse, J., Chang, X. J., Boodhoo, A., Potvin, B., and Cumming, D. A. (1998). Molecular cloning of human GDP-mannose 4,6-dehydratase and reconstitution of GDP-fucose biosynthesis *in vitro*. *J. Biol. Chem.* **273**, 8193–8202.

Warnock, D. E., Roberts, C., Lutz, M. S., Blackburn, W. A., Young, W. W., Jr., and Baenziger, J. U. (1993). Determination of plasma membrane lipid mass and composition in cultured Chinese hamster ovary cells using high gradient maGnetic affinity chromatography. *J. Biol. Chem.* **268**, 10145–10153.

Warren, C. E., Krizus, A., Roy, P. J., Culotti, J. G., and Dennis, J. W. (2002). The Caenorhabditis elegans gene, gly-2, can rescue the N-acetylglucosaminyltransferase V mutation of Lec4 cells. *J. Biol. Chem.* **277**, 22829–22838.

Weinstein, J., Sundaram, S., Wang, X., Delgado, D., Basu, R., and Stanley, P. (1996). A point mutation causes mistargeting of Golgi GlcNAc-TV in the Lec4A Chinese hamster ovary glycosylation mutant. *J. Biol. Chem.* **271**, 27462–27469.

Worton, R. G. (1978). Karyotypic heterogeneity in CHO cell lines. *Cytogenet. Cell Genet.* **21**, 105–110.

Zhang, A., Potvin, B., Zaiman, A., Chen, W., Kumar, R., Phillips, L., and Stanley, P. (1999). The gain-of-function Chinese hamster ovary mutant LEC11B expresses one of two Chinese hamster FUT6 genes due to the loss of a negative regulatory factor. *J. Biol. Chem.* **274**, 10439–10450.

## [12]   CHO Glycosylation Mutants: GPI Anchor

*By* Yusuke Maeda, Hisashi Ashida, and Taroh Kinoshita

### Abstract

Glycosylphosphatidylinositol (GPI) is used for anchoring many cell surface proteins to the plasma membrane. Biosynthesis of GPI anchor, its attachment to proteins, and modification of GPI-anchored proteins (GPI-APs) en route to the plasma membrane are complex processes (Ferguson, 1999; Kinoshita and Inoue, 2000). GPI-AP–defective mutant cell lines derived from CHO and other cells have been very useful in elucidating GPI biosynthetic pathway and cloning genes involved in these processes. In this chapter, we overview GPI-AP biosynthesis, establishment and characterization of GPI-AP–defective mutant cell lines, expression cloning using those mutant cells, and characteristics of GPI-AP–defective mutant cell lines.

### Overview of GPI-AP Biosynthesis

*Biosynthesis of GPI*

GPI is synthesized by the sequential additions of sugars and other components to phosphatidylinositol (PI) in the endoplasmic reticulum (ER) (Fig. 1) (Ferguson, 1999; Kinoshita and Inoue, 2000). At least eight reaction steps are required for generation of GPI that is competent for attachment to proteins (Fig. 2). The biosynthetic pathway is initiated by the transfer of N-acetylglucosamine (GlcNAc) from UDP-GlcNAc to PI on the

METHODS IN ENZYMOLOGY, VOL. 416
0076-6879/06 $35.00
DOI: 10.1016/S0076-6879(06)16012-7

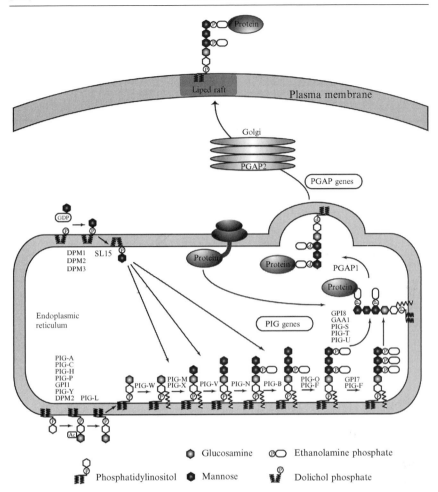

FIG. 1. Biosynthesis and transport of GPI-APs. GPI-anchor is synthesized in the endoplasmic reticulum and is en bloc transferred to proteins. GPI-APs are then tranported to the plasma membrane via the Golgi apparatus.

cytoplasmic side of the ER, generating the first intermediate, GlcNAc-PI (step 1). GPI-*N*-acetylglucosaminyltransferase (GPI-GnT) that catalyzes this reaction is a multisubunit enzyme consisting of at least seven components, PIG-A (Miyata *et al.*, 1993), PIG-C (Inoue *et al.*, 1996), PIG-H (Kamitani *et al.*, 1993), PIG-P (Watanabe *et al.*, 2000), GPI1 (also termed PIG-Q) (Hong *et al.*, 1999b), PIG-Y (Murakami *et al.*, 2005), and DPM2 (Maeda *et al.*, 1998; Watanabe *et al.*, 2000).

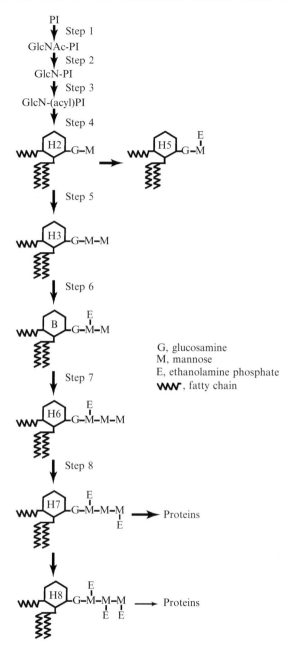

FIG. 2. Biosynthesis of GPI-anchor. GPI-anchor is synthesized from PI through sequential reaction steps. H7 and H8 are major and minor forms of mature GPI-anchors, respectively.

The step 2 in the pathway is de-$N$-acetylation of GlcNAc-PI to generate the second intermediate, glucosaminyl (GlcN)-PI. This reaction, which occurs still on the cytoplasmic side of the ER, is mediated by deacetylase PIG-L (Nakamura *et al.*, 1997). GlcN-PI is then translocated into the luminal side of the ER where all the later reactions occur. The luminal translocation of GlcN-PI is likely mediated by some enzyme with flippase or scramblase activity; however, it has not been identified or cell lines defective in this translocation have not been isolated.

The step 3 is the addition of acyl-chain, usually palmitoyl chain, from acyl-coenzyme A (acyl-CoA) to position 2 of inositol (inositol acylation) to generate GlcN-(acyl)PI, the third intermediate. This reaction is mediated by acyltransferase PIG-W (Murakami *et al.*, 2003).

The fourth step is the transfer of the first mannose (Man1) to generate the fourth intermediate, Man-GlcN-(acyl)PI or H2 (Fig. 2). Man1, linked to GlcN via $\alpha1$–4 linkage, is transferred from dolichol-phosphate-mannose (Dol-P-Man) by GPI-mannosyltransferase I that is a complex of a catalytic subunit PIG-M (Maeda *et al.*, 2001) and a regulatory subunit PIG-X (Ashida *et al.*, 2005).

The fifth step is the transfer of the second mannose (Man2) from Dol-P-Man to generate the fifth intermediate, Man-Man-GlcN-(acyl)PI or H3. Man2, linked to Man1 via $\alpha1$–6 linkage, is transferred by PIG-V, GPI-mannosyltransferase II (Kang *et al.*, 2005).

The step 6 is the addition of ethanolamine-phosphate (EtNP) side-chain to position 2 of Man1, generating the sixth intermediate, Man-(EtNP)Man-GlcN-(acyl)PI or B. This reaction is mediated by PIG-N, GPI EtNP transferase I (Hong *et al.*, 1999a), and phosphatidylethanolamine (PE) is the donor substrate. This EtNP-side-chain is essential in *Saccharomyces cerevisiae* (Imhof *et al.*, 2004), whereas it is not an obligatory component in mammalian cells. YW3548/BE49385A is a terpenoid lactone that selectively inhibits GPI EtNP transferase I (Hong *et al.*, 2002; Sutterlin *et al.*, 1997, 1998).

The step 7 is the transfer of the third mannose (Man3) to Man2 from Dol-P-Man by PIG-B, GPI-mannosyltransferase III (Takahashi *et al.*, 1996), generating the seventh intermediate, Man-Man-(EtNP)Man-GlcN-(acyl)PI or H6. Man3 is linked to Man2 via $\alpha1$–2 linkage. Sometimes the fourth mannose (Man4) is added to Man3 in $\alpha1$–2 linkage as a side-chain. Man4 is transferred from Dol-P-Man by SMP3, GPI-mannosyltransferase IV (Taron *et al.*, 2004). Man4 is essential in *S. cerevisiae*, whereas it is not a common component in mammalian GPI-APs and its biological requirement is yet to be clarified.

The eighth step is the addition of EtNP from PE to position 6 of Man3, generating EtNP-Man-Man-(EtNP)Man-GlcN-(acyl)PI or H7, one of the so-called "mature" GPI precursors that are competent for attachment to proteins. This EtNP (bridging EtNP) links GPI to proteins. The transfer of this EtNP is mediated by GPI EtNP transferase II that is a complex of a catalytic subunit PIG-O (Hong *et al.*, 2000) and a regulatory subunit PIG-F (Inoue *et al.*, 1993).

The "mature" GPI precursor H7 can be further modified by the second EtNP-side-chain linked to position 6 of Man2, generating EtNP-Man-(EtNP)Man-(EtNP)Man-GlcN-(acyl)PI or H8. This reaction is mediated by GPI EtNP transferase III consisting of catalytic GPI7 (also termed PIG-G) and regulatory PIG-F (Shishioh *et al.*, 2005). The modification by the second EtNP-side-chain is not essential at the cell level and the biological significance is yet to be determined.

Dol-P-Man, a donor of all four mannoses in GPI, is synthesized on the cytoplasmic side of the ER by Dol-P-Man synthase and then translocated into the ER lumen for usage in GPI (Fig. 1). Mammalian Dol-P-Man synthase consists of a catalytic subunit DPM1 (Tomita *et al.*, 1998) and two regulatory subunits DPM2 (Maeda *et al.*, 1998) and DPM3 (Ashida *et al.*, 2006; Maeda *et al.*, 2000). DPM2 is essential for Dol-P-Man synthase and is found in a fraction of GPI GnT complex. DPM2 enhances GPI GnT three to fourfold (Watanabe *et al.*, 2000). For luminal translocation of Dol-P-Man, SL15 (the product of MPDU1 gene) (Ware and Lehrman, 1996), which is probably involved in translocation reaction, is required (Anand *et al.*, 2001).

*Attachment of GPI to Proteins*

Attachment of "mature" GPI precursors, H7 or H8, to proteins is mediated by GPI transamidase (Fig. 1). Precursor proteins that are to be GPI anchored have carboxyl-terminal signal sequence for GPI attachment. GPI transamidase recognizes and cleaves off the GPI attachment signal peptide, forming enzyme-substrate intermediate linked via thioester. The intermediate is dissolved by a nucleophilic attack by the terminal amino group in GPI, generating GPI-APs. GPI transamidase consists of five subunits, GPI8 (also termed PIG-K) (Benghezal *et al.*, 1996; Yu *et al.*, 1997), GAA1 (Hiroi *et al.*, 1998), PIG-S (Ohishi *et al.*, 2001), PIG-T (Ohishi *et al.*, 2001), and PIG-U (Hong *et al.*, 2003).

*Post–GPI-Attachment Processing*

The first event after attachment of GPI to proteins is inositol-deacylation (i.e., removal of palmitoyl chain from the inositol ring) within the ER (Fig. 1). The inositol-deacylation is mediated by an ER-resident deacylase PGAP1 (for post-GPI attachment to proteins 1) (Tanaka *et al.*,

2004). After this reaction, GPI-APs become sensitive to cleavage by PI-specific phospholipase C (PI-PLC).

Inositol-deacylated GPI-APs are then transported to the cell surface through the Golgi apparatus via the secretary pathway (Fig. 1). One protein termed PGAP2 that is mainly found in the Golgi was identified by Tashima *et al.* (2006). Although the exact function of PGAP2 is yet to be determined, PGAP2 is very likely involved in lipid remodeling of GPI. It is known that fatty chains of PI within GPI-APs are usually saturated chains, which are compatible with liquid ordered phase of the membranes thought to exist in lipid rafts. Because typical cellular PIs have unsaturated fatty acid, the lipid portion of GPIs may be remodeled to have only saturated chains (see later discussion of the characteristics of PGAP2-deficient cells).

## Establishment of GPI-AP–Defective CHO Cells

In this section, we describe methods for the establishment of CHO mutant cells with defective surface expressions of GPI-APs.

### *Establishment of Parent CHO Cells Expressing GPI-APs*

Parent cells are required for the expression of GPI-APs as reporters against which antibodies are available. Because of the difficulty associated with obtaining antibodies against endogenous GPI-APs expressed on CHO cells, 3B2A parent CHO cells have been established by stable transfection of a plasmid encoding the human DAF (CD55), CD59, and neomycin-resistant genes into CHO-K1 cells under selection with 600 $\mu$g/ml G418 (Nakamura *et al.*, 1997). The use of two marker proteins, namely DAF and CD59, eliminates the incidence of picking up incorrect clones in which the marker gene itself is mutated irrelevantly toward GPI-anchor biosynthesis. DAF and CD59, which are both complement regulatory molecules, are widely expressed on human cell lines, and these antibodies are also useful for analyzing human mutant cells. After selection for 2 wk with 600 $\mu$g/ml G418, cells expressing high levels of DAF and CD59 are sorted using a cell sorter (FACS-Vantage) and subjected to limiting dilution to establish clones. Mouse monoclonal antibodies IA10 against DAF and 5H8 against CD59 are used to stain the cells for cell sorting.

Because mutagenesis of CHO cells with one mutagen show the strong unfavorable tendency that only a few kinds of complementary groups of mutants are obtained, presumably due to the existence of hot spots for mutations, we need devices to create more mutant groups from a second round of mutagenesis with the same mutagen. One strategy for eliminating this tendency is to use a new parent cell line stably transfected with the genes responsible for the mutants derived from the first round of mutagenesis. Expression vectors containing the antibiotic-resistance genes for puromycin (5–6 $\mu$g/ml), hygromycin (500–800 $\mu$g/ml), and blasticidin are useful for the selection and establishment of such new parent cell lines. CHO F21 (Ashida et al., 2005) and CHO D9 (Kang et al., 2005) (Table I) are such parent cells. Another strategy is to use a different mutagen, such as $N$-methyl-$N'$-nitro-$N$-nitrosoguanidine.

## Mutagenesis and Selection of Mutant Cells

At the logarithmic phase, the aforementioned parent CHO cells are treated with the mutagen ethyl methanesulfonate (EMS; 0.4 mg/ml) for 18–24 h. Since EMS is both volatile and a strong mutagen, it must be handled carefully. Because 30–70% of the cells are typically dying after 4–5 days if the EMS treatment is optimal, mutant cell selections are performed at 5–7 days after the EMS treatment. Two methods are useful for the selection of GPI-negative cells. One is to use a cell sorter (FACS-Vantage) and the other is to use the toxins Aeromonas hydrophila proaerolysin (Protox Biotech) and Clostridium septicum $\alpha$-toxin, which form pores and kill the cells by binding to the surface GPIs. These selection techniques are based on the fact that defects in genes involved in the biosynthesis of GPI strongly decrease or abrogate the surface expression of GPI-APs. For cell sorting, the cells are stained with the 5H8 antibody (10 $\mu$g/ml in phosphate-buffered saline [PBS] containing 1% bovine serum albumin [BSA]) against CD59 on ice for 20 min, washed twice, and stained with fluorescein-isothiocyanate (FITC)- or PE-conjugated secondary antibodies against mouse immunoglobulin G (IgG) on ice for 20 min. Next, cells with low or null surface expression of CD59 are collected. For typical selection of GPI-APs–negative cells with toxins, the cells are incubated in normal culture medium containing 1–5 n$M$ proaerolysin for 1–2 days or 0.1–1.0 n$M$ $\alpha$-toxin for 0.5–2.0 days, although the optimal conditions must be established for each parent cell line. The enrichment of GPI-APs–negative mutant cells is confirmed by FACS, and if the percentage of negative cells is not sufficiently high, repeated selection is necessary. If the percentage is sufficiently high, limiting dilution is performed in 96-well plates.

Mutagenesis with chemical mutagens followed by the aforementioned selections can be applied to other cell lines in addition to CHO cells. A method has been reported for ES cells in which expression of the Bloom's syndrome gene is regulated and a variety of GPI mutants are created efficiently (Yusa et al., 2004), although very specific techniques and materials are required.

TABLE I

List of Mutant Cells Defective in GPI-Anchor Biosynthetic Genes

| Gene | Mutant | Original cell | Reference |
|---|---|---|---|
| PIG-A | JY5 | Human B lymphoblastoid JY | Hollander et al., 1988 |
| | S49-Thy-1$^-$ class A | Mouse T lymphoma S49 | Hyman, 1988 |
| PIG-C | T1M1-Thy-1$^-$ class C | Mouse T lymphoma T1M1 | Inoue et al., 1996 |
| PIG-H | Ltk$^-$ | L929 cells | Kamitani et al., 1993 |
| | S49-Thy-1$^-$ class H | Mouse T lymphoma S49 | Hyman, 1988 |
| PIG-P | 2.10(GPI$^-$) | Mouse T cell | Watanabe et al., 2000 |
| GPI1 (PIG-Q) | GPI1 KO | Mouse embryonal carcinoma F9 | Hong et al., 1999b |
| PIG-Y | Daudi | Human Burkitt lymphoma Daudi | Murakami et al., 2005 |
| PIG-L | M2S2 | CHO IIIB2A | Nakamura et al., 1997 |
| | G9PLAP.85 | CHO | Stevens et al., 1996 |
| | LA1 | CHO | Abrami et al., 2001 |
| PIG-W | CHOPA10.14 | CHO | Murakami et al., 2003 |
| | Molt4-1D10 | Human T-lymphoma Molt4 | Murakami et al., 2003 |
| PIG-M | Ramos517 | Human Burkitt lymphoma Ramos | Maeda et al., 2001 |
| PIG-X | CHO2.46 | CHO F21 | Ashida et al., 2005 |
| PIG-V | D9PA15.6, 25.2, 72.1 | CHO D9 | Kang et al., 2005 |
| PIG-B | S1A-Thy-1$^-$ class B | Murine T-lymphoma S1A | Hyman, 1988 |
| SMP3 | None | | |

(continued)

TABLE I (*continued*)

| Gene | Mutant | Original cell | Reference |
|------|--------|---------------|-----------|
| DPM1 | BW5147-Thy1⁻ class E | Mouse T-lymphoma BW5147 | Trowbridge *et al.*, 1978 |
| DPM2 | Lec15 | CHO | Maeda *et al.*, 1998 |
|      | B4-2-1 | CHO | Stoll *et al.*, 1982 |
| DPM3 | CHO2.38 | CHO F21 | Ashida *et al.*, 2006 |
| SL15 (MPDU1) | Lec35 | CHO | Ware and Lehrman, 1996 |
| PIG-N | PIG-N KO | Mouse embryonal carcinoma F9 | Hong *et al.*, 1999 |
| PIG-O | PIG-O KO | Mouse embryonal carcinoma F9 | Hong *et al.*, 2000 |
|       | No name | CHO | Hong *et al.*, 2000 |
| PIG-F | EL4-Thy1⁻ class F | Murine T lymphoma EL4 | Conzelmann *et al.*, 1988 |
|       | CHO30.5 | CHO F21 | Shishioh *et al.*, 2005 |
| GPI7 (PIG-G) | None | | |
| GPI8 (PIG-K) | Class K | Human K562 | Yu *et al.*, 1997 |
| GAA1 | GAA KO | Mouse embryonal carcinoma F9 | Ohishi *et al.*, 2000 |
| PIG-S | PIG-S KO | Mouse embryonal carcinoma F9 | Ohishi *et al.*, 2001 |
| PIG-T | PIG-T KO | Mouse embryonal carcinoma F9 | Ohishi *et al.*, 2001 |
| PIG-U | PA9.1, 12.1, 16.1 | CHO C311 | Hong *et al.*, 2003 |
| PGAP1 | C10 | CHO IIIB2A | Tanaka *et al.*, 2004 |
| PGAP2 | C84 | CHO GD3S-C37 | Tashima *et al.*, 2006 |
|       | AM-B | CHO F21 | Tashima *et al.*, 2006 |

*Elimination of Mutant Cells Belonging to Known Mutant Classes*

Mutant cell lines newly obtained by limiting dilution are examined to evaluate whether they belong to new complementary groups of mutants or known groups. A feasible method for achieving this aim is to transfect cells with genes known to be involved in GPI biosynthesis, followed by FACS analysis to assess whether the surface expressions of CD59 and DAF are restored. Although more than 20 genes are involved in GPI biosynthesis, the transfection efficiency of either electroporation (Gene-Pulser; Bio-Rad) or lipofection (Lipofectamine 2000; Invitrogen) is fairly good and more than five genes can be mixed for one transfection assay. A total of $4–10 \times 10^6$ cells in 0.4 ml Opti-MEM (Invitrogen) are electroporated with 20–25 $\mu$g plasmid DNA at 960 $\mu$F and 250–260 V. Lipofection is performed according to the manufacturer's recommendations. If none of the genes can restore the surface expressions of DAF and CD59, the mutant cells are considered to belong to a new class. If the defective step is identified by metabolic labeling as described later in this chapter, the genes related to the step may be sufficient for transfection. Cell fusion analysis using polyethylene glycol (PEG), as described in a previous report (Fuller *et al.*, 1987), is useful for further classifying mutants among the newly obtained cell lines.

Characterization of Mutants

The best way to characterize mutant cells is to label GPI intermediates with radioisotopes and analyze the labeled molecules by thin-layer chromatography (TLC). $^3$H-labeled UDP-GlcNAc for steps 1–3 and GDP-mannose for the later steps and biosynthesis of Dol-P-Man are commonly used for *in vitro* labeling of GPI intermediates in microsomes. $^3$H-labeled mannose is also very useful for the later steps of *in vivo* labeling (Fig. 3). Radiolabeled inositol is used for *in vivo* labeling of the former steps and GPI-APs (Fig. 3A), while radiolabeled ethanolamine is used for *in vivo* labeling of GPI-APs. In some cases, the labeled glycolipids are treated with the GPI-specific phospholipase D (GPI-PLD) or jack bean $\alpha$-mannosidase (Sigma) before application to TLC as described later in this chapter. Any defect in the biosynthesis reaction causes abnormal accumulation of the intermediate that is the usual substrate for the affected step, although this is not always the case for PIG-W mutants, in which accumulation of intermediates for the later steps is observed. Many abnormal accumulations are sometimes observed due to the traffic jam in the biosynthetic pathway, as observed in the case of defects in GPI transamidase (Fig. 3C).

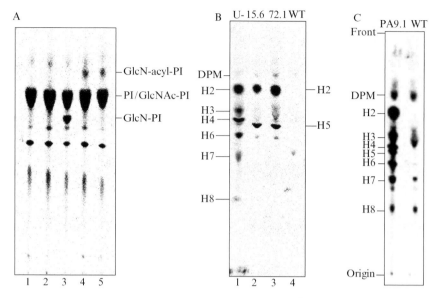

FIG. 3. (A) Cells were metabolically labeled with *myo*-[³H]inositol. The glycolipids were applied onto a silica gel thin-layer chromatography (TLC) plate and developed with chloroform:methanol:1 N NH4OH (10:10:3, v/v/v). *Lane 1*, CHO F21 (wild-type cells); *lane 2*, PIG-L–defective CHO 1.46; *lane 3*, PIG-W–defective CHO 10.14; *lane 4*, PIG-X– defective CHO 2.46; *lane 5*, Dol-P-Man synthase (DPM3)–defective CHO 2.38. (B, C) Cells were metabolically labeled with [³H]mannose. The glycolipids were applied onto a silica gel TLC plate and developed with chloroform/methanol/H₂O (10:10:3, v/v/v). *DPM*, dolichol-phosphate-mannose; *H2*, Man-GlcN-(acyl)PI; *H3*, Man-Man-GlcN-(acyl)PI; *H4*, Man-Man-Man-GlcN-(acyl)PI; *H5*, (EtNP)Man-GlcN-(acyl)PI; *H6*, Man-Man-(EtNP)-Man-GlcN-(acyl) PI; *H7*, EtNP-Man-Man-(EtNP)Man-GlcN-(acyl)PI; *H8*, EtNP-Man-(EtNP)Man-(EtNP) Man-GlcN-(acyl)PI. (B) *Lane 1*, PIG-U–defective CHO C311PA16; *lane 2*, PIG-V–defective CHO D9PA15.6; *lane 3*, PIG-V–defective CHO D9PA72.1; *lane 4*, CHOD9 (wild-type cells). (C) *Lane 1*, PIG-U–defective CHO PA9.1; *lane 2*, CHO-K1 (wild-type cells). (A, modified, with permission, from Ashida *et al.*, 2005; B, modified, with permission, from Kang *et al.*, 2005; C, modified, with permission, from Hong *et al.*, 2003.)

If no abnormal accumulation of intermediates is observed by metabolic labeling, the defective gene may be involved in processes other than GPI biosynthesis. PGAP2 mutant cells do not show any spots of accumulated molecules upon TLC, and PGAP2 is considered to be involved in the lipid remodeling of GPI-APs that occurs during their transport from the ER to the Golgi. Examining the fates of GPI-APs by pulse-chase metabolic labeling with ³⁵S-methionine or immunofluorescence microscopy may be useful for understanding which process beyond the biosynthesis in the ER is affected in the mutant cells.

## In Vivo *Labeling of GPI Intermediates*

For labeling with mannose, cells ($2 \times 10^6$) are briefly washed twice with glucose-free medium and incubated in 1 ml of glucose-free medium containing 20 m$M$ HEPES–NaOH (pH 7.4), 100 $\mu$g/ml glucose, 10% dialyzed fetal calf serum (FCS), and 10 $\mu$g/ml tunicamycin (Sigma-Aldrich) for 1 h. Next, 40 $\mu$Ci of D-[2-$^3$H]mannose (Amersham Pharmacia Biotech) is added, and the cells are incubated in the same medium for a further 45 min (Hirose *et al.*, 1992).

For labeling with inositol, $1 \times 10^6$ cells are washed with inositol-free Dulbecco modified eagle medium (DMEM) (Gibco-BRL) and then cultured in 1 ml of the same medium supplemented with 10% dialyzed FCS and 40 $\mu$Ci of *myo*-(2-$^3$H[N])inositol (American Radiolabeled Chemicals) for 1–2 days.

After washing with PBS and centrifugation, glycolipids in the cell pellet are extracted twice with 0.5 ml chloroform/methanol/water (10:10:3, v/v/v) and the combined extracts are evaporated. The dried lipids are then extracted twice with 0.3 ml of water-saturated *n*-butanol, and the combined butanol extracts are back-washed with 0.3 ml of butanol-saturated water and evaporated. The dried glycolipids are suspended in 20 $\mu$l of chloroform: methanol (2:1, v/v) and separated by TLC on Kieselgel 60 (Merck) using a solvent of chloroform/methanol/1$M$ NH$_4$OH (10:10:3, v/v/v) for the first three steps and chloroform/methanol/H$_2$O (10:10:3, v/v/v) for the later steps. The radiolabeled glycolipids are analyzed using an Image Analyzer BAS 1500 (Fuji Film Co.) after exposure for 1–2 days (Fig. 3).

## In Vitro *Labeling of GPI Intermediates*

After preincubation in culture medium containing 5 $\mu$g/ml tunicamycin (Sigma) for 2 h, CHO cells ($3 \times 10^7$) are washed with PBS and suspended in 1 ml of hypotonic buffer (20 m$M$ Tris–HCl pH 7.4, 2 $\mu$g/ml leupeptin, and 0.1 m$M$ 1-chloro-3-tosylamido-7-1-2-heptanone [TLCK]) on ice for 10 min. A cell lysate is obtained by disrupting the cells as gently as possible with a proper homogenizer. For the labeling of the first three steps, better results may be obtained from the cell lysate than from microsomes. After removal of the cell debris and nuclei by centrifugation at 10,000$g$ for 10 min, microsomes are collected by centrifugation at 100,000$g$ for 1 h, suspended in a buffer (50 m$M$ HEPES–NaOH pH 7.4, 25 m$M$ KCl, 2 $\mu$g/ml leupeptin, and 0.1 m$M$ TLCK), and stored at $-80°$ for the labeling of the later steps.

For the first three steps, cell lysates ($3 \times 10^6$) are incubated in 0.15 ml of buffer (2 $\mu$Ci UDP-[6-$^3$H]GlcNAc [American Radiolabeled Chemicals], 50 m$M$ HEPES–NaOH pH 7.4, 25 m$M$ KCl, 5 m$M$ MgCl$_2$, 5 m$M$ MnCl$_2$,

1 m$M$ ATP, 0.5 m$M$ dithiothreitol [DTT], 1 m$M$ GTP, 2 $\mu M$ CoA, 0.5 $\mu$g/ml tunicamycin, 0.1 m$M$ TLCK, and 2 $\mu$g/ml leupeptin) for 1 h at 37°.

For assays of DPM synthase and the later steps of GPI biosynthesis, microsomes (3 × 10$^6$) are incubated in 0.15 ml of buffer (2 $\mu$Ci GDP-[$^3$H] mannose [American Radiolabeled Chemicals], 50 m$M$ HEPES–NaOH pH 7.4, 25 m$M$ KCl, 5 m$M$ MgCl$_2$, 5 m$M$ MnCl$_2$, 10 $\mu$g dolichol phosphate, 0.5 $\mu$g/ml tunicamycin, 0.1 m$M$ TLCK and 2 $\mu$g/ml leupeptin) at 37° for 10 min to 1 h.

Both reactions are terminated by adding 1 ml of chloroform/methanol (1:1, v/v). After centrifugation at 18,000$g$ for 10 min at 4°, the supernatant is transferred to a new tube and dried. The glycolipids are extracted with $n$-butanol and analyzed by TLC as described earlier.

## Enzyme Treatments of Labeled Lipids

Glycolipids are dissolved in 0.15 ml of one of the following three buffers (100 m$M$ Tris–HCl pH 7.4, 0.1% Triton X-100, and 1 $\mu$l of PI-PLC [Invitrogen]; 100 m$M$ sodium acetate pH 5.0, 1 m$M$ ZnCl$_2$, 0.1% sodium taurodeoxycholate, and 15 $\mu$l [1.7 U] of jack bean $\alpha$-mannosidase [Sigma]; or 50 m$M$ Tris–HCl pH 7.4, 10 m$M$ NaCl, 2.5 m$M$ CaCl$_2$, 0.1% Triton X-100, and 15 $\mu$l of human serum as a source of GPI-PLD). After incubation for 12–16 h at 37°, the reactions are terminated by adding 1 ml of chloroform/methanol (1:1, v/v), and the glycolipids are extracted and analyzed by TLC as described earlier.

## Expression Cloning of the Gene Responsible for the Defects

GPI biosynthesis is essential for the cell surface expressions of GPI-APs, so a defect in GPI biosynthesis abrogates these surface expressions. Because of this nature, expression cloning can be applied to identify the gene responsible for the defect in mutant cells based on restoration of the surface expressions of GPI-APs. The methods for expression cloning are described in detail in a report by Kinoshita et al. (1995), but the procedure for transient expression cloning, which is suitable for CHO cells, is introduced here. Similar to the combination of the large T antigen and replication origin of the SV40 virus used in primate cells, plasmids containing the replication origin of the polyoma virus are replicated to several thousands of copies in rodent cells expressing the polyoma large T antigen, which enables recovery of the plasmids from the cells. Therefore, the plasmid for construction of a cDNA library has to contain a polyoma virus replication origin, and the cDNA library and polyoma large T-antigen expression plasmids are co-transfected into mutant cells. CHO mutant cells ($\sim 2 \times 10^8$) are suspended in 4 ml of HEPES–buffered saline (HeBS; 20 m$M$ HEPES, 137 m$M$ NaCl, 5 m$M$ KCl, 0.7 m$M$ Na$_2$HPO$_4$, and 6 m$M$ glucose, adjusted to pH 7.05 with 10 N NaOH) or OptiMEM I (Invitrogen) containing 200 $\mu$g each of a rat C6 glioma cDNA library and pcDNA-PyT (ori-) plasmids (Nakamura et al., 1997), divided into 10 cuvettes and electroporated at 260–280 V and 960 $\mu$F using a Gene Pulser (Bio-Rad). At 2–3 days after the transfection, the cells are stained with anti-CD59 antibody 5H8 and a phycoerythrin-conjugated secondary antibody against mouse IgG, and cells with restored surface expression of CD59 are collected using a cell sorter (FACS-Vantage). Usually less than 0.01% of the starting cells are sorted out as a positive population. Plasmids are recovered from these cells by the Hirt method (Hirt, 1967). Since the pcDNA-PyT (ori-) plasmid does not have an ampicillin-resistance gene, only plasmids derived from the cDNA library are rescued under ampicillin selection. The efficiency of the Escherichia coli strain used for electroporation is important and should be more than $10^9$ cfu/$\mu$g, which leads to the recovery of more than a few colonies per sorted cell. After another round of electroporation of the recovered plasmids and pcDNA-PyT (ori-) into mutant cells, cell sorting and plasmid recovery by the Hirt method, the recovered plasmids are transfected into the mutant cells in order to check whether there is a population of CD59-positive cells, which are normally observed in success-ful cases. Next, the recovered plasmids are transformed into E. coli and the transformants are plated on ampicillin plates to form isolated colonies. More than 1000 colonies are picked up from the plates and independently cultured in 96-well plates. Plasmid DNA purified from each plate (contain-ing 96 independent clones) is transfected into the mutant cells and the surface expression of CD59 is analyzed by FACS to examine whether the recovered plasmids can complement the defect in the mutant cells. If a positive plate is obtained, each clone in the plate is analyzed in the same way.

A retrovirus cDNA library is also useful for expression cloning. Although CHO cells are rodent cells, ecotropic viruses cannot be used because of the lack of appropriate receptors. As an alternative, the appro-priate receptor encoded by a CAT1 gene must be expressed on CHO cells before infection with the viruses or amphotropic viruses must be used (Albritton et al., 1989). A few cycles of sorting of the infected cells enrich the CD59-positive population. The integrated cDNA is amplified by poly-merase chain reaction (PCR) from the genomic DNA of the CD59-positive cells and is confirmed to be the gene responsible for the defect in the mutant cells by transfection.

*Recovery of Plasmids from CHO Cells (Hirt's Method)*

1. Suspend the sorted cell pellet in 300 μl of 0.6% sodium dodecylsulfate (SDS) and 10 mM ethylenediaminetetraacetic acid (EDTA), and leave it for 20 min at room temperature.
2. Add 75 μl of 5 M NaCl and incubate the mixture on ice for 15 min.
3. Centrifuge the mixture at 15,000 rpm for 15 min at 4° and transfer the supernatant to a new tube.
4. Add 400 μl of phenol/chloroform (1:1), vortex well, and centrifuge the mixture at 15,000 rpm for 5 min at room temperature.
5. Transfer the upper phase to a new tube, add 5 μl of polyacrylamide carrier and 1 ml of cold ethanol, and incubate the tube at −80° for 1 h.
6. After centrifugation, wash the DNA pellet with 1 ml of 70% ethanol, dry it, and resuspend it in 10 μl TE (10 mM Tris–HCl pH 8.0 and 1 mM EDTA).

## Characteristics of Mutant CHO and Other Mammalian Cells Defective in GPI-APs

Properties of mutant cells in each step (Table I) are described generally in order of reaction steps.

1. GPI-*N*-acetylglucosaminyltransferase (GPI-GnT)

   *Reaction.* PI + UDP-GlcNAc → GlcNAc-PI + UDP

   GPI-GnT that catalyzes this reaction consists of seven different components, PIG-A, PIG-C, PIG-H, PIG-P, GPI1, PIG-Y, and DPM2

   *PIG-A mutant.* PIG-A is thought to be a catalytic subunit of GPI-GnT because it was a significant sequence homology with the members of glycosyltransferase (GT) family 4 in the CAZy database and because it binds UDP analog (Kostova *et al.*, 2000). PIG-A mutant JY5 cells were derived from a human B lymphoblastoid JY cells (Hollander *et al.*, 1988) (Table I). The surface expression of GPI-APs (CD59 and DAF) on JY5 cells was completely negative. The microsome from JY5 cells did not show GPI-GnT activity. Complementation class A cells of Thy-1– defective mutants derived from mouse T-lymphoma cells, such as S49 (Hyman, 1988) are also PIG-A mutants.

   *PIG-C mutant.* Class C mutant derived from mouse T-cell lymphoma T1M1 cells had almost no surface expression of Thy-1 (Inoue *et al.*, 1996).

*PIG-H mutant.* Ltk⁻ cell line was established from L929 cells in continuous culture. These cells did not express Ly-6E, a GPI-AP, on the cell surface (Kamitani *et al.*, 1993).

*PIG-P mutant.* Mouse T-cell clone 2.10 GPI(−) is GPI-APs (Thy-1, CD48, and Sca1) negative (Marmor *et al.*, 1999; Watanabe *et al.*, 2000).

*GPI1 (PIG-Q) mutant.* GPI1 knockout F9 cells showed severe but not complete defect in the expression of Thy-1 (Hong *et al.*, 1999b).

*PIG-Y mutant.* Burkitt lymphoma–derived Daudi cell was reported as prion protein–negative cell (Morelon *et al.*, 2001). The Daudi cell was later found to be a mutant of PIG-Y (Murakami *et al.*, 2005). PIG-Y is not essential for complex formation of the other components of GPI-GnT but is important for efficient activity of the enzyme. Thus, Daudi cells showed leaky phenotype of GPI-APs expression.

*DPM2 mutant.* DPM2 was originally identified as a component of Dol-P-Man synthase complex (Maeda *et al.*, 1998) by expression cloning using CHO Lec15 cells (Beck *et al.*, 1990), which did not express GPI-APs. DPM2 is not essential for GPI-GnT; however, the enzyme activity is enhanced three to fourfold in the presence of DPM2 (Watanabe *et al.*, 1998).

2. GlcNAc-PI de-*N*-acetylase

*Reaction.* GlcNAc-PI → GlcN-PI + acetate

GlcNAc moiety of GlcNAc-PI is de-*N*-acetylated by PIG-L to generate GlcN-PI. PIG-L is the ER membrane protein with a large cytosolic domain.

*PIG-L mutant.* PIG-L–defective CHO M2S2 cells were generated by chemical mutagenesis from CHO IIIB2A cells stably transfected with GPI-AP markers, human CD59 and DAF. CHO M2S2 cells were GPI-AP negative and accumulated GlcNAc-PI (Nakamura *et al.*, 1997). PIG-L mutant is the most frequently isolated mutant from CHO cells. CHO G9PLAP.85, first GPI-AP-defective CHO mutant to be reported (Stevens *et al.*, 1996), was later found to be PIG-L mutant. Another PIG-L mutant CHO cell line, CHO-LA1, was established by resistance to bacterial pore-forming toxin aerolysin, which specifically binds to GPI-APs (Abrami *et al.*, 2001).

3. GPI inositol acyltransferase

*Reaction.* GlcN-PI + acyl-CoA → GlcN-(acyl)PI + CoA

The first reaction in the ER lumen, the transfer of acyl-chain, usually palmitoyl group, from acyl-CoA to the position 2 of inositol, is catalyzed by PIG-W. GPI inositol acylation is important but not essential for effective mannosylation reactions, whereas it is essential for step 8 (Figs. 1 and 2) to occur.

*PIG-W mutant.* Molt4 1D10 and CHO PA10.14 cells showed severely reduced expression of GPI-APs on the cell surface. The cells accumulate GlcN-PI and Man-containing GPI intermediates without inositol acylation (Murakami *et al.*, 2003).

4. GPI-mannosyltransferase I, II, and III

*Reaction.* GlcN-(acyl)PI + 3 × Dol-P-Man → Manα1-2Manα1-6Manα1-4GlcN-(acyl)PI + 3 × Dol-P

Three Man with different glycosidic bonds, namely α1-4, α1-6, and α1-2, is transferred sequentially from Dol-P-Man to GlcN-(acyl) PI. GPI-mannosyltransferase I, II, and III were identified to be PIG-M/PIG-X, PIG-V, and PIG-B, respectively. PIG-M, PIG-V, and PIG-B show some structural similarity to one another, namely multi-transmembrane topology, and to the other Dol-P-Man–dependent mannosyltransferases acting in the ER lumen, although the primary sequences show only very weak similarity. PIG-M, PIG-V, and PIG-B are classified into GT family 50, 76, and 22, respectively. PIG-X is a unique subcomponent for GPI-mannosyltransferase I.

*PIG-M mutant.* Ramos 517 cells were isolated from culture of original Ramos cells. Ramos 517 cells accumulated GlcN-(acyl)PI and did not express GPI-APs on the surface (Maeda *et al.*, 2001).

*PIG-X mutant.* CHO2.46 cells were generated by chemical mutagenesis of the parental CHO F21 cells that stably transfected with 12 PIG-genes and GPI-AP markers. Because PIG-M becomes highly unstable in the absence of PIG-X, CHO2.46 cells shows phenotype closely similar to Ramos 517 cells (Ashida *et al.*, 2005).

*PIG-V mutant.* CHO PA15.6, PA25.2, and PA72.1 were generated from CHO D9 cells that previously transfected with seven PIG-genes and GPI-AP markers. All of these mutant cells accumulated H2 and H5. CHO PA15.6 and PA72.1 cells did not express GPI-APs on their cell surface, whereas CHO PA25.2 cells exhibited leaky phenotype (Kang *et al.*, 2005) (Fig. 3).

*PIG-B mutant.* Murine thymoma S1A-b cells accumulated a GPI-intermediate B and did not express GPI-APs (Hyman, 1988; Takahashi *et al.*, 1996).

5. Synthesis and usage of Dol-P-Man

*Reaction.* Dol-P + GDP-Man → Dol-P-Man + GDP

Dol-P-Man is synthesized on the cytosolic surface of the ER by Dol-P-Man synthase, flipped into the ER lumen, and used as a donor. Dol-P-Man synthase consists of three components, DPM1,

DPM2, and DPM3. SL15 is required for usage of both Dol-P-Man and Dol-P-Glc in the ER lumen.

*DPM1 mutant.* DPM1 is the catalytic subunit of this enzyme complex. Class E mutant of mouse T-cell lymphoma was found to be a DPM1 mutant. These cells did not express Thy-1 (Tomita *et al.*, 1998; Trowbridge *et al.*, 1978).

*DPM2 mutant.* DPM2 binds and stabilizes DPM3 to regulate Dol-P-Man synthase. CHO Lec15 cells (Beck *et al.*, 1990) and B4-2-1 cells (Stoll *et al.*, 1982) are defective in Dol-P-Man synthase activity and GPI-AP expression.

*DPM3 mutant.* DPM3 directly tethers DPM1 on the ER membrane. CHO2.38 cells were generated from CHO F21 parental cells (see "PIG-X mutant") by chemical mutagenesis. CHO2.38 cells did not express GPI-APs and accumulated GlcN-(acyl)PI (Ashida *et al.*, 2006).

*SL15 mutant.* CHO Lec35 cells are defective in general usage of both Dol-P-Man and Dol-P-Glc (Ware and Lehrman, 1996). These cells showed reduced expression of GPI-APs and accumulation of Dol-P-Man and GlcN-(acyl)PI. This mutant also accumulated an immature $N$-glycan precursor Dol-PP-GlcNAc$_2$-Man$_5$ and was impaired in C-mannosylation (Anand *et al.*, 2001).

6. Transfer of ethanolamine phosphate (EtNP) to Man residues

*Reaction.* Man$\alpha$1-2Man$\alpha$1-6Man$\alpha$1-4GlcN-(acyl)PI + 3 × PE → EtNP-6Man$\alpha$1-2(EtNP-6)Man$\alpha$1-6(EtNP-2)Man$\alpha$1-4GlcN-(acyl) PI + 3 × diacylglycerol

The addition of EtNP to the position 2 of Man1 and the position 6 of Man2, as side-chain modifications, is catalyzed by PIG-N and GPI7/ PIG-F, respectively. The addition of EtNP to the position 6 of Man3 is catalyzed by PIG-O/PIG-F. This EtNP is used for linkage to proteins. PIG-N, GPI7, and PIG-O are the catalytic subunit of the enzymes. PIG-F is a common subunit of 6-EtNP transferases.

*PIG-N mutant.* PIG-N knockout F9 cells showed reduced expression of Thy-1 on the cell surface and accumulated several GPI precursors without EtNP modification to Man1, such as H3, H4, and H7 (Hong *et al.*, 1999a). PIG-N mutant cells have not been known.

*GPI7 (PIG-G) mutant.* Neither mutant nor knockout cells have been known. Knockdown of GPI7 by RNAi caused increase of H7 and concomitant decrease of H8 but had no effect on GPI-APs expression (Shishioh, 2005).

*PIG-O mutant.* PIG-O knockout F9 cells expressed very little GPI-APs and mainly accumulated H6 (Hong, 2000). PIG-O mutant CHO cells were later established (Hong, 2002).

*PIG-F mutant.* Mouse T-cell lymphoma–derived Thy-1⁻f cells (Conzelmann *et al.*, 1988) showed strongly reduced expression of Thy-1 and accumulated H6 (Inoue, 1993). PIG-F mutant CHO30.5 cells were later generated from CHO F21 parental cells (Shishioh *et al.*, 2005).

7. GPI transamidase

*Reaction*: Mature GPI (H7 or H8) + pro-protein → GPI-anchored protein + signal peptide

GPI transamidase consists of five components, GPI8 (PIG-K), GAA1, PIG-S, PIG-T, and PIG-U.

*GPI8 (PIG-K) mutant.* GPI8 is a catalytic component of GPI-transamidase and has significant homology with cysteine proteases. GPI8-defective class K mutant derived from human erythroleukemia K562 cells accumulated various GPI intermediates, H2 to H8, and devoid of surface expression of GPI-APs (Yu *et al.*, 1997).

*GAA1 mutant.* GAA1 knockout F9 cells mainly accumulated H7 and H8, and expressed no Thy-1 on their cell surface (Ohishi *et al.*, 2000).

*PIG-S mutant.* PIG-S knockout F9 cells showed similar phenotype to GAA1 knockout cells (Ohishi *et al.*, 2000).

*PIG-T mutant.* PIG-T knockout F9 cells showed similar phenotype to GAA1 knockout cells (Ohishi *et al.*, 2000).

*PIG-U mutant.* CHO PA9.1, 12.1, and 16.1 cells were generated from parental cells and stably express PIG-L, DPM2, SL15, PIG-A, and GPI-AP markers. These cells accumulated various GPI intermediates H2 to H8 and expressed almost no GPI-APs (Hong *et al.*, 2003) (Fig. 3).

8. GPI inositol deacylase

*Reaction.* Inositol-acylated GPI-APs → Inositol-deacylated GPI-APs + fatty acid

In most cells, acyl-chain on the inositol moiety of newly generated GPI-anchored proteins is removed before the departure from the ER. GPI inositol deacylation is catalyzed by PGAP1 bearing a lipase motif with an essential Ser residue. The removal of acyl-chain on inositol is important for the ER-to-Golgi transport of GPI-anchored protein but are not essential.

*PGAP1 mutant.* Inositol-acylated GPI is resistant to PI-PLC. CHO C10 cells whose surface GPI-APs were not released by PI-PLC were derived from CHO IIIB2A cells. The steady state levels of surface GPI-APs of C10 cells were similar to those of the wild type, although the rate of transport from ER to Golgi was slower (Tanaka *et al.*, 2004).

9. Processing of GPI in the secretary pathway

GPI-APs are transported to the cell surface through the secretary pathway. During the course, GPI-APs receive several modifications such as side-chain addition and lipid remodeling and are finally incorporated into lipid raft. PGAP2 is the first identified molecule involved in this process. PGAP2 is a membrane protein mainly localized in the Golgi and is essential for correct processing of GPI-APs. PGAP2 might function in the lipid modification of GPI, but the detail is yet to be determined.

*PGAP2 mutant.* Two PGAP2 mutants were established. Surface expression of GPI-APs on CHO C84 cells was partially defective, and that of CHO AM-B cells was nearly negative. The biosynthesis of GPI in the ER and the ER-to-Golgi transport of GPI-APs are normal in both cells. Because of the rapid secretion of GPI-APs to the medium, the levels of GPI-APs on the cell surface were low. GPI-APs were modified/cleaved by two reaction steps in the mutant cells. First, the GPI anchor was converted to lyso-GPI before exiting the trans-Golgi network. Second, lyso-GPI-APs were cleaved by an unknown phospholipase D after transport to the plasma membrane. Therefore, PGAP2 deficiency caused transport to the cell surface of lyso-GPI-APs that were sensitive to a phospholipase D (Tashima *et al.*, 2006).

## References

Abrami, L., Fivaz, M., Kobayashi, T., Kinoshita, T., Parton, R. G., and van der Goot, F. G. (2001). Cross-talk between caveolae and glycosylphosphatidylinositol-rich domains. *J. Biol. Chem.* **276,** 30729–30736.

Albritton, L. M., Tseng, L., Scadden, D., and Cunningham, J. M. (1989). A putative murine ecotropic retrovirus receptor gene encodes a multiple membrane-spanning protein and confers susceptibility to virus infection. *Cell* **57,** 659–666.

Anand, M., Rush, J. S., Ray, S., Doucey, M. A., Weik, J., Ware, F. E., Hofsteenge, J., Waechter, C. J., and Lehrman, M. A. (2001). Requirement of the Lec35 gene for all known classes of monosaccharide-P-dolichol-dependent glycosyltransferase reactions in mammals. *Mol. Biol. Cell* **12,** 487–501.

Ashida, H., Hong, Y., Murakami, Y., Shishioh, N., Sugimoto, N., Kim, Y. U., Maeda, Y., and Kinoshita, T. (2005). Mammalian PIG-X and yeast Pbn1p are the essential components of glycosylphosphatidylinositol-mannosyltransferase I. *Mol. Biol. Cell* **16,** 1439–1448.

Ashida, H., Maeda, Y., and Kinoshita, T. (2006). DPM1, the catalytic subunit of dolichol-phosphate mannose synthase, is tethered to and stabilized on the endoplasmic reticulum membrane by DPM3. *J. Biol. Chem.* **281,** 896–904.

Beck, P. J., Gething, M. J., Sambrook, J., and Lehrman, M. A. (1990). Complementing mutant alleles define three loci involved in mannosylation of Man5-GlcNAc2-P-P-dolichol in Chinese hamster ovary cells. *Somat. Cell. Mol. Genet.* **16,** 539–548.

Benghezal, M., Benachour, A., Rusconi, S., Aebi, M., and Conzelmann, A. (1996). Yeast Gpi8p is essential for GPI anchor attachment onto proteins. *EMBO J.* **15**, 6575–6583.

Conzelmann, A., Spiazzi, A., Bron, C., and Hyman, R. (1988). No glycolipid anchors are added to Thy-1 glycoprotein in Thy-1 negative mutant thymoma cells of four different complementation classes. *Mol. Cell. Biol.* **8**, 674–678.

Ferguson, M. A. (1999). The structure, biosynthesis and functions of glycosylphosphatidylinositol anchors, and the contributions of trypanosome research. *J. Cell Sci.* **112**, 2799–2809.

Fuller, S. A., Takahashi, M., and Hurrell, G. R. (1987). Fusion of myeloma cells with immune spleen cells. *In* "Current Protocols in Molecular Biology" (F. M. Ausbel *et al.*, eds.), pp. 11.7.1–11.7.4. John Wiley & Sons, Inc.

Hiroi, Y., Komuro, I., Chen, R., Hosoda, T., Mizuno, T., Kudoh, S., Georgescu, S. P., Medof, M. E., and Yazaki, Y. (1998). Molecular cloning of human homolog of yeast *GAA1* which is required for attachment of glycosylphosphatidylinositols to proteins. *FEBS Lett.* **421**, 252–258.

Hirose, S., Prince, G. M., Sevlever, D., Ravi, L., Rosenberry, T. L., Ueda, E., and Medof, M. E. (1992). Characterization of putative glycoinositol phospholipid anchor precursors in mammalian cells. Localization of phosphoethanolamine. *J. Biol. Chem.* **267**, 16968–16974.

Hirt, B. (1967). Selective extraction of Polyoma DNA from infected mouse cell cultures. *J. Mol. Biol.* **26**, 365–369.

Hollander, N., Selvaraj, P., and Springer, T. A. (1988). Biosynthesis and function of LFA-3 in human mutant cells deficient in phosphatidylinositol-anchored proteins. *J. Immunol.* **141**, 4283–4290.

Hong, Y., Maeda, Y., Watanabe, R., Inoue, N., Ohishi, K., and Kinoshita, T. (2000). Requirement of PIG-F and PIG-O for transferring phosphoethanolamine to the third mannose in glycosylphosphatidylinositol. *J. Biol. Chem.* **275**, 20911–20919.

Hong, Y., Maeda, Y., Watanabe, R., Ohishi, K., Mishkind, M., Riezman, H., and Kinoshita, T. (1999a). Pig-N, a mammalian homologue of yeast Mcd4p, is involved in transferring phosphoethanolamine to the first mannose of the glycosylphosphatidylinositol. *J. Biol. Chem.* **274**, 35099–35106.

Hong, Y., Ohishi, K., Inoue, N., Kang, J. Y., Shime, H., Horiguchi, Y., van der Goot, F. G., Sugimoto, N., and Kinoshita, T. (2002). Requirement of *N*-glycan on GPI-anchored proteins for efficient binding of aerolysin but not *Clostridium septicum* α-toxin. *EMBO J.* **21**, 5047–5056.

Hong, Y., Ohishi, K., Kang, J. Y., Tanaka, S., Inoue, N., Nishimura, J., Maeda, Y., and Kinoshita, T. (2003). Human PIG-U and yeast Cdc91p are the fifth subunit of GPI transamidase that attaches GPI-anchors to proteins. *Mol. Biol. Cell* **14**, 1780–1789.

Hong, Y., Ohishi, K., Watanabe, R., Endo, Y., Maeda, Y., and Kinoshita, T. (1999b). GPI1 stabilizes an enzyme essential in the first step of glycosylphosphatidylinositol biosynthesis. *J. Biol. Chem.* **274**, 18582–18588.

Hyman, R. (1988). Somatic genetic analysis of the expression of cell surface molecules. *Trends Genet.* **4**, 5–8.

Imhof, I., Flury, I., Vionnet, C., Roubaty, C., Egger, D., and Conzelmann, A. (2004). Glycosylphosphatidylinositol (GPI) proteins of *Saccharomyces cerevisiae* contain ethanolamine phosphate groups on the α1,4-linked mannose of the GPI anchor. *J. Biol. Chem.* **279**, 19614–19627.

Inoue, N., Kinoshita, T., Orii, T., and Takeda, J. (1993). Cloning of a human gene, PIG-F, a component of glycosylphosphatidylinositol anchor biosynthesis, by a novel expression cloning strategy. *J. Biol. Chem.* **268**, 6882–6885.

Inoue, N., Watanabe, R., Takeda, J., and Kinoshita, T. (1996). PIG-C, one of the three human genes involved in the first step of glycosylphosphatidylinositol biosynthesis is a homologue of *Saccharomyces cerevisiae* GPI2. *Biochem. Biophys. Res. Comm.* **226**, 193–199.

Kamitani, T., Chang, H. M., Rollins, C., Waneck, G. L., and Yeh, E. T. H. (1993). Correction of the class H defect in glycosylphosphatidylinositol anchor biosynthesis in Ltk-cells by a human cDNA clone. *J. Biol. Chem.* **268**, 20733–20736.

Kang, J. Y., Hong, Y., Ashida, H., Shishioh, N., Murakami, Y., Morita, Y. S., Maeda, Y., and Kinoshita, T. (2005). PIG-V involved in transferring the second mannose in glycosyl-phosphatidylinositol. *J. Biol. Chem.* **280**, 9489–9497.

Kinoshita, T., Miyata, T., Inoue, N., and Takeda, J. (1995). Expression cloning strategies for glycosylphosphatidylinositol-anchor biosynthesis enzymes and regulators. *Methods Enzymol.* **250**, 547–560.

Kinoshita, T., and Inoue, N. (2000). Dissecting and manipulating the pathway for glycosylphos-phatidylinositol-anchor biosynthesis. *Curr. Opin. Chem. Biol.* **4**, 632–638.

Kostova, Z., Rancour, D. M., Menon, A. K., and Orlean, P. (2000). Photoaffinity labelling with P³-(4-azidoanilido)uridine 5′-triphosphate identifies Gpi3p as the UDP-GlcNAc-binding subunit of the enzyme that catalyses formation of GlcNAc-phosphatidylinositol, the first glycolipid intermediate in glycosylphosphatidylinositol synthesis. *Biochem. J.* **350**, 815–822.

Maeda, Y., Tanaka, S., Hino, J., Kangawa, K., and Kinoshita, T. (2000). Human dolichol-phosphate-mannose synthase consists of three subunits, DPM1, DPM2 and DPM3. *EMBO J.* **19**, 2475–2482.

Maeda, Y., Tomita, S., Watanabe, R., Ohishi, K., and Kinoshita, T. (1998). DPM2 regulates biosynthesis of dolichol phosphate-mannose in mammalian cells: Correct subcellular localization and stabilization of DPM1, and binding of dolichol phosphate. *EMBO J.* **17**, 4920–4929.

Maeda, Y., Watanabe, R., Harris, C. L., Hong, Y., Ohishi, K., Kinoshita, K., and Kinoshita, T. (2001). PIG-M transfers the first mannose to glycosylphosphatidylinositol on the luminal side of the ER. *EMBO J.* **20**, 250–261.

Marmor, M. D., Bachmann, M. F., Ohashi, P. S., Malek, T. R., and Julius, M. (1999). Immobilization of glycosylphosphatidylinositol-anchored proteins inhibits T cell growth but not function. *Int. Immunol.* **11**, 1381–1393.

Miyata, T., Takeda, J., Iida, Y., Yamada, N., Inoue, N., Takahashi, M., Maeda, K., Kitani, T., and Kinoshita, T. (1993). Cloning of PIG-A, a component in the early step of GPI-anchor biosynthesis. *Science* **259**, 1318–1320.

Morelon, E., Dodelet, V., Lavery, P., Cashman, N. R., and Loertscher, R. (2001). The failure of Daudi cells to express the cellular prion protein is caused by a lack of glycosyl-phosphatidylinositol anchor formation. *Immunology* **102**, 242–247.

Murakami, Y., Siripanyaphinyo, U., Hong, Y., Tashima, Y., Maeda, Y., and Kinoshita, T. (2005). The initial enzyme for glycosylphosphatidylinositol biosynthesis requires PIG-Y, a seventh component. *Mol. Biol. Cell* **16**, 5236–5246.

Murakami, Y., Siripanyapinyo, U., Hong, Y., Kang, J. Y., Ishihara, S., Nakakuma, H., Maeda, Y., and Kinoshita, T. (2003). PIG-W is critical for inositol acylation but not for flipping of glycosylphosphatidylinositol-anchor. *Mol. Biol. Cell* **14**, 4285–4295.

Nakamura, N., Inoue, N., Watanabe, R., Takahashi, M., Takeda, J., Stevens, V. L., and Kinoshita, T. (1997). Expression cloning of PIG-L, a candidate *N*-acetylglucosaminyl-phosphatidylinositol deacetylase. *J. Biol. Chem.* **272**, 15834–15840.

Ohishi, K., Inoue, N., and Kinoshita, T. (2001). PIG-S and PIG-T, essential for GPI anchor attachment to proteins, form a complex with GAA1 and GPI8. *EMBO J.* **20,** 4088–4098.

Ohishi, K., Inoue, N., Maeda, Y., Takeda, J., Riezman, H., and Kinoshita, T. (2000). Gaa1p and gpi8p are components of a glycosylphosphatidylinositol (GPI) transamidase that mediates attachment of GPI to proteins. *Mol. Biol. Cell* **11,** 1523–1533.

Shishioh, N., Hong, Y., Ohishi, K., Ashida, H., Maeda, Y., and Kinoshita, T. (2005). GPI7 is the second partner of PIG-F and involved in modification of glycosylphosphatidylinositol. *J. Biol. Chem.* **280,** 9728–9734.

Stevens, V. L., Zhang, H., and Harreman, M. (1996). Isolation and characterization of a CHO mutant defective in the second step in glycosylphosphatidylinositol biosynthesis. *Biochem. J.* **313,** 253–258.

Stoll, J., Robbins, A. R., and Krag, S. S. (1982). Mutant of Chinese hamster ovary cells with altered mannose 6-phosphate receptor activity is unable to synthesize mannosylphosphoryldolichol. *Proc. Natl. Acad. Sci. USA* **79,** 2296–2300.

Sutterlin, C., Escribano, M. V., Gerold, P., Maeda, Y., Mazon, M. J., Kinoshita, T., Schwarz, R. T., and Riezman, H. (1998). *Saccharomyces cerevisiae* GPI10, the functional homologue of human PIG-B, is required for glycosylphosphatidylinositol-anchor synthesis. *Biochem. J.* **332,** 153–159.

Sutterlin, C., Horvath, A., Gerold, P., Schwarz, R. T., Wang, Y., Dreyfuss, M., and Riezman, H. (1997). Identification of a species-specific inhibitor of glycosylphosphatidylinositol synthesis. *EMBO J.* **16,** 6374–6383.

Takahashi, M., Inoue, N., Ohishi, K., Maeda, Y., Nakamura, N., Endo, Y., Fujita, T., Takeda, J., and Kinoshita, T. (1996). PIG-B, a membrane protein of the endoplasmic reticulum with a large luminal domain, is involved in transferring the third mannose of the GPI anchor. *EMBO J.* **15,** 4254–4261.

Tanaka, S., Maeda, Y., Tashima, Y., and Kinoshita, T. (2004). Inositol deacylation of glycosylphosphatidylinositol-anchored proteins is mediated by mammalian PGAP1 and yeast Bst1p. *J. Biol. Chem.* **279,** 14256–14263.

Taron, B. W., Colussi, P. A., Grimme, J. M., Orlean, P., and Taron, C. H. (2004). Human Smp3p adds a fourth mannose to yeast and human glycosylphosphatidylinositol precursors *in vivo. J. Biol. Chem.* **279,** 36083–36092.

Tashima, Y., Taguchi, R., Murata, C., Ashida, H., Kinoshita, T., and Maeda, Y. (2006). PGAP2 is essential for correct processing and stable expression of GPI-anchored proteins. *Mol. Biol. Cell* **17,** 1410–1420.

Tomita, S., Inoue, N., Maeda, Y., Ohishi, K., Takeda, J., and Kinoshita, T. (1998). A homologue of *Saccharomyces cerevisiae* Dpm1p is not sufficient for synthesis of dolichol-phosphate-mannose in mammalian cells. *J. Biol. Chem.* **273,** 9249–9254.

Trowbridge, I. S., Hyman, R., and Mazauskas, C. (1978). The synthesis and properties of T25 glycoprotein in Thy-1-negative mutant lymphoma cells. *Cell* **14,** 21–32.

Ware, F. E., and Lehrman, M. A. (1996). Expression cloning of a novel suppressor of the Lec15 and Lec35 glycosylation mutations of chinese hamster ovary cells. *J. Biol. Chem.* **271,** 13935–13938.

Watanabe, R., Inoue, N., Westfall, B., Taron, C. H., Orlean, P., Takeda, J., and Kinoshita, T. (1998). The first step of glycosylphosphatidylinositol biosynthesis is mediated by a complex of PIG-A, PIG-H, PIG-C and GPI1. *EMBO J.* **17,** 877–885.

Watanabe, R., Murakami, Y., Marmor, M. D., Inoue, N., Maeda, Y., Hino, J., Kangawa, K., Julius, M., and Kinoshita, T. (2000). Initial enzyme for glycosylphosphatidylinositol biosynthesis requires PIG-P and is regulated by DPM2. *EMBO J.* **19,** 4402–4411.

Yu, J., Nagarajan, S., Knez, J. J., Udenfriend, S., Chen, R., and Medof, M. E. (1997). The affected gene underlying the class K glycosylphosphatidylinositol (GPI) surface protein defect codes for the GPI transamidase. *Proc. Natl. Acad. Sci. USA* **94,** 12580–12585.

Yusa, K., Horie, K., Kondoh, G., Kouno, M., Maeda, Y., Kinoshita, T., and Takeda, J. (2004). Genome-wide phenotype analysis in ES cells by regulated disruption of Bloom's syndrome gene. *Nature* **429,** 896–899.

## [13] CHO Glycosylation Mutants: Proteoglycans

*By* Lijuan Zhang, Roger Lawrence,
Beth A. Frazier, and Jeffrey D. Esko

### Abstract

Most glycosaminoglycan (GAG)-defective mutants have been isolated and characterized from Chinese hamster ovary (CHO) cells. Wild-type and GAG-defective CHO cells have been used by several hundreds of laboratories to study how altering the GAG structure of proteoglycans affects fundamental properties of cells, such as bacterial/viral infection, signaling, protein degradation, and cell adhesion. This chapter describes methods used to construct and characterize new CHO cell lines with gain-of-function GAG structures. These novel CHO cell lines allow herpes simplex virus (HSV) entry or have anticoagulant properties that are not possessed by wild-type CHO cells. The method used to study GAG biosynthetic mechanisms that control specific GAG sequence assembly is also described.

### Overview

GAGs are among the most complex sugar chains. Three major types of GAGs are heparan sulfate (HS), chondroitin sulfate (CS), and dermatan sulfate (DS). These linear and highly sulfated polysaccharides are abundantly expressed on mammalian cell surface and in the extracellular matrix in the forms of HS, CS, DS, or HS/CS(DS) hybrid proteoglycans (Esko and Zhang, 1996). More than 50 core proteins encoded by specific genes have been cloned from vertebrates, and in some cases, homologs exist in invertebrates. Protein and proteoglycan interactions in the matrix and on the cell surface depend in part on the arrangement of sulfated sugar residues and uronic acid epimers in limited segments of the chain. These GAG sequences are not directly encoded by genes but are assembled in the *Golgi*

METHODS IN ENZYMOLOGY, VOL. 416                                    0076-6879/06 $35.00
DOI: 10.1016/S0076-6879(06)16013-9

by enzymes encoded by more than 40 genes, thus generating a great variety of GAG sequences in a cell- and tissue-specific fashion.

HS and CS interact with hundreds of ligands, including growth factors, cytokines, chemokines, extracellular matrix components, proteases, protease inhibitors, lipoproteins, and lipolytic enzymes (Conrad, 1998). HS and CS proteoglycans have been credited with controlling many multicellular processes, including blood coagulation (Rosenberg et al., 1997), wound healing, axon guidance (Inatani et al., 2003), lipid metabolism, metastasis, angiogenesis (Vlodavsky and Friedmann, 2001), morphogenesis (Selleck, 2000), inflammation (Wang et al., 2005), matrix assembly, and bacterial and viral entry (Shukla et al., 1999). The diversified functions of HS and CS are mainly attributed to the complex saccharide sequence and sulfation patterns of the polysaccharides, along with the unique properties and dynamic expressions of the core proteins to which they are attached (Iozzo, 2001). GAG-defective mutant cell lines provide a powerful tool for studying the biosynthesis, structure, and function of proteoglycans under controlled cell culture conditions.

Most GAG-defective mutants have been obtained from CHO cells and characterized both genetically and biochemically (Table I). CHO cells are hypodiploid and have substantial functional hemizygosity at many loci, which allows obtaining recessive mutants after single hits on certain GAG assembling genes. The methodologies used to generate recessive mutants by replica plating and direct selection have been reviewed (Bai et al., 2001; Esko, 1989).

However, wild-type CHO cells make rather simple HS and CS because they express a limited set of GAG modification enzymes and proteoglycan core proteins. For example, CHO cells express two out of four HS $N$-sulfotransferases, one out of three HS 6-$O$-sulfotransferases, and none of the seven HS 3-$O$-sulfotransferases found in both human and mouse. Moreover, CHO cells synthesize less GAG compared to mast cells and chondrocytes, which make copious amounts of heparin and chondroitin sulfate, respectively, and moderately less than fibroblasts, osteoblasts, and certain tumor cell lines (L. Zhang et al., unpublished results). Because of these limitations, CHO cell GAGs may not show all of the biological activities possible for GAGs and proteoglycans.

To overcome the limited GAG biosynthetic enzyme expression repertoire in CHO cells, different strategies have been developed to make a CHO cell library where each cell line expresses a unique set of GAG assembly enzymes and expresses unique GAG structures. This chapter describes the methods used to generate novel CHO cell lines that bind to antithrombin III (ATIII) (Zhang et al., 2001b) and heparin cofactor II or have a unique HS motif that serves as a HSV entry receptor (O'Donnell et al., 2006; Shukla et al., 1999; Xia

TABLE I
CHO PROTEOGLYCAN MUTANTS

| Definition | Biochemical characteristics | Phenotype |
|---|---|---|
| pgsA; CHO-745 (Esko et al., 1985) | Xylosyltransferase | GAG deficient |
| pgsB; CHO-761 (Esko et al., 1987) | Galactosyltransferase I | GAG deficient |
| pgsC (Esko et al., 1986) | Sulfate transporter | Normal GAG; deficient labeling with $^{35}$SO4 |
| pgsD: CHO-677 (Lidholt et al., 1992; Wei et al., 2000) | N-acetylglucosaminyl/glucuronosyltransferase (EXT-1) | HS deficient |
| pgsE; CHO-606 (Bame and Esko, 1989) | N-deacetylase/N-sulfotransferase 1 | Undersulfated HS |
| pgsF; CHO-F17 (Bai and Esko, 1996) | HS 2-O-sulfotransferase | 2-O-sulfation–deficient HS |
| pgsG (Bai et al., 1999) | Glucuronosyltransferase I | GAG deficient |
| ldlD (Esko et al., 1988; Kingsley et al., 1986) | UDP-glucose/galactose and UDP-GlcNAc/GalNAc 4-epimerase | CS deficient when starved for GalNAc; GAG-deficient when starved for galactose |
| Resistant to Sindbis virus infection (Jan et al., 1999) | Unknown | HS deficient and reduced CS synthesis |
| Gain of antithrombin binding (Zhang et al., 2001a,b) | HS 3-O-sulfotransferase 1 expression | HS has high affinity for antithrombin |
| Gain of heparin cofactor II binding | CS 2-O-sulfotransferase expression | CS has high affinity for heparin cofactor II |
| Gain of herpes simplex virus entry | HS 3-O-sulfotransferase 2, 3, 4, 5, or 6 expression (O'Donnell et al., 2006; Shukla et al., 1999; Tiwari et al., 2005; Xia et al., 2002; Xu et al., 2005) | HS serves as herpes simplex viral entry receptor |
| Loss of antithrombin binding (Zhang et al., 2001a) | HS 3-O-sulfotransferase 1 expression and reduced 6-O-sulfotransferase activity | Reduced HS 6-O-sulfation and diminished antithrombin binding |

*et al.*, 2002). Methods used to study early GAG assembling enzymes that limit specific GAG sequence synthesis are also described (Zhang *et al.*, 2001a).

## Making Gain-of-Function CHO Cell Mutants

### Principle

The general principal guiding the development of gain-of-function mutants is to introduce genes involved in GAG biosynthesis into a cell line that does not normally express the desired enzyme or GAG-binding sequence. For example, ATIII requires a special 3-*O* and 6-*O*-sulfated HS structural motif for high affinity binding, which is generated by 3-*O*-sulfotransferase 1 or 5 and 6-*O*-sulfotransferase isoforms on nascent HS chains. HSV requires a 3-*O*-sulfated HS motif, which can be generated by 3-OST2, 3A, 3B, 4, 5, or 6, but not by 3-OST1. Because wild-type CHO cells do not express any of the HS 3-*O*-sulfotransferases, expressing each 3-OST isoform in wild-type or mutant CHO cells has successfully produced a new CHO cell phenotype (O'Donnell *et al.*, 2006; Shukla *et al.*, 1999; Tiwari *et al.*, 2005; Xia *et al.*, 2002; Xu *et al.*, 2005; Zhang *et al.*, 2001a,b). In theory, any CHO or mammalian cell line can be used to express GAG assembling enzymes by transient or stable cDNA transfection or retroviral transduction. The cell lines generated by expressing a GAG synthetic enzyme not present or present in a limited amount in the parental cells are expected to have different GAG structures.

### Cell Culture

Parental CHO K1 cells require exogenous proline for growth and are normally maintained in Ham's F-12 medium supplemented with 10% fetal bovine serum (FBS), penicillin G (100 U/ml), and streptomycin sulfate (100 $\mu$g/ml) at 37° under an atmosphere of 5% $CO_2$ in air and 100% relative humidity. The cells should be passaged every 3–4 days with 0.125% (w/v) trypsin with or without 1 m$M$ ethylenediaminetetraacetic acid (EDTA), and after 20–25 cycles, fresh cells should be revived from stocks stored under liquid nitrogen.

### Transient and Stable Transfection Methods and Retroviral Transduction Methods

Transient and stable transfection (e.g., LipofectAMINE 2000, Invitrogen) has proven effective for introducing genes in CHO cells (O'Donnell *et al.*,

2006; Shukla *et al.*, 1999; Tiwari *et al.*, 2005; Xia *et al.*, 2002; Xu *et al.*, 2005; Zhang *et al.*, 2001a) and other types of mammalian cell lines (McCormick *et al.*, 2000; Pikas *et al.*, 2000).

Retroviral transduction has the advantage over DNA transfection methods by increasing the probability of obtaining a single copy of the desired gene integrated into the genome of the target cells. The murine stem cell virus pMSCV vector used for retroviral transduction derived from the murine embryonic stem cell virus (MESV) and the LN retroviral vectors has proven useful (Grez *et al.*, 1990; Lieu *et al.*, 1997; Miller and Rosman, 1989) and can be purchased from Clontech (Lieu *et al.*, 1997). The specificity of infection depends on the presence of appropriate viral receptors on the cell surface. Murine cells express ecotropic receptors (*MCAT1*). Ecotropic receptor can be transfected into nonecotropic expressing cells. Introduction of pMSCV into an ecotropic packaging cell line results in production of high-titer, replication-incompetent infectious virus particles. These particles can infect ecotropic receptor expressing target cells and integrate the desired complementary DNA (cDNA) into different chromosomal locations. Pantropic viruses can also be produced by pseudo-typing the virus with VSV-G protein in the suitable packaging line (Chen *et al.*, 2005).

Multiple rounds of retroviral transduction result in insertion of multiple copies of the gene. Increasing gene dosage in this way is highly desirable for searching novel genes that affect GAG assembly while avoiding mutations in the transduced gene when using a chemical mutagenesis approach to detect other genes required for a specific GAG sequence assembly.

### Materials for Retroviral Transduction

*CHO cells expressing ecotropic receptor.* Both CHO K1 (Gu *et al.*, 2000) and CHO GAG-defective mutant CHO pgsF17 (Zhang *et al.*, 2001b) have been generated by stable transfection with *MCAT1* cDNA. These CHO ecotropic expressing cell lines are obtainable through the authors. Other ecotropic expressing cell lines can be made by following the protocol. In brief, cells are transfected with a mammalian expression vector containing *MCAT1* (ecotropic receptor) cDNA and a selection marker cDNA (e.g., Hygromycin resistance gene cDNA), from separate constitutive promoters. The transfected cells are selected for drug resistance (e.g., with 200 μg/ml Hygromycin B). Each stable drug-resistant clone is assayed for its ability to be infected by reporter virus, for example, pMSCVPLAP containing alkaline phosphatase (Fields-Berry *et al.*, 1992).

1. *The PHOENIX ecotropic retroviral packaging cell line.* This cell line can be purchased from ATCC (SD 3443).

2. *Virion.* The retrovirus plasmid pMSCV, which contains cDNA of puromycin-resistant gene, can be purchased from Clontech and used as a vector for different GAG enzymes. For example, the retroviral expressing vector for 3-OST1 is constructed (Shworak *et al.*, 1997) by digestion with *BglII* and *XhoI* to release the wild-type murine 3-OST1 cDNA. The fragment (1623 bp) is then cloned into the *BglII* + *XhoI* sites in pMSCVpac (Lieu *et al.*, 1997). Infectious virions can be produced by programming ecotropic PHOENIX packaging cells with recombinant provirus plasmids such as the 3-OST1 construct by standard transient transfection protocol, for example, FuGENE 6 (Roche) or LipofectAMINE 2000 (Invitrogen) following manufacturer's protocol. Viral supernatants are collected, either flash-frozen in liquid nitrogen and stored at $-80°$ or used directly after low-speed centrifugation. Viron titer is determined by infection of NIH/3T3 fibroblasts or CHO ecotropic expressing cells with serial dilutions of virus followed by counting antibiotic resistant colonies or colonies expressing desired activity (AP, green fluorescent protein [GFP], specific epitope-positive cells). Alternatively, flow cytometry may also be used where GFP expression or immunofluorescence is possible.

3. Institutional approval for working with retrovirus (BL2 level).

*Transduction Protocol*

Day 1. CHO cells containing ecotropic receptors are plated in a six-well plate.

Day 2. Cells at 60–70% confluence are transduced by overnight incubation with viral supernatants containing 5 $\mu$g/ml polybrene surfactant.

Day 3. The virus containing media is replaced with fresh growth media.

Day 4. Cells are passaged into a T-25 flask and maintained in 7.5 $\mu$g/ml puromycin containing culture media or effective concentration of other selective agent appropriate for the type of retroviral used to select for transduced cells. Repeat Days 1–4 once or twice if high percentage and multiple copies of cDNA expression in the cells are wanted.

Gain-of-Function Mutant Characterization

The introduction of 3-OST1 into CHO cells results in presentation of HS on the cell surface that can bind to antithrombin. In contrast, introduction of 3-OST3a increases binding of HSV glycoprotein D (gD) and renders the cells susceptible to viral infection. Characterizing the gain-of-function phenotype depends on the nature of the ligand or biological process under study.

The following section describes the characterization of cell lines containing 3-OSTs and can be used as a model for characterizing cell lines transfected with other genes.

## CHO Cell HSV Entry Assay

CHO cells are exposed to recombinant HSV-1 (HSV-1[KOS]gL86) that expresses $\beta$-galactosidase upon viral entry. The HSV entry assay is based on the quantification of the activity of $\beta$-galactosidase. Six hours post-infection, the cells are fixed in PBS containing 2% formaldehyde and 0.2% glutaraldehyde, permeabilized in 2 m$M$ MgCl$_2$ containing 0.01% deoxycholate and 0.02% Nonidet P-40, and incubated with buffered X-gal (0.5 mg/ml). Three hours later, the infected cells are stained blue due to the action of $\beta$-galactosidase on X-gal. Wild-type CHO cells and the CHO cells expressing an HS-independent HSV-1 entry receptor (nectin-1) can be used as positive and negative controls, respectively (O'Donnell et al., 2006; Xia et al., 2002; Xu et al., 2005).

GAGs are involved in many types of viral infections (Liu and Thorp, 2002). The strategy that works for HSV might work for other types of viruses as well.

## Specific CHO Cell Surface GAG and Protein Interaction by FACS Analysis

FLUORESCENT LABELING ATIII OR HEPARIN COFACTOR II. Human ATIII and heparin cofactor II are fluorescently labeled on the oligosaccharide side-chains. The aldehyde groups are generated by removing terminal sialic acid and oxidizing galactose residues by galactose oxidase. The standard reaction mixture contains 20 m$M$ NaH$_2$PO$_4$ (pH 7.0), 0.3 m$M$ CaCl$_2$, 25 $\mu$g of human ATIII or heparin cofactor II, 4 mU neuraminidase, 4 mU galactose oxidase in a final volume of 270 $\mu$l. The mixture is incubated at 37° for 1 h before 10 $\mu$l of 12.5 mg/ml fluorescein hydrazide is added and then incubated at 37° for another hour. Phosphate-buffered saline (PBS) (1 ml) and a 50% slurry of heparin–Sepharose CL-6B in PBS (100 $\mu$l) is added and mixed end-over-end from 20 min to 16 h. After centrifugation, the heparin–Sepharose CL-6B is washed four times with PBS (1 ml). Labeled protein is eluted with four 0.25-ml aliquots of 10× concentrated PBS and desalted by centrifugation for 35 min at 5000$g$ through two Microcon-10 column (Millipore). The labeled protein is combined and diluted into 0.5 ml 10% FBS in PBS. Inclusion of [125]I-ATIII showed that the overall recovery of material was 20–30%. Therefore, the labeled proteins have a final concentration around 10–15 $\mu$g/ml.

The previous protocol works well for fluorescein hydrazide. However, it is not successful for Alexa-488 hydrazide and other hydrazide compounds in which lower amounts of labeling reagents are used because of the high cost of these reagents. Because mouse anti-human ATIII antibody can be purchased from Sigma-Aldrich (A5816), quantitating binding by fluorescence-activated cell sorting (FACS) using anti-ATIII antibody followed by Phycolink goat anti-mouse IgG (Fc)-RPE (Prozyme) or other commercially available fluorescent-labeled anti-mouse IgG (Fc) is recommended.

*Materials*

Human ATIII or human heparin cofactor II can be purchased from
   Sigma or purified from normal human serum (Lian *et al.*, 2001).
Neuraminidase (Worthington Biochemical Corp., 4520)
Galactose oxidase (Worthington Biochemical Corp., 4759)
Fluorescein hydrazide (Molecular Probes, C-356)
Heparin–Sepharose CL-6B (Amersham Pharmacia Biotech)
Microcon-10 column (Millipore)

FIBROBLAST GROWTH FACTOR-2 (FGF-2) LABELING. Fluorescent-labeled FGF-2 is prepared by mixing 50 $\mu$l of 1 $M$ sodium bicarbonate with 0.5 ml of PBS containing 2 mg/ml bovine serum albumin (BSA) and 25 $\mu$g FGF-2. The mixture is transferred to a vial of reactive dye (Alexa-594, Molecular Probes) and stirred at room temperature for 1 h. The isolation of the labeled FGF-2 is the same as described earlier for ATIII and heparin cofactor II, except 2 $M$ NaCl in 1× PBS is used for labeled FGF-2 elution.

Biotinated FGF-2 is prepared by protecting 0.4 mg of FGF-2 with heparin (0.4 mg) in 0.2 $M$ HEPES buffer (pH 8.4) and mixed with biotin hydrazide (Pierce Chemical, long arm, water-soluble, 40 $\mu$g) in a final volume of 0.2 ml. After 2 h at room temperature, 40 $\mu$l of 10 mg/ml glycine solution is added to stop the reaction. The sample is diluted with 30 ml of 20 m$M$ HEPES buffer (pH 7.4) containing 0.5 $M$ NaCl and 0.2% BSA and loaded onto a 1-ml column of heparin–Sepharose CL-6B (Amersham Pharmacia Biotech). The column is washed with 30 ml of buffer and eluted with 2.5 ml of solution adjusted to 3 $M$ NaCl. The sample is desalted on a PD-10 column (Amersham Pharmacia Biotech) equilibrated with 20 m$M$ HEPES buffer (pH 7.4) containing 0.2% BSA.

FACS ANALYSIS. Both cell sorting and FACS analysis depend on the integrity of cell surface proteoglycans, so trypsin must be avoided during the release of the cells. For this purpose, cells are detached with 5 m$M$ EDTA in PBS, washed with 10 ml of PBS, and suspended in 1 ml of 10% FBS in PBS. The cells are kept on ice all the time.

The cell surface GAG and fluorescent-labeled FGF-2 binding is performed by suspending cells in 0.1 ml 10% FBS in PBS containing about 50 ng of fluorescent-labeled protein per $10^6$ cells. After 30 min, the cells are washed once and resuspended in 1 ml of 10% FBS in PBS and are ready for FACS analysis.

The cell surface GAG and biotinated FGF-2 binding is performed by incubating cells with about 0.5 $\mu$g/ml biotin–FGF-2 and 1% BSA for 1 h. The cells are washed twice with 0.5 ml of cold 20 m$M$ NaH$_2$PO$_4$ (pH 7.4) buffer containing 150 m$M$ NaCl, and resuspended in 0.2 ml of buffer containing 5 $\mu$g/ml fluorescein/avidin/DCS (Vector Laboratories). After shaking the cells in the dark for 20 min, they are washed once and resuspended in cold phosphate buffer and are ready for FACS analysis.

FACS and cell sorting are performed on FACScan and FACStar (Becton Dickinson) or similar instruments following the manufacturer's protocol.

## Acquiring Loss-of-Function Mutants from Gain-of-Function Cells

### Identifying GAG Biosynthetic Enzymes Other Than the Transduced One that Limit Specific GAG Motif Assembly

Gain-of-function mutants described in the last section are very useful in identifying novel GAG-binding proteins that require special sulfation patterns of GAGs for high-affinity binding or special biological function. However, other GAG assembly enzymes also contribute to the special GAG sequence assembly. To search for such genes, chemical mutagenesis can be used to affect specific GAG sequence motif assembly while avoiding mutations in the transduced gene where multiple copies have been introduced. The following section describes a strategy that can be used as a model for searching critical genes in other GAG-involved systems.

### Mutant Strategy

3-OST1 expression gives rise to CHO cells with the ability to bind to fluorescent-labeled ATIII on their cell surface. To search for other HS synthetic enzymes that are also responsible for ATIII binding, a CHO cell line having three copies of *3-OST1* is chemically mutagenized and selected by cell sorting. The mutant scheme is depicted in Fig. 1. The advantage of having multiple copies of 3-OST1 is that other enzymes that are responsible for generating specific HS precursor structures can be sought after chemical mutagenesis without concern for the complete loss of 3-OST1 activity. FGF-2 selection is employed to detect the mutant cells that still make HS. By using this scheme, two major types of mutants are obtained: (1) mutants that do not bind to both FGF-2 and ATIII, and (2) mutants

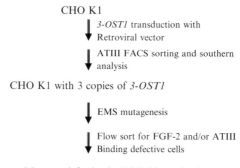

CHO K1

3-OST1 transduction with
Retroviral vector

ATIII FACS sorting and southern
analysis

CHO K1 with 3 copies of 3-OST1

EMS mutagenesis

Flow sort for FGF-2 and/or ATIII
Binding defective cells

Mutants defective in GAG biosynthesis

FIG. 1. Mutant scheme. Using recombinant retroviral transduction, the human heparan sulfate (HS) 3-O-sulfotransferase 1 (3-OST1) gene is transduced into CHO cells. 3-OST1 expression gives rise to CHO cells with the ability to produce anticoagulant HS. A cell line that has three copies of 3-OST1 was chosen by Southern analysis. After chemical mutagenesis of this cell line, fibroblast growth factor-2 (FGF-2) binding–positive and antithrombin III (ATIII) binding–negative mutant cells are FACS-sorted and cloned. The advantage of having three copies of 3-OST1 is that upstream genes that are responsible for generating specific HS precursor structures can be sought by chemical mutagenesis without losing 3-OST1. FGF-2 selection is employed to select mutant cells that still make HS.

that bind to FGF-2 and have reduced binding to ATIII. The mutant generation and characterization are described next.

*Determination of Gene Copy Numbers in 3-OST1–Transduced Mutant Clones*

A CHO cell line expressing multiple copies of 3-OST1 is needed for the cell sorting–based mutant scheme. 3-OST1 cDNA is transduced into CHO wild-type cells as described earlier. Transduced cells that bind to fluorescent-labeled ATIII are single-cell sorted into a 96-well plate. Typically 20–30 wells will have a single-cell colony that can be visualized within 10 days. The colonies that have medium size are expanded and frozen under liquid nitrogen.

Genomic DNA is isolated from each clone ($\sim 10^7$ cells) and quantitated. Genomic DNA (10 $\mu$g) is digested with 40 U of *Eco*RI overnight at 37°, electrophoresed on a 0.7% (w/v) agarose gel, transferred to GeneScreen Plus (PerkinElmer Life Sciences), and probed with 3-OST1 cDNA labeled with the Megaprimer labeling kit (Amersham Pharmacia Biotech). The blot is hybridized with ExpressHyb solution (Clontech) containing the 3-OST1 probe ($2 \times 10^6$ cpm/ml), followed by autoradiography (Fig. 2). All the clones have an endogenous copy of 3-OST1. No enzyme activity was found in non-transduced CHO cells, which indicates that the endogenous gene was inactive. The 12 clones analyzed have one to five inserted copies of the gene in different chromosomal positions.

Mutagenesis: Typically, loss-of-function mutations in CHO cells are detected after chemical mutagenesis, which increases the incidence of mutants in the population. Chemical mutagenesis can increase $10^2$- to $10^4$-fold the incidence of the desired mutants. However, mutagenesis also increases the probability of inducing DNA damage and many of the cells can contain multiple mutations. To circumvent this problem, cells can be analyzed by FACS, which has the capacity of about $10^9$ cells/day. The collected cells can be sorted multiple times, thus obviating the need for mutagen treatment to find the desired mutants. By combining low-dose chemical mutagenesis with cell sorting, the best attributes of both approaches can be applied to finding desirable mutants (Esko, 1989). Other mutagenesis methods (e.g., viral insertion and gene trap procedures) might prove useful as well. These latter methods have the advantage of marking chromosomal sites at the position of the relevant gene.

Cell sorting–based mutant selection: A portion of mutagenized cells are thawed and propagated in a T-175 flask for 3 days. The cells are

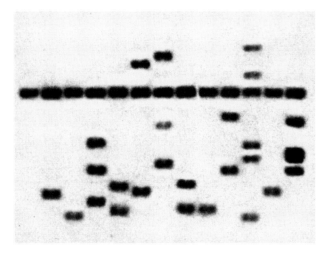

FIG. 2. Determination of 3-OST1 gene copy numbers in twelve 3-OST1–transduced CHO cell clones. EcoRI-digested genomic DNA was hybridized to a 3-OST1 probe. 0, control; 1–12, 3-OST1–transduced CHO cell clones.

detached and labeled with both Alexa-594–FGF-2 and fluorescein–ATIII when they are approaching confluence. The labeled cells are sorted. Both FGF-2–positive and ATIII-negative cells and double-negative cells are collected. The double-negative cells are propagated once. The cells are labeled with Alexa-594–FGF-2 and fluorescein–ATIII and are then single-cell sorted into a 96-well plate with a stringent criterion that only the double-negative cells are collected. The single cell clones are expanded and frozen away for further analysis. Approximately $1 \times 10^4$ FGF-2 binding–positive and ATIII binding–negative cells are collected into 1 ml of complete F-12 Ham's media and then plated in a T-75 flask. Sorted cell populations are maintained in complete F-12 Ham's medium for 1 wk, and then the cells are labeled and sorted again. After five rounds of sorting, FGF-2 binding–positive and ATIII binding–negative cells are single-cell sorted into a 96-well plate. The single-cell clones are expanded and frozen away for further analysis.

*FGF-2 and ATIII binding–negative mutants:* The double-negative mutants named *CHO M1* (Broekelmann *et al.*, 2005) have a point mutation in EXT-1 that converts Gln179 to a premature stop codon, resulting in expression of a truncated protein missing the majority of the enzyme and is, therefore, inactive. CHO M1 cells do not make HS but synthesize increased amounts of CS compared to CHO wild-type cells. CS at CHO M1 cells surface does not bind to fluorescein-labeled heparin cofactor II (Fig. 3, upper panel). Transducing CS 2-OST in CHO M1 augments heparin cofactor II binding (Fig. 3, lower panel), which is consistent with the literature that both 2-$O$ and 4-$O$-sulfates in CS are required for high-affinity heparin cofactor II binding (Maimone and Tollefsen, 1990).

*FGF-2 binding–positive and ATIII binding–negative mutants:* Twenty-two clones of CHO mutant that have normal FGF-2 binding and reduced ATIII binding have been obtained. All the mutants make HS as predicted by their FGF-2 binding ability. HS composition analysis indicates that most of the mutants have a decreased amount of 6-$O$-sulfate containing disaccharides. One of the mutants has decreased 6-$O$-sulfotransferase activity. Transducing 6-OST1 into the mutant cells fully restored the ATIII binding. The profiles of 3-OST1 expression in CHO K1, the mutant, and the 6-OST1 transduced correctant are shown by dual-color (Alexa-594 FGF-2 and fluorescein–ATIII) FACS analysis (Fig. 4). This result indicates that not only 3-OST1 but also 6-OST are critical for ATIII binding HS sequence assembly. Both enzymes can limit ATIII binding HS biosynthesis in a cell line.

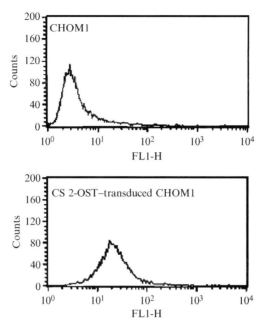

FIG. 3. Flow cytometric analysis of the binding of fluorescein-labeled heparin cofactor II to CHOM1 mutant (upper panel) and CS 2-OST–transduced CHOM1 (lower panel). CHOM1 and CS 2-OST–transduced CHOM1 cells are labeled with fluorescein heparin cofactor II and analyzed by FACS.

The advantage of cell sorting–based mutant selection is that many mutants with similar properties can be obtained simultaneously during the selection process. We suspect the cell-sorting strategy may work in other diploid mammalian cell lines. The cell sorting has a high capacity ($10^9$ cells) and the repeated sorting can enrich the rare mutants until the desired mutant is obtained. This technique may also constitute a general approach for defining and characterizing the components of GAG assembly machinery once the terminal biosynthetic enzyme has been obtained and the proteins that recognize a specific GAG structural motif are identified.

In summary, we have described the method of generating gain-of-function CHO cell mutants. This method is straightforward and should be applicable to other types of mammalian cells. The strategy used to generate the loss-of-function mutants from the gain-of-function cells takes advantage of the combined power of cell sorting and low-dose chemical mutagenesis. This strategy might work for other mammalian cells in which GAG-defective

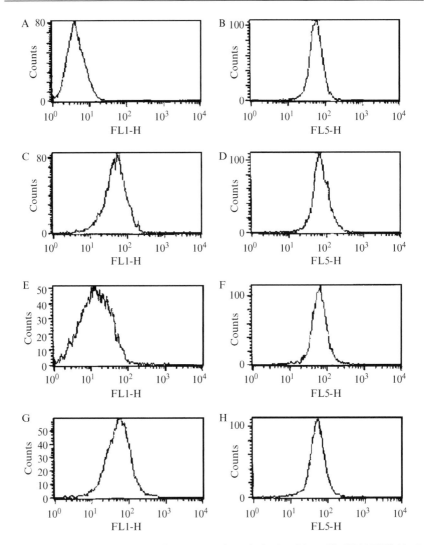

FIG. 4. Dual-color fluorescence flow cytometric analysis of antithrombin III (ATIII) (A, C, E, and G) and fibroblast growth factor-2 (FGF-2) (B, D, F, and H) binding to wild-type, mutant, and 6-OST1–transduced correctant. CHO wild-type (A and B); wild-type CHO cell clone with three copies of 3-OST1 (C and D), mutant cell clone with three copies of 3-OST1 (E and F), and 6-OST1 correctant of the mutant (G and H) are double-labeled with fluorescein–ATIII (A, C, E, and G) and Alexa-594–FGF-2 (B, D, F, and H) and subjected to dual-color FACS analysis.

mutants have not been obtained with other strategies. GAGs bind not only to hundreds of mammalian proteins but also to many lectins. The gain-of-function mutants could be used to screen for novel protein ligands or lectins that require special sulfated GAG sequence for high-affinity binding and to generate novel GAG-defective mutants to understand the GAG sequence assembly mechanism.

## References

Bai, X., and Esko, J. D. (1996). An animal cell mutant defective in heparan sulfate hexuronic acid 2-O-sulfation. *J. Biol. Chem.* **271,** 17711–17717.

Bai, X., Crawford, B., and Esko, J. D. (2001). Selection of glycosaminoglycan-deficient mutants. *Methods Mol. Biol.* **171,** 309–316.

Bai, X., Wei, G., Sinha, A., and Esko, J. D. (1999). Chinese hamster ovary cell mutants defective in glycosaminoglycan assembly and glucuronosyltransferase I. *J. Biol. Chem.* **274,** 13017–13024.

Bame, K. J., and Esko, J. D. (1989). Undersulfated heparan sulfate in a Chinese hamster ovary cell mutant defective in heparan sulfate N-sulfotransferase. *J. Biol. Chem.* **264,** 8059–8065.

Broekelmann, T. J., Kozel, B. A., Ishibashi, H., Werneck, C. C., Keeley, F. W., Zhang, L., and Mecham, R. P. (2005). Tropoelastin interacts with cell-surface glycosaminoglycans via its COOH-terminal domain. *J. Biol. Chem.* **280,** 40939–40947.

Chen, Y., Miller, W. M., and Aiyar, A. (2005). Transduction efficiency of pantropic retroviral vectors is controlled by the envelope plasmid to vector plasmid ratio. *Biotechnol. Prog.* **21,** 274–282.

Conrad, H. E. (1998). "Heparin-Binding Proteins." Academic Press, San Diego.

Esko, J. D. (1989). Replica plating of animal cells. *Methods Cell. Biol.* **32,** 387–422.

Esko, J. D., and Zhang, L. (1996). Influence of core protein sequence on glycosaminoglycan assembly. *Curr. Opin. Struct. Biol.* **6,** 663–670.

Esko, J. D., Elgavish, A., Prasthofer, T., Taylor, W. H., and Weinke, J. L. (1986). Sulfate transport-deficient mutants of Chinese hamster ovary cells. Sulfation of glycosaminoglycans dependent on cysteine. *J. Biol. Chem.* **261,** 15725–15733.

Esko, J. D., Rostand, K. S., and Weinke, J. L. (1988). Tumor formation dependent on proteoglycan biosynthesis. *Science* **241,** 1092–1096.

Esko, J. D., Stewart, T. E., and Taylor, W. H. (1985). Animal cell mutants defective in glycosaminoglycan biosynthesis. *Proc. Natl. Acad. Sci. USA* **82,** 3197–3201.

Esko, J. D., Weinke, J. L., Taylor, W. H., Ekborg, G., Rodén, L., Anantharamaiah, G., and Gawish, A. (1987). Inhibition of chondroitin and heparan sulfate biosynthesis in Chinese hamster ovary cell mutants defective in galactosyltransferase I. *J. Biol. Chem.* **262,** 12189–13195.

Fields-Berry, S., Halliday, A., and Cepko, C. (1992). A recombinant retrovirus encoding alkaline phosphatase confirms clonal boundary assignment in lineage analysis of murine retina. *PNAS* **89,** 693–697.

Grez, M., Akgun, E., Hilberg, F., and Ostertag, W. (1990). Embryonic stem cell virus, a recombinant murine retrovirus with expression in embryonic stem cells. *PNAS* **87,** 9202–9206.

Gu, X., Lawrence, R., and Krieger, M. (2000). Dissociation of the high density lipoprotein and low density lipoprotein binding activities of murine scavenger receptor class B type I (mSR-BI) using retrovirus library-based activity dissection. *J. Biol. Chem.* **275**, 9120–9130.

Inatani, M., Irie, F., Plump, A. S., Tessier-Lavigne, M., and Yamaguchi, Y. (2003). Mammalian brain morphogenesis and midline axon guidance require heparan sulfate. *Science* **302**, 1044–1046.

Iozzo, R. V. (2001). Heparan sulfate proteoglycans: Intricate molecules with intriguing functions. *J. Clin. Invest.* **108**, 165–167.

Jan, J. T., Byrnes, A. P., and Griffin, D. E. (1999). Characterization of a Chinese hamster ovary cell line developed by retroviral insertional mutagenesis that is resistant to Sindbis virus infection. *J. Virol.* **73**, 4919–4924.

Kingsley, D. M., Kozarsky, K. F., Hobbie, L., and Krieger, M. (1986). Reversible defects in O-linked glycosylation and LDL receptor expression in a UDP-Gal/UDP-GalNAc 4-epimerase deficient mutant. *Cell* **44**, 749–759.

Lian, F., He, L., Colwell, N. S., Lollar, P., and Tollefsen, D. M. (2001). Anticoagulant activities of a monoclonal antibody that binds to exosite II of thrombin. *Biochemistry* **40**, 8508–8513.

Lidholt, K., Weinke, J. L., Kiser, C. S., Lugemwa, F. N., Bame, K. J., Cheifetz, S., Massague, J., Lindahl, U., and Esko, J. D. (1992). A single mutation affects both N-acetylglucosaminyl-transferase and glucoronosyltransferase activities in a Chinese hamster ovary cell mutant defective in heparan sulfate biosynthesis. *Proc. Natl. Acad. Sci. USA* **89**, 2267.

Lieu, F. H., Hawley, T. S., Fong, A. Z., and Hawley, R. G. (1997). Transmissibility of murine stem cell virus-based retroviral vectors carrying both interleukin-12 cDNAs and a third gene: Implications for immune gene therapy. *Cancer Gene Ther.* **4**, 167–175.

Liu, J., and Thorp, S. C. (2002). Cell surface heparan sulfate and its roles in assisting viral infections. *Med. Res. Rev.* **22**, 1–25.

Maimone, M. M., and Tollefsen, D. M. (1990). Structure of a dermatan sulfate hexasaccharide that binds to heparin cofactor II with high affinity. *J. Biol. Chem.* **265**, 18263–18271.

McCormick, C., Duncan, G., Goutsos, K. T., and Tufaro, F. (2000). The putative tumor suppressors EXT1 and EXT2 form a stable complex that accumulates in the Golgi apparatus and catalyzes the synthesis of heparan sulfate. *PNAS* **97**, 668–673.

Miller, A. D., and Rosman, G. J. (1989). Improved retroviral vectors for gene transfer and expression. *Biotechniques* **7**, 980–982, 984–986, 989–990.

O'Donnell, C. D., Tiwari, V., Oh, M. J., and Shukla, D. (2006). A role for heparan sulfate 3-O-sulfotransferase isoform 2 in herpes simplex virus type 1 entry and spread. *Virology* **346**, 452–459.

Pikas, D. S., Eriksson, I., and Kjellen, L. (2000). Overexpression of different isoforms of glucosaminyl *N*-deacetylase/*N*-sulfotransferase results in distinct heparan sulfate N-sulfation patterns. *Biochemistry* **39**, 4552–4558.

Rosenberg, R. D., Shworak, N. W., Liu, J., Schwartz, J. J., and Zhang, L. (1997). Heparan sulfate proteoglycans of the cardiovascular system. Specific structures emerge but how is synthesis regulated? *J. Clin. Invest.* **100**, S67–S75.

Selleck, S. B. (2000). Proteoglycans and pattern formation: Sugar biochemistry meets developmental genetics. *Trends Genet.* **16**, 206–212.

Shukla, D., Liu, J., Blaiklock, P., Shworak, N. W., Bai, X., Esko, J. D., Cohen, G. H., Eisenberg, R. J., Rosenberg, R. D., and Spear, P. G. (1999). A novel role for 3-O-sulfated heparan sulfate in herpes simplex virus 1 entry. *Cell* **99**, 13–22.

Shworak, N. W., Liu, J., Fritze, L. M., Schwartz, J. J., Zhang, L., Logeart, D., and Rosenberg, R. D. (1997). Molecular cloning and expression of mouse and human cDNAs encoding heparan sulfate D-glucosaminyl 3-O-sulfotransferase. *J. Biol. Chem.* **272**, 28008–28019.

Tiwari, V., O'Donnell, C. D., Oh, M. J., Valyi-Nagy, T., and Shukla, D. (2005). A role for 3-O-sulfotransferase isoform-4 in assisting HSV-1 entry and spread. *Biochem. Biophys. Res. Commun.* **338,** 930–937.

Vlodavsky, I., and Friedmann, Y. (2001). Molecular properties and involvement of heparanase in cancer metastasis and angiogenesis. *J. Clin. Invest.* **108,** 341–347.

Wang, L., Fuster, M., Sriramarao, P., and Esko, J. D. (2005). Endothelial heparan sulfate deficiency impairs L-selectin- and chemokine-mediated neutrophil trafficking during inflammatory responses. *Nat. Immunol.* **6,** 902–910.

Wei, G., Bai, X., Gabb, M. M., Bame, K. J., Koshy, T. I., Spear, P. G., and Esko, J. D. (2000). Location of the glucuronosyltransferase domain in the heparan sulfate copolymerase EXT1 by analysis of Chinese hamster ovary cell mutants. *J. Biol. Chem.* **275,** 27733–27740.

Xia, G., Chen, J., Tiwari, V., Ju, W., Li, J. P., Malmstrom, A., Shukla, D., and Liu, J. (2002). Heparan sulfate 3-O-sulfotransferase isoform 5 generates both an antithrombin-binding site and an entry receptor for herpes simplex virus, type 1. *J. Biol. Chem.* **277,** 37912–37919.

Xu, D., Tiwari, V., Xia, G., Clement, C., Shukla, D., and Liu, J. (2005). Characterization of heparan sulphate 3-O-sulphotransferase isoform 6 and its role in assisting the entry of herpes simplex virus type 1. *Biochem. J.* **385,** 451–459.

Zhang, L., Beeler, D. L., Lawrence, R., Lech, M., Liu, J., Davis, J. C., Shriver, Z., Sasisekharan, R., and Rosenberg, R. D. (2001a). 6-O-sulfotransferase-1 represents a critical enzyme in the anticoagulant heparan sulfate biosynthetic pathway. *J. Biol. Chem.* **276,** 42311–42321.

Zhang, L., Lawrence, R., Schwartz, J. J., Bai, X., Wei, G., Esko, J. D., and Rosenberg, R. D. (2001b). The effect of precursor structures on the action of 3-OST-1 and the biosynthesis of anticoagulant heparan sulfate. *J. Biol. Chem.* **24,** 24.

## Further Reading

Esko, J. D., and Selleck, S. B. (2002). Order out of chaos: Assembly of ligand binding sites in heparan sulfate. *Annu. Rev. Biochem.* **71,** 435–471.

# Section V

# Function of Sulfation

# [14]   Determination of Substrate Specificity of Sulfotransferases and Glycosyltransferases (Proteoglycans)

By Hiroko Habuchi, Osami Habuchi, Kenji Uchimura, Koji Kimata, and Takashi Muramatsu

## Abstract

Proteoglycans have sulfated linear polysaccharide chains, that is, heparan sulfate, heparin, chondroitin sulfates, dermatan sulfate, and keratan sulfate. Many glycosyltransferases and sulfotransferases are involved in biosynthesis of the polysaccharides. Specificities of these enzymes have been mainly determined by evaluating their activities to various acceptor carbohydrates and by analyzing the structure of the products. For the latter purpose, enzymatic hydrolysis using heparitinases, heparinase, and chondroitinases or chemical degradation employing nitrous acid deamination has been effectively used in combination with high-performance liquid chromatography (HPLC) of the degraded products. As examples, we describe methods for assays and product characterization of sulfotransferases involved in biosynthesis of these polysaccharides, namely heparan sulfate 2-sulfotransferase, heparan sulfate 6-sulfotransferases, chondroitin 4-sulfotransferases, chondroitin 6-sulfotransferase, N-acetylgalactosamine 4-sulfate 6-sulfotransferase, and N-acetylglucosamine 6-sulfotransferases.

## Overview

Heparan sulfate, heparin, chondroitin sulfates, dermatan sulfate, and keratan sulfate are polysaccharidic components of mammalian proteoglycans. Together with hyaluronan, which is not linked to proteins, these polysaccharides are collectively called *glycosaminoglycans*. Polysaccharides of proteoglycans are mainly formed by concerted action of glycosyltransferases and sulfotransferases, which transfer a sulfate group from 3′-phosphoadenosine 5′-phosphosulfate (PAPS) to the acceptor. Before describing methods of specificity determination, we provide an overall view of the enzymes (Fig. 1).

### Heparan Sulfate and Heparin

Heparan sulfate has repeating units composed of glucosamine (GlcN) derivatives and uronic acids, which are either glucuronic acid (GlcA) or

METHODS IN ENZYMOLOGY, VOL. 416
0076-6879/06 $35.00
DOI: 10.1016/S0076-6879(06)16014-0

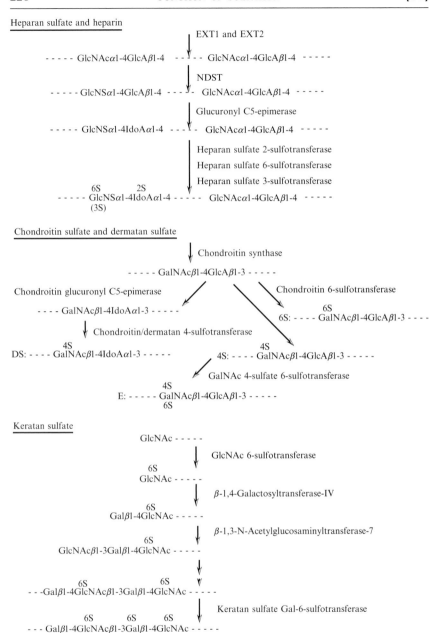

Fig. 1. The proposed major pathway of the synthesis of polysaccharide chains in proteoglycans. In heparan sulfate, both a (left) highly modified portion and a (right) relatively unmodified portion are present. DS, dermatan sulfate; E, chondroitin sulfate E; 6S, chondroitin 6-sulfate; 4S, chondroitin 4-sulfate; NDST, heparan sulfate/heparin $N$-deacetylase $N$-sulfotransferase.

L-iduronic acid (IdoA). The polysaccharide chain is formed by a molecular complex of two proteins called *EXT1* and *EXT2* (Lind *et al.*, 1998; McCormick *et al.*, 1998, 2000). Both of them have two domains of glycosyl-transferases: One catalyzes the transfer of an *N*-acetylglucosamine (GlcNAc) residue and the other catalyzes the transfer of a GlcA residue. The synthesized chain has a repeating unit of GlcNAc$\alpha$1-4 GlcA$\beta$1-4. The polysaccharide is then modified by sulfotransferases and an epimerase. Heparan sulfate is a heterogeneous molecule with various degrees of modification. When the modification is extensive, the molecule is called *heparin*, the fundamental repeating unit of which is GlcNS (6S) $\alpha$1-4 IdoA(2S) $\alpha$1-4, in which S is an SO$_3$ group attached to N or O atom of monosaccharides.

The key enzyme of the modification is a heparan sulfate/heparin *N*-deacetylase *N*-sulfotransferase (NDST), which catalyzes both de-*N*-acetylation and *N*-sulfation of a GlcN residue, because other modifying enzymes either require or prefer *N*-sulfated structures in the substrate. Four species of NDSTs are present: NDST-1 is involved in the synthesis of heparan sulfate, and NDST-2 in the synthesis of heparin in mast cells (Aikawa *et al.*, 2001; Eriksson *et al.*, 1994; Kusche-Gullberg *et al.*, 1998). Glucuronyl C-5-epimerase catalyzes the epimerization of C-5 of GlcA to IdoA (Li *et al.*, 1997). The enzyme requires an adjacent *N*-sulfated GlcN (GlcNS) toward the non-reducing end-side. Heparan sulfate 2-sulfotransferase sulfates C-2 of uronic acids and prefers IdoA rather than GlcA (Kobayashi *et al.*, 1997). There is only one species of 2-sulfotransferase.

Heparan sulfate 6-sulfotransferase (HS6ST) sulfates the 6-position of GlcN, and three molecular species of the enzyme are present (Habuchi *et al.*, 1998, 2000; Jemth *et al.*, 2003; Smeds *et al.*, 2003). All three enzymes prefer a GlcNS residue for action but exhibit a low level of activity toward structures with a GlcNAc residue. Each isoform shows different specificity toward the isomeric uronic acid adjacent to the targeted GlcNS; HS6ST-1 prefers the IdoA-GlcNS while HS6ST-2 has a different preference, depending on the substrate concentration, and HS6ST-3 acts on either substrate, as long as a GlcA-GlcNS repeating polysaccharide and a IdoA-GlcNS repeating polysaccharide are used as acceptor substrates. However, when heparan sulfate is a substrate, all of them have a preference for IdoA-containing targets with or without 2-sulfation. HS6ST-1 shows relatively higher activity toward target sequences lacking 2-sulfation, and HS6ST-2 and -3 show a preference for 2-sulfated substrates.

Finally, the C-3 position of a GlcN residue is sulfated by heparan sulfate 3-sulfotransferase; this sulfated residue is essential for the formation of an antithrombin III (ATIII)–binding site but is present only in the restricted portion of the polysaccharide chain. There are six molecular species of 3-sulfotransferases with different specificities (Chen *et al.*, 2003; Liu *et al.*, 1999a,b; Mochizuki *et al.*, 2003; Shworak *et al.*, 1997).

The general order of modification in heparan sulfate is $N$-deacetylation/$N$-sulfation, followed by C-5 epimerization and finally 2-, 3-, and 6-sulfation. As can be seen from the *in vitro* specificities of the enzymes, this sequence of events is not strict, and in ES cells lacking both NDST-1 and -2, a 6-sulfated structure without $N$-sulfation is found (Holmborn *et al.*, 2004).

## Chondroitin Sulfates and Dermatan Sulfate

Chondroitin sulfates and dermatan sulfate have building blocks composed of uronic acids and $N$-acetylgalactosamine (GalNAc). The polysaccharide backbone of chondroitin sulfates is chondroitin with repeating units of GalNAc$\beta$1-4GlcA$\beta$1-3 and is formed by chondroitin synthase (DeAngelis and Padgett-McCue, 2000; Kitagawa *et al.*, 2001; Yada *et al.*, 2003a,b). Chondroitin glucuronyl C5-epimerase changes GlcA to IdoA in the chain, forming dermatan with repeating units of GalNAc$\beta$1-4IdoA$\alpha$1-3, which is the backbone of dermatan sulfate.

Chondroitin 6-sulfate is formed as the result of 6-sulfation of chondroitin by chondroitin 6-sulfotransferase (Fukuta *et al.*, 1995). Chondroitin 4-sulfate (also called *chondroitin sulfate A*) and dermatan sulfate are formed by 4-sulfation of the GalNAc residue by chondroitin/dermatan 4-sulfotransferase. Three molecular species of the 4-sulfotransferase are present, and they have different specificities in terms of uronic acids in the substrates (Hiraoka *et al.*, 2000; Mikami *et al.*, 2003; Yamada *et al.*, 2004; Yamauchi *et al.*, 2000). One enzyme preferentially uses chondroitin and yields chondroitin 4-sulfate, the second one preferentially uses dermatan and yields dermatan sulfate, and the last one uses both substrates with similar efficiency. Chondroitin sulfate E, which has GalNAc (4S, 6S)$\beta$1-4GlcA$\beta$1-3 unit, is formed by 6-sulfation of chondroitin 4-sulfate by GalNAc 4-sulfate 6-sulfotransferase (Ohtake *et al.*, 2001). Chondroitin sulfate D, which has GalNAc(6S)$\beta$1-4GlcA(2S)$\beta$1-3 units, is suggested to be formed by 2-sulfation of GlcA in chondroitin 6-sulfate. Uronosyl 2-sulfotransferase sulfates C-2 of IdoA in dermatan sulfate and C-2 of GlcA in chondroitin sulfate with lesser activity (Kobayashi *et al.*, 1999). This enzyme is probably involved in the synthesis of chondroitin sulfate D.

## Keratan Sulfate

Keratan sulfate has repeating units of Gal$\beta$1-4GlcNAc$\beta$1-3, in which GlcNAc is always 6-sulfated and Gal is occasionally sulfated. The polysaccharide chain is believed to be extended by the alternative actions of a $\beta$-1,

4-galactosyltransferase ($\beta$4GalT) and a $\beta$-1,3-$N$-acetyl glucosaminyltransferase ($\beta$3GnT). The sequence of biosynthesis is $N$-acetylglucosaminylation, 6-sulfation of a GlcNAc residue exposed at the non-reducing end, and galactosylation. After formation of the polysaccharide chain, a part of Gal residue is sulfated. Using appropriate oligosaccharides as substrates, $\beta$4GalT-IV was shown to effectively act on substrates with exposed 6-sulfated GlcNAc and is suggested to be the enzyme performing galactosylation (Seko et al., 2003). Similarly, $\beta$3GnT-7 is suggested to be the enzyme responsible for $N$-acetylglucosaminylation (Seko and Yamashita, 2004).

GlcNAc 6-sulfation is catalyzed by GlcNAc 6-sulfotransferases (GlcNAc6STs). GlcNAc6ST-5 is involved in the biosynthesis of keratan sulfate in the cornea, and its null mutation in humans leads to macular corneal dystrophy (Akama et al., 2000). GlcNAc6ST-1 and -3 (Akama et al., 2001; Lee et al., 2000; Uchimura et al., 1998, 2002) also participate in the synthesis of keratan sulfate. 6-Sulfation of Gal is performed by keratan sulfate Gal-6-sulfotransferase (Fukuta et al., 1997).

Determination of Enzyme Specificities

The specificities of these enzymes were mainly determined by the enzymatic actions toward different substrates and analysis of the products. Specifically, desulfated heparins were useful in determining the specificity of sulfotransferases involved in the synthesis of heparan sulfate and heparin. Oligosaccharide substrates were also used to study the enzymes forming keratan sulfate. Oligosaccharide libraries have been applied to determine the specificities of sulfotransferases (Jemth et al., 2003). Digestion with specific enzymes such as heparitinases, heparinase, and chondroitinases and analysis of the products by HPLC gave precise information on the structure of the synthesized polysaccharides. These enzymes cleave the polysaccharide at hexosamine residues (GlcNS, GlcNAc, and GalNAc). Because they are eliminases, the products have an unsaturated bond between C-4 and C-5 of uronic acids at the non-reducing ends. Deamination by nitrous acid at pH levels of 1.5 and 4.0 is also used to cleave heparan sulfate/heparin chain at GlcNS and GlcN, respectively. As examples, we describe here methods for determination of enzymatic activities and analysis of the product structure of representative sulfotransferases that are involved in synthesis of heparan sulfate/heparin, chondroitin sulfates/dermatan sulfate, and keratan sulfate. The described method is applicable to determination of specificities of other sulfotransferases and glycosyltransferases involved in proteoglycan biosynthesis.

*Heparan Sulfate 2-Sulfotransferase and Heparan Sulfate*
*6-Sulfotransferase-1, -2, and -3*

Methods described in this section have been used for characterization of heparan sulfate 2-sulfotransferase (HS2ST) as well as HS6ST-1, HS6ST-2, and HS6ST-3 (Ashikari-Hada, 2004; Habuchi *et al.*, 1995, 1998, 2000, 2003; Kobayashi *et al.*, 1996, 1997; Nogami *et al.*, 2004; Rong *et al.*, 2000; Smeds *et al.*, 2003).

*Materials*

YMC-Pack Polyamine II column (4.6 mm × 25 cm) from YMC (Kyoto, Japan).

Partisil-10 SAX column (4.6 mm × 25 cm) from Whatman.

Superdex 30 HR 16/60 pg and Fast Desalting Column HR 10/10 from Amersham Bioscience (see Note 1, later in this section).

Heparitinase I (EC 4.2.2.8), heparitinase II, heparinase (EC 4.2.2.7), completely desulfated and *N*-resulfated heparin (CDSNS-heparin), *N*-sulfated heparosan (NS-heparosan, a GlcA-GlcNS repeating polysaccharide), chondroitin sulfate A, and unsaturated heparan sulfate/heparin disaccharide mixture (H mix) from Seikagaku Corporation (Tokyo, Japan).

pH 1.5 Nitrous acid reagent is freshly prepared as follows. Solution of 0.5 $M$ $H_2SO_4$ and 0.5 $M$ $Ba(NO_2)_2$ are cooled separately to 0°. 0.5 ml of each solution is mixed and centrifuged at 10,000$g$ for 1 min. The supernatant should be used immediately.

[$^{35}$S]PAPS is prepared from $^{35}$SO$_4$ and ATP using PAPS synthetase prepared from chick embryo chondrocytes. Commercially available [$^{35}$S]PAPS (PerkinElmer NEG010) is also suitable for the assay.

*Solution*

Heparitinase digestion buffer (5× enriched buffer): 250 m$M$ Tris–HCl, 5 m$M$ CaCl$_2$, and 0.1 mg/ml of bovine serum albumin (BSA).

*Preparation of Recombinant HS2ST, HS6ST-1, -2, and -3*

1. Cos-7 cells (5 × 10$^5$) precultured for 48 h in a 60-mm culture dish are transfected with 4 $\mu$g of pFLAG-CMV-2-HS2ST, -HS6ST-1, -2, or -3.

2. After incubation in Dulbecco Modified Eagle Medium (DMEM) for 72 h, the spent media are collected, and the cell layers are washed with phosphate-buffered saline (PBS) (−), and extracted with 1 ml of buffer A (10 m$M$ Tris–HCl, pH 7.2, 0.5% (w/v) Triton X-100, 0.15 $M$ NaCl, 10 m$M$

MgCl$_2$, 2 m$M$ CaCl$_2$, and 20% glycerol) for 1 h on a rotatory shaker at 4°. The extracts are centrifuged at 10,000$g$ for 30 min (crude extracts).

3. Crude extracts are applied to an anti-FLAG M2 antibody-conjugated agarose column (bed volume 0.5 ml) (Sigma). The absorbed materials are eluted with 1.5 ml of buffer A containing 118 $\mu$g FLAG peptide (Sigma).

### Assay of HS2ST, HS6ST-1, -2, and -3 Activities

1. The standard reaction mixture (50 $\mu$l) contains 2.5 $\mu$mol of imidazole–HCl, pH 5.6 (for HS2ST) or pH 6.3 (for HS6STs), 3.75 $\mu$g of protamine chloride, 2 nmol (as hexuronic acid) of CDSNS-heparin or other glycosaminoglycans, 50 pmol [$^{35}$S]PAPS ($\sim$5 × 10$^5$ cpm), 5 $\mu$l of 1 $M$ NaCl in buffer A, and enzyme (see Notes 2 and 3, later in this section).

2. After incubation for 20 min at 37°, the reaction is stopped by heating at 100° for 1 min. Chondroitin sulfate A as carrier (0.1 $\mu$mol as GlcA) is added to the reaction mixture. After cooling, 2.5 volumes of cold 95% ethanol/1.3% (w/v) potassium acetate/0.5 m$M$ ethylenediaminetetraacetic acid (EDTA) is added to the reaction mixture, and stood at −20° for 30 min. $^{35}$S-labeled polysaccharides are precipitated by centrifugation at 10,000$g$ for 30 min. The precipitate is dissolved in 70 $\mu$l of distilled water.

3. Fifty microliters of the solution is injected to the Fast Desalting Column equilibrated with 0.2 $M$ NH$_4$HCO$_3$ at a flow rate of 2 ml/min to remove [$^{35}$S]PAPS and its degradation products. $^{35}$S-labeled products are recovered in the void volume and mixed with 3 ml of Ready Safe Scintillator, and the radioactivity is determined (see Note 4, later in this section).

### Determination of Sulfated Positions in the Products

#### ANALYSIS OF HEPARITINASE/HEPARINASE DIGESTS

1. $^{35}$S-labeled products are digested with a mixture of 5 milliunits of heparitinase I, 2.5 milliunits of heparitinase II, and 5 milliunits of heparinase in 50 $\mu$l of 50 m$M$ Tris–HCl, pH 7.2, 1 m$M$ CaCl$_2$, and 5 $\mu$g of BSA at 37° for 2 h.

2. The reaction is stopped by heating at 100° for 1 min.

3. The digests are injected together with 1 nmol of standard unsaturated disaccharides into a column of YMC-Pack Polyamine II.

4. The elution is performed by a linear gradient from 40 m$M$- to 550 m$M$-KH$_2$PO$_4$ at a flow rate of 1.2 ml/min at 40°.

5. Fractions of 0.6 ml are collected and mixed with 3 ml of Ready Safe Scintillator, and the radioactivity is determined. The separation pattern of standard materials is indicated in Fig. 2A. Under the conditions, GlcNS

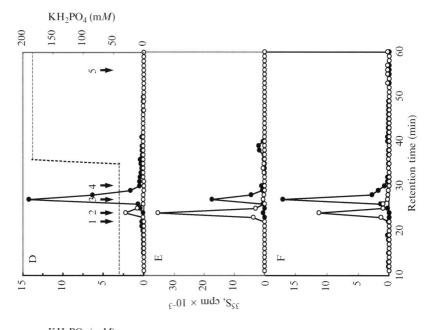

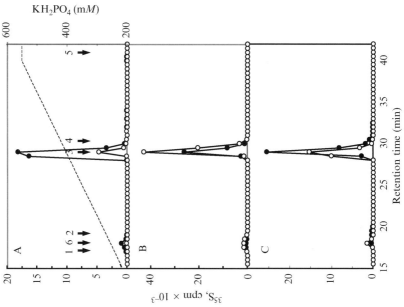

(6S) (see Note 5, later in this section), which are derived from the non-reducing terminal, and $\Delta$Di-6S, $\Delta$Di-NS, $\Delta$Di-(N,6)diS, $\Delta$Di-(N,2)diS, and $\Delta$Di-(N,6,2)triS, which are derived from the internal portion or reducing terminal, are separated. When CDSNS-heparin is used as the acceptor, radioactivity derived from three isoforms of HS6ST is detected at the positions $\Delta$Di-(N,6)diS and GlcNS(6S). In the case of HS2ST, radioactivity is detected at the position $\Delta$Di-(N,2)diS.

ANALYSIS OF PRODUCTS DEGRADED BY NITROUS ACID AT pH 1.5
(SHIVELY AND CONRAD, 1976)

1. Eighty microliters of 0.5 $M$ nitrous acid is added to 20 $\mu$l of $^{35}$S-labeled products.

2. The mixture is incubated at room temperature for 10 min, and then adjusted to pH 8.5 by 1 $M$ Na$_2$CO$_3$.

3. The samples are reduced by adding 40 $\mu$l of 0.5 $M$ NaBH$_4$, followed by incubation at 50° for 30 min.

4. Excess NaBH$_4$ is destroyed by addition of 4 $M$ acetic acid to give slightly acidic pH, and then neutralized with NaOH.

5. The samples are treated with 25 m$M$ H$_2$SO$_4$ (final concentration) at 80° for 30 min to cleavage anomalous deamination products.

6. The final samples are subjected to gel chromatography on a Superdex 30-pg column equilibrated with 0.2 $M$ ammonium acetate at a flow rate of 1 ml/min. The $^{35}$S-labeled disaccharide fractions are injected to a Partisil-10 SAX column and separated with isocratic elution of 40 m$M$ KH$_2$PO$_4$ up to

---

FIG. 2. High-performance liquid chromatography (HPLC) analysis of $^{35}$S-labeled disaccharides derived from completely desulfated and $N$-resulfated (CDSNS) heparin (CDSNS-heparin) and $N$-sulfated heparosan (NS-heparosan). $^{35}$S-labeled products are synthesized from CDSNS-heparin and NS-heparosan by incubation with [$^{35}$S]PAPS and the recombinant purified mouse HS6ST-1 (A and D), HS6ST-2 (B and E), and HS6ST-3 (C and F), respectively. (A–C) The labeled products are digested with a mixture of heparitinases and applied to a YMC-Pack Polyamine II column. (D–F) The disaccharide fractions of the degraded products with nitrous acid are applied to a Partisil 10-SAX column. The closed circle and the open circle show the elution patterns of the products derived from CDSNS-heparin and NS-heparosan, respectively. The arrows in Fig. 2A indicate the elution position: 1, $\Delta$Di-6S; 2, $\Delta$Di-NS; 3, $\Delta$Di-(N,6)diS; 4, $\Delta$Di-(N,2)diS; 5, $\Delta$Di-(N,6,2)triS; 6, GlcNS(6SO$_4$). The arrows in Fig. 2D show the elution position: 1, GlcA(2S)-AMan$_R$; 2, GlcA-AMan(6S)$_R$; 3, IdoA-AMan(6S)$_R$; 4, IdoA(2S)-AMan$_R$; 5, IdoA(2S)-AMan(6S$_4$)$_R$, respectively derived from GlcA(2S)-GlcNS, GlcA-GlcNS (6S), IdoA-GlcNS(6S), IdoA(2S)-GlcNS, and IdoA(2S)-GlcNS(6S). Permission granted to reproduce from Habuchi *et al.* (2000). Abbreviations: $\Delta$Di, 2-acetamide-2-deoxy-4-$O$-(4-deoxy-$\alpha$-L-*threo*-hex-4-enepyranosyluronic acid)-D-glucose; AMan$_R$, 2,5-anhydro-D-mannitol. $\Delta$Di-6S, $\Delta$Di-NS, $\Delta$Di-(N,6)diS, $\Delta$Di-(N,2)diS, and $\Delta$Di-(N,6,2)triS indicate the position and number of sulfate residues in the $\Delta$Di unsaturated disaccharide.

35 min and 185 m$M$ KH$_2$PO$_4$ from 36 to 75 min. The separation pattern of standard materials is indicated in Fig. 2D. Under the conditions, GlcA(2S)-AMan$_R$, GlcA-AMan(6S)$_R$, IdoA-AMan(6S)$_R$, IdoA(2S)-AMan$_R$, and IdoA(2S)-AMan(6S)$_R$, respectively derived from GlcA(2S)-GlcNS, GlcA-GlcNS(6S), IdoA-GlcNS(6S), and IdoA(2S)-GlcNS, are well separated. When CDSNS-heparin is used as the acceptor, radioactivity derived from three isoforms of HS6ST is detected at the position IdoA-AMan(6S)$_R$. As to HS2ST, radioactivity is detected at the position IdoA(2S)-AMan$_R$.

*Notes*

1. Instead of Fast Desalting Column, a spin column can be used. The spin column (bed volume 0.9 ml) is made by packing Sephadex G-50 medium (Amersham Bioscience) suspended in 0.1 $M$ NH$_4$HCO$_3$ into a 1-ml syringe under centrifugation at 2000 rpm for 4 min. Samples (100 $\mu$l) are loaded to the top of the gel and centrifuged at 2000 rpm for 4 min. [$^{35}$S] Glycosaminoglycans are recovered in the flow-through fractions.

2. When crude extracts are used as enzyme, 1 $\mu$mol NaF and 0.1 $\mu$mol AMP are added to the reaction mixture to prevent degradation of PAPS.

3. HS6ST isoforms are inhibited by dithiothreitol (DTT) but not HS2ST. HS6ST-1 activity is decreased to 19% of control in 10 m$M$ DTT.

4. As to the sensitivity of this assay, crude extract from 1 $\times$ 10$^5$ fibroblasts is enough to determine HS2ST and HS6ST activities.

5. Standard GlcNS(6S) is able to be prepared from $\Delta$Di-(N,6)diS by mercuric acetate treatment. $\Delta$Di-(N,6)diS is dissolved in 100 $\mu$l of 70 m$M$ mercuric acetate adjusted to pH 5.0 and the mixture is left at room temperature for 30 min. To remove mercuric acetate, 60 $\mu$l of a 50% slurry of Dowex 50 H$^+$ resin is added to the reaction mixture. The resin is removed by filtration and washed with water. The flow-through fractions and the washings are combined and neutralized with 2 $M$ NaOH.

*Chondroitin 6-Sulfotransferase, Chondroitin 4-Sulfotransferase, and GalNAc 4-Sulfate 6-O-Sulfotransferase*

Activities of chondroitin 6-sulfotransferase (C6ST), chondroitin 4-sulfotransferase (C4ST), and GalNAc 4-sulfate 6-$O$-sulfotransferase (GalNAc4S-6ST) are determined by measuring the rate of transfer of radioactive sulfate from [$^{35}$S]PAPS to chondroitin (for C6ST and C4ST) or chondroitin sulfate A (for GalNAc4S-6ST) (Fukuta *et al.*, 1995; Habuchi *et al.*, 1993; Ito *et al.*, 2000; Ohtake *et al.*, 2001, 2003; Yamada *et al.*, 2004; Yamauchi *et al.*, 1999, 2000). The key step of this assay is complete separation of the $^{35}$S-labeled products from [$^{35}$S]PAPS and its degradation materials such as $^{35}$SO$_4^{2-}$ that are included in the reaction mixtures in large excess.

We cleared this step by gel filtration in a solution containing volatile salt, $NH_4HCO_3$.

*Materials*

Chondroitin sulfate A from whale cartilage (Seikagaku Corporation, Tokyo, Japan).

Chondroitin prepared from squid skin. Commercially available chondroitin (Seikagaku Corporation, Tokyo, Japan) is also suitable for the assay.

[$^{35}$S]PAPS prepared from $^{35}$SO$_4$ and ATP using PAPS synthetase prepared from chick embryo chondrocytes. Commercially available [$^{35}$S]PAPS (PerkinElmer NEG010) is also suitable for the assay.

Fast Desalting Column HR 10/10 (Amersham Bioscience) equilibrated with 0.1 $M$ $NH_4HCO_3$ and run at a flow rate of 2 ml/min (see Note 1, later in this section).

Superdex 30 HR 16/60 pg (Amersham Bioscience) equilibrated with 0.2 $M$ $NH_4HCO_3$ and run at a flow rate of 2 ml/min.

Partisil-10 SAX column (4.6 mm × 25 cm) from Whatman.

Chondroitinase ACII, chondroitinase ABC, chondro-6-sulfatase, chondro-4-sulfatase, and unsaturated chondro-disaccharide kit (C kit) from Seikagaku Corporation.

Monosaccharide standards, GalNAc(4S) and GalNAc(6S) from Funakoshi, Tokyo, Japan (see Note 2, later in this section).

*Enzyme Preparation*

The recombinant C4ST, C6ST, and GalNAc4S-6ST are expressed in Cos-7 cells as fusion proteins with a FLAG peptide. From the transfected cells, each sulfotransferase is extracted with buffer A (0.15 $M$ NaCl, 10 m$M$ Tris–HCl, pH 7.2, 10 m$M$ MgCl$_2$, 2 m$M$ CaCl$_2$, 0.5% Triton X-100, and 20% glycerol) for 30 min on a rotary shaker at 4°. The extracts are centrifuged at 10,000$g$ for 10 min. The supernatant fractions are applied to an anti-FLAG M2 antibody-conjugated agarose column (bed volume 0.5 ml) (Sigma). The absorbed materials are eluted with 1.5 ml of buffer A containing 118 $\mu$g FLAG peptide (Sigma).

*Assay of C6ST, C4ST, and GalNAc4S-6ST*

1. The standard reaction mixture contains 2.5 $\mu$mol imidazole–HCl, pH 6.8, 1.25 $\mu$g protamine chloride (for C6ST and C4ST) or 0.5 $\mu$mol CaCl$_2$ (for GalNAc4S-6ST), 0.1 $\mu$mol DTT (for C6ST or C4ST) or 1 $\mu$mol reduced glutathione (for GalNAc4S-6ST) (see Note 3, later in this section), 25 nmol (as galactosamine) chondroitin (for C6ST or C4ST) or chondroitin

sulfate A (for GalNAc4S-6ST), 50 pmol [$^{35}$S]PAPS ($\sim$5.0 × 10$^5$ cpm), and enzyme in a final volume of 50 $\mu$l. For determining the activity toward various glycosaminoglycans or oligosaccharides, chondroitin or chondroitin sulfate A is replaced with 25 nmol (as galactosamine) of glycosaminoglycans or oligosaccharides.

2. The reaction mixtures are incubated at 37° for 20 min and the reaction is stopped by immersing the reaction tubes in a boiling waterbath for 1 min.

3. After the reaction is stopped, add 50 nmol chondroitin sulfate A as carrier and three volumes of ethanol containing 1.3% potassium acetate to the reaction mixture. The mixtures are stirred well and placed on ice for 30 min.

4. $^{35}$S-labeled glycosaminoglycans are precipitated with centrifugation (10,000$g$ × 10 min) and dissolved in 70 $\mu$l water.

5. Fifty microliters of the solution is injected to the Fast Desalting Column. $^{35}$S-labeled glycosaminoglycans are recovered in the void volume (1.0–1.5 min). The samples are injected to the column at a 5-min interval.

6. When oligosaccharides are used as acceptors, the reaction mixtures are applied directly to the Superdex 30 column, and the enzymatic reaction products are separated from [$^{35}$S]PAPS and inorganic sulfate.

*Determination of Sulfated Positions in the Products*

1. Solutions of $^{35}$S-labeled glycosaminoglycans or oligosaccharides are lyophilized to remove NH$_4$HCO$_3$ and digested with chondroitinase ACII (see Note 4, later in this section) or chondroitinase ABC in the combination with chondro-6-sulfatase or chondro-4-sulfatase (see Note 5, later in this section).

2. The digests are injected to Partisil-10 SAX column equilibrated with 10 m$M$ KH$_2$PO$_4$ after filtration with a spin filter (Ultrafree MC, HV 0.45 $\mu$m, Millipore). The column is developed with 10 m$M$ KH$_2$PO$_4$ for 10 min, followed by a 45-min linear gradient from 10 to 500 m$M$ KH$_2$PO$_4$. After the gradient, the column is washed with 500 m$M$ KH$_2$PO$_4$ for 10 min. A column temperature is 40°. Fractions (0.5 ml) are collected at a flow rate of 1 ml/min (see Note 6, later in this section). Each fraction is mixed with 2 ml of Clearsol I (Nakarai Tesque, Kyoto, Japan) and the radioactivity is measured. An example of the separation pattern is shown in Fig. 3 in which $^{35}$S-labeled product from chondroitin sulfate A after the reaction with GalNAc4S-6ST is analyzed. Radioactivity is detected at the position of GalNAc(4,6S), $\Delta$Di-diS$_E$ and a trisaccharide derived from the non-reducing terminal (see Note 7, later in this section). After chondro-6-sulfatase digestion, radioactivity of GalNAc(4,6S) and $\Delta$Di-diS$_E$ is shifted to the position of inorganic sulfate.

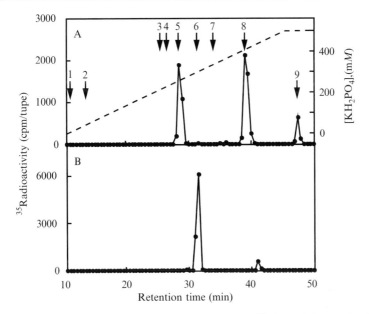

FIG. 3. Partisil 10-SAX HPLC of disaccharides derived from $^{35}$S-labeled glycosaminoglycans formed from chondroitin sulfate A after reaction with GalNAc4S-6ST. The labeled glycosaminoglycans were digested with chondroitinase ACII (A) or chondroitinase ACII plus chondro-6-sulfatase (B). Arrows indicate the elution position of: 1, GalNAc(6S); 2, GalNAc(4S); 3, $\Delta$Di-6S; 4, $\Delta$Di-4S; 5, GalNAc(4, 6S); 6, $SO_4^{2-}$; 7, $\Delta$Di-diS$_D$; 8, $\Delta$Di-diS$_E$ and 9, a trisaccharide derived from non-reducing terminal of the glycosaminoglycan (see Note 7). Abbreviations: $\Delta$Di-6S, 2-acetamide-2-deoxy-3-$O$-($\beta$-D-gluco-4-enepyranosyluroic acid)-6-$O$-sulfo-D-galactose; $\Delta$Di-4S, 2-acetamide-2-deoxy-3-$O$-($\beta$-D-gluco-4-enepyranosyluroic acid)-4-$O$-sulfo-D-galactose; $\Delta$Di-diS$_D$, 2-acetamide-2-deoxy-3-$O$-(2-$O$-sulfo-$\beta$-D-gluco-4-enepyranosyluroic acid)-6-$O$-sulfo-D-galactose; $\Delta$Di-diS$_E$, 2-acetamide-2-deoxy-3-$O$-($\beta$-D-gluco-4-enepyranosyluroic acid)-4,6-bis-$O$-sulfo-D-galactose.

From these results, GalNAc4S-6ST is demonstrated to transfer sulfate to position 6 of both non-reducing terminal and internal GalNAc(4S) residues.

*Notes*

1. If Fast Desalting Column is not available, Sephadex G-25 superfine (Amersham Bioscience) suspended in 0.1 $M$ NaCl is packed to a column (10 mm × 10 cm) at a flow rate of 6 ml/min. This column is equilibrated with 0.1 $M$ NH$_4$HCO$_3$ and run at a flow rate of 2 ml/min.

2. If commercial GalNAc(4,6S) is not available, it can be prepared from $\Delta$Di-diS$_E$ by mercuric acetate treatment. $\Delta$Di-diS$_E$ is dissolved in 1 ml of 35 m$M$ mercuric acetate in 25 m$M$ Tris/25 m$M$ sodium acetate, pH 5.0. The reaction is carried out for 1 h at room temperature. After the reaction is

over, the sample is applied to Dowex 50 (H$^+$) column (bed volume of 1 ml). The column is washed with 3 ml of water. The flow-through fractions and the washings are combined and lyophilized. The lyophilized materials are further purified with Superdex 30.

3. Glutathione is neutralized with NaOH before use.

4. Digestion with chondroitinase ACII or chondroitinase ABC under the standard conditions is carried out for 4 h at 37° in the reaction mixture containing, in a final volume of 25 μl, $^{35}$S-labeled glycosaminoglycans or $^{35}$S-labeled oligosaccharides, 1.25 μmol of Tris–acetate buffer, pH 7.5, 2.5 μg BSA, and 30 milliunits of chondroitinase ACII or chondroitinase ABC. For complete degradation of GlcA(2S)-containing sequence by chondroitinase ACII, stronger conditions are required (Ohtake et al., 2003).

5. For complete degradation of GalNAc(4,6S), digestion with chondro-6-sulfatase is carried out for 5 h at 37° in the reaction mixture containing, in a final volume of 25 μl, 1.25 μmol of Tris–acetate buffer, pH 7.5, 2.5 μg BSA, and 100 milliunits of chondro-6-sulfatase, and the reaction is continued for further 5 h in the presence of 100 milliunits of the fresh enzyme.

6. Standard unsaturated disaccharides (5 nmol each) and monosaccharides (18 nmol each) are detected simultaneously at 210 nm. Precise retention times of the standard materials may vary when different lots of the column are used.

7. This trisaccharide is deduced to be GalNAc(4,6S)-GlcA(2S)-GalNAc(6S).

### GlcNAc 6-Sulfotransferases

GlcNAc 6-sulfotransferases (GlcNAc6STs) are involved in synthesis of both keratan sulfate and glycoprotein-bound glycans. Five molecular species of GlcNAc6STs so far identified have different specificities (Akama et al., 2000, 2001; Kadomatsu et al., 2004; Lee et al., 2000; Uchimura et al., 1998, 2000, 2002, 2004). An enzymatic assay employing various oligosaccharides derived from N-linked or O-linked glycoprotein-bound glycans and keratan sulfate is applicable to compare substrate specificity of GlcNAc6STs and to elucidate possible contribution of each enzyme to extend keratan sulfate chain and/or initiate addition of keratan sulfate to different core oligosaccharides bound to scaffold proteins. The assay described here employing thin-layer chromatography (TLC) can handle a large number of samples.

### Materials

Oligosaccharide substrates: GlcNAcβ1-3Galβ1-4GlcNAc prepared from Galβ1-4GlcNAcβ1-3Galβ1-4GlcNAc in combination with β-galactosidase digestion and purification with Superdex 30

size-fractionating column (Uchimura *et al.*, 1998). GlcNAc$\beta$1–6Man-
O-methyl from Sigma. GlcNAc$\beta$1–2Man is from Dextra Laboratories
(Reading, UK). GlcNAc$\beta$1–6[Gal$\beta$1–3]GalNAc-p-nitrophenyl (core
2-p-nitrophenyl) and GlcNAc$\beta$1–3GalNAc-p-nitrophenyl (core 3-p-
nitrophenyl) from Toronto Research Chemicals (Ontario, Canada).
[35][S]PAPS obtained from PerkinElmer NEG010. After dilution with
water, the aliquots are stored as frozen.
TLC aluminum sheet precoated with cellulose (0.1-mm thick) from
Merck (Whitehouse Station, NJ).

*Solutions*

Tris–HCl, pH 7.5 (0.5 *M*), 0.1 *M* MgCl$_2$, and 1% Triton X-100 in water
is prepared and kept at room temperature.
50 m*M* AMP (Sigma) and 1 *M* NaF (Sigma) in water are prepared and
stored in freezer (see Note 1, later in this section).
Developing buffer: ethanol/pyridine/n-butyl alcohol/water/acetic acid
(100:10:10:30:3, by volume).

*Assay Methods*

1. Enzymes are extracted from Cos-7 cells transfected with an
expression plasmid encoding GlcNAc6STs (Uchimura *et al.*, 1998, 2000,
2002). If a plasmid encoding a soluble protein A–fused GlcNAc6ST-1 or
other GlcNAc6ST is transfected, enzymes are purified with IgG–Sepharose
beads (Uchimura *et al.*, 2002). The standard reaction mixture contains
1 $\mu$mol Tris–HCl, pH 7.5, 0.2 $\mu$mol MnCl$_2$, 0.04 $\mu$mol AMP, 2 $\mu$mol NaF,
20 $\mu$mol oligosaccharide, 150 pmol [35S]PAPS (1.5 × 10$^6$ cpm), 0.05% of
Triton X, and 5 $\mu$l of suspension of IgG–Sepharose beads with the enzyme
or the extract from Cos-7 cells in a final volume of 20 $\mu$l. The reaction
mixture is incubated at 30° for 1 h.
2. Then, aliquots of 2 $\mu$l of the reaction mixture are applied to TLC
plates, which are developed with developing buffer. When the solvent
front reaches to the top, the development is ended. The [35]S-labeled
products migrate faster than [35S]PAPS.

The radioactivity of the [35]S-labeled products is visualized and measured
with a BAS2000 bioimaging analyzer (Uchimura *et al.*, 2002).

*Notes*

1. The remaining aliquots of reaction mixture can be analyzed by
Superdex 30 gel chromatography as described on page 233 to confirm the
result of TLC.

2. Sulfation of GlcNAc in oligosaccharides results in unsusceptibility to Jack bean $\beta$-N-acetylhexosaminidase. Digestion of $^{35}$S-labeled products with the enzyme may be used to confirm the sulfation on GlcNAc.

## References

Aikawa, J., Grobe, K., Tsujimoto, M., and Esko, J. D. (2001). Multiple isozymes of heparan sulfate/heparin GlcNAc N-deacetylase/GlcN N-sulfotransferase. Structure and activity of the fourth member, NDST4. *J. Biol. Chem.* **276,** 5876–5882.

Akama, T. O., Nakayama, J., Nishida, K., Hiraoka, N., Suzuki, M., McAuliffe, J., Hindsgaul, O., Fukuda, M., and Fukuda, M. N. (2001). Human corneal GlcNAc 6-O-sulfotransferase and mouse intestinal GlcNAc 6-O-sulfotransferase both produce keratan sulfate. *J. Biol. Chem.* **276,** 16271–16278.

Akama, T. O., Nakayama, J., Nishida, K., Watanabe, H., Ozaki, K, Nakamura, T., Dota, A., Kawasaki, S., Inoue, Y., Maeda, N., Yamamoto, S., Fujiwara, T., Thonar, E. J., Shimomura, Y., Kinoshita, S., Tanigami, A., and Fukuda, M. N. (2000). Macular corneal dystrophy type I and type II are caused by distinct mutations in a new sulphotransferase gene. *Nat. Genet.* **26,** 237–241.

Ashikari-Hada, S., Habuchi, H., Itoh, N., Reddi, A. H., and Kimata, K. (2004). Characterization of growth factor-binding structures in heparin/heparan sulfate using an octasaccharide library. *J. Biol. Chem.* **279,** 12346–12354.

Chen, J., Duncan, M. B., Carrick, K., Pope, R. M., and Liu, J. (2003). Biosynthesis of 3-O-sulfated heparan sulfate: Unique substrate specificity of heparan sulfate 3-O-sulfotransferase isoform 5. *Glycobiology* **13,** 785–794.

DeAngelis, P. L., and Padgett-McCue, A. J. (2000). Identification and molecular cloning of a chondroitin synthase from *Pasteurella multocida* type F. *J. Biol. Chem.* 275, **24,** 124–129.

Eriksson, I., Sandback, D., Ek, B., Lindahl, U., and Kjellen, L. (1994). cDNA cloning and sequencing of mouse mastocytoma glucosaminyl N-deacetylase/N-sulfotransferase, an enzyme involved in the biosynthesis of heparin. *J. Biol. Chem.* **269,** 10438–10443.

Fukuta, M., Inazawa, J., Torii, T., Tsuzuki, K., Shimada, E., and Habuchi, O. (1997). Molecular cloning and characterization of human keratan sulfate Gal-6-sulfotransferase. *J. Biol. Chem.* **272,** 32321–32328.

Fukuta, M., Uchimura, K., Nakashima, K., Kato, M., Kimata, K., Shinomura, T., and Habuchi, O. (1995). Molecular cloning and expression of chick chondrocyte chondroitin 6-sulfotransferase. *J. Biol. Chem.* **270,** 18575–18580.

Habuchi, H., Habuchi, O., and Kimata, K. (1995). Purification and characterization of heparan sulfate 6-sulfotransferase from the culture medium of chinese hamster ovary cells. *J. Biol. Chem.* **270,** 4172–4179.

Habuchi, H., Kobayashi, M., and Kimata, K. (1998). Molecular characterization and expression of heparan-sulfate 6-sulfotransferase. Complete cDNA cloning in human and partial cloning in Chinese hamster ovary cells. *J. Biol. Chem.* **273,** 9208–9213.

Habuchi, O., Matsui, Y., Kotoya, Y., Aoyama, Y., Yasuda, Y., and Noda, M. (1993). Purification of chondroitin 6-sulfotransferase secreted from cultured chick embryo chondrocytes. *J. Biol. Chem.* **268,** 21968–21974.

Habuchi, H., Miyake, G., Nogami, K., Kuroiwa, A., Matsuda, Y., Kusche-Gullberg, M., Habuchi, O., Tanaka, M., and Kimata, K. (2003). Biosynthesis of heparan sulphate with diverse structures and functions: Two alternatively spliced forms of human heparan sulphate 6-O-sulphotransferase-2 having different expression patterns and properties. *Biochem. J.* **371,** 131–142.

Habuchi, H., Tanaka, M., Habuchi, O., Yoshida, K., Suzuki, H., Ban, K., and Kimata, K. (2000). The occurrence of three isoforms of heparan sulfate 6-$O$-sulfotransferase having different specificities for hexuronic acid adjacent to the targeted $N$-sulfoglucosamine. *J. Biol. Chem.* **275**, 2859–2868.

Hiraoka, N., Nakagawa, H., Ong, E., Akama, T. O., Fukuda, M. N., and Fukuda, M. (2000). Molecular cloning and expression of two distinct human chondroitin 4-$O$-sulfotransferases that belong to the HNK-1 sulfotransferase gene family. *J. Biol. Chem.* **275**, 20188–20196.

Holmborn, K., Ledin, J., Smeds, E., Eriksson, I., Kusche-Gullberg, M., and Kjellen, L. (2004). Heparan sulfate synthesized by mouse embryonic stem cells deficient in NDST1 and NDST2 is 6-$O$-sulfated but contains no $N$-sulfate groups. *J. Biol. Chem.* **279**, 42355–42358.

Ito, Y., and Habuchi, O. (2000). Purification and characterization of N-acetylgalactosamine 4-sulfate 6-$O$-sulfotransferase from the squid cartilage. *J. Biol. Chem.* **275**, 34728–34736.

Jemth, P., Smeds, E., Do, A. T., Habuchi, H., Kimata, K., Lindahl, U, and Kusche-Gullberg, M. (2003). Oligosaccharide library-based assessment of heparan sulfate 6-$O$-sulfotransferase substrate specificity. *J. Biol. Chem.* **278**, 24371–24376.

Kadomatsu, K., Zhang, H., Uchimura, K., and Muramatsu, T. (2004). $N$-acetylglucosamine-6-$O$-sulfotransferase-1 deficiency causes loss of keratan sulfate in the developing brain and injured brain. *Glycobiology,* **14**, 1065 (Abstract).

Kitagawa, H., Uyama, T., and Sugahara, K. (2001). Molecular cloning and expression of a human chondroitin synthase. *J. Biol. Chem.* **276**, 38721–38726.

Kobayashi, M., Habuchi, H., Habuchi, O., Saito, M., and Kimata, K. (1996). Purification and characterization of heparan sulfate 2-sulfotransferase from cultured chinese hamster ovary cells. *J. Biol. Chem.* **271**, 7645–7653.

Kobayashi, M., Habuchi, H., Yoneda, M., Habuchi, O., and Kimata, K. (1997). Molecular cloning and expression of Chinese hamster ovary cell heparan-sulfate 2-sulfotransferase. *J. Biol. Chem.* **272**, 13980–13985.

Kobayashi, M., Sugumaran, G., Liu, J., Shworak, N. W., Silbert, J. E., and Rosenberg, R. D. (1999). Molecular cloning and characterization of a human uronyl 2-sulfotransferase that sulfates iduronyl and glucuronyl residues in dermatan/chondroitin sulfate. *J. Biol. Chem.* **274**, 10474–10480.

Kusche-Gullberg, M., Eriksson, I., Pikas, D. S., and Kjellen, L. (1998). Identification and expression in mouse of two heparan sulfate glucosaminyl $N$-deacetylase/$N$-sulfotransferase genes. *J. Biol. Chem.* **273**, 11902–11907.

Lee, J. K., Bhakta, S., Rosen, S. D., and Hemmerich, S. (2000). Cloning and characterization of a mammalian $N$-acetylglucosamine-6-sulfotransferase that is highly restricted to intestinal tissue. *Biochem. Biophys. Res. Commun.* **263**, 543–549.

Li, J., Hagner-McWhirter, A., Kjellen, L., Palgi, J., Jalkanen, M., and Lindahl, U. (1997). Biosynthesis of heparin/heparan sulfate. cDNA cloning and expression of D-glucuronyl C5-epimerase from bovine lung. *J. Biol. Chem.* **272**, 28158–28163.

Lind, T., Tufaro, F., McCormick, C., Lindahl, U., and Lidholt, K. (1998). The putative tumor suppressors EXT1 and EXT2 are glycosyltransferases required for the biosynthesis of heparan sulfate. *J. Biol. Chem.* **273**, 26265–26268.

Liu, J., Shriver, Z., Blaiklock, P., Yoshida, K., Sasisekharan, R., and Rosenberg, R. D. (1999a). Heparan sulfate D-glucosaminyl 3-$O$-sulfotransferase-3A sulfates $N$-unsubstituted glucosamine residues. *J. Biol. Chem.* **274**, 38155–38162.

Liu, J., Shworak, N. W., Sinay, P., Schwartz, J. J., Zhang, L., Fritze, L. M., and Rosenberg, R. D. (1999b). Expression of heparan sulfate D-glucosaminyl 3-$O$-sulfotransferase isoforms reveals novel substrate specificities. *J. Biol. Chem.* **274**, 5185–5192.

McCormick, C., Duncan, G., Goutsos, K. T., and Tufaro, F. (2000). The putative tumor suppressors EXT1 and EXT2 form a stable complex that accumulates in the Golgi apparatus and catalyzes the synthesis of heparan sulfate. *Proc. Natl. Acad. Sci. USA* **97**, 668–673.

McCormick, C., Leduc, Y., Martindale, D., Mattison, K., Esford, L. E., Dyer, A. P., *et al.* (1998). The putative tumour suppressor EXT1 alters the expression of cell-surface heparan sulfate. *Nat. Genet.* **19**, 158–161.

Mikami, T., Mizumoto, S., Kago, N., Kitagawa, H., and Sugahara, K. (2003). Specificities of three distinct human chondroitin/dermatan N-acetylgalactosamine 4-*O*-sulfotransferases demonstrated using partially desulfated dermatan sulfate as an acceptor: Implication of differential roles in dermatan sulfate biosynthesis. *J. Biol. Chem.* **278**, 36115–36127.

Mochizuki, H., Yoshida, K., Gotoh, M., Sugioka, S., Kikuchi, N., Kwon, Y. D., Tawada, A., Maeyama, K., Inaba, N., Hiruma, T., Kimata, K., and Narimatsu, H. (2003). Characterization of a heparan sulfate 3-*O*-sulfotransferase-5, an enzyme synthesizing a tetrasulfated disaccharide. *J. Biol. Chem.* **278**, 26780–26787.

Nogami, K., Suzuki, H., Habuchi, H., Ishiguro, N., Iwata, H., and Kimata, K. (2004). Distinctive expression patterns of heparan sulfate *O*-sulfotransferases and regional differences in heparan sulfate structure in chick limb buds. *J. Biol. Chem.* **279**, 8219–8229.

Ohtake, S., Ito, Y., Fukuta, M., and Habuchi, O. (2001). Human N-acetylgalactosamine 4-sulfate 6-*O*-sulfotransferase cDNA is related to human B cell recombination activating gene-associated gene. *J. Biol. Chem.* **276**, 43894–43900.

Ohtake, S., Kimata, K., and Habuchi, O. (2003). A unique nonreducing terminal modification of chondroitin sulfate by N-acetylgalactosamine 4-sulfate 6-*O*-sulfotransferase. *J. Biol. Chem.* **278**, 38443–38452.

Rong, J., Habuchi, H., Kimata, K., Lindahl, U., and Kusche-Gullberg, M. (2000). Expression of heparan sulphate L-iduronyl 2-*O*-sulphotransferase in human kidney 293 cells results in increased D-glucuronyl 2-*O*-sulphation. *Biochem. J.* **346**, 463–468.

Seko, A., Dohmae, N., Takio, K., and Yamashita, K. (2003). β1,4-galactosyltransferase (β4GalT)-IV is specific for GlcNAc 6-*O*-sulfate. β4GalT-IV acts on keratan sulfate-related glycans and a precursor glycan of 6-sulfosialyl-Lewis X. *J. Biol. Chem.* **278**, 9150–9158.

Seko, A., and Yamashita, K. (2004). β1,3-N-acetylglucosaminyltransferase-7 (β3Gn-T7) acts efficiently on keratan sulfate-related glycans. *FEBS Lett.* **556**, 216–220.

Shively, J. E., and Conrad, H. E. (1976). Formation of anhydrosugars in the chemical depolymerization of heparin. *Biochemistry* **15**, 3932–3942.

Shworak, N. W., Liu, J., Fritze, L. M., Schwartz, J. J., Zhang, L., Logeart, D., and Rosenberg, R. D. (1997). Molecular cloning and expression of mouse and human cDNAs encoding heparan sulfate D-glucosaminyl 3-*O*-sulfotransferase. *J. Biol. Chem.* **272**, 28008–28019.

Smeds, E., Habuchi, H., Do, A. T., Hjertson, E., Grundberg, H., Kimata, K., Lindahl, U., and Kusche-Gullberg, M. (2003). Substrate specificities of mouse heparan sulphate glucosaminyl 6-*O*-sulphotransferases. *Biochem. J.* **372**, 371–380.

Uchimura, K., El-Fasakhany, F. M., Hori, M., Hemmerich, S., Blink, S. E., Kansas, G. S., Kanamori, A., Kumamoto, K., Kannagi, R., and Muramatsu, T. (2002). Specificities of N-acetylglucosamine-6-*O*-sulfotransferases in relation to L-selectin ligand synthesis and tumor-associated enzyme expression. *J. Biol. Chem.* **277**, 3979–3984.

Uchimura, K., Fasakhany, F., Kadomatsu, K., Matsukawa, T., Yamakawa, T., Kurosawa, N., and Muramatsu, T. (2000). Diversity of N-acetylglucosamine-6-*O*-sulfotransferases: Molecular cloning of a novel enzyme with different distribution and specificities. *Biochem. Biophys. Res. Commun.* **274**, 291–296.

Uchimura, K., Kadomatsu, K., El-Fasakhany, F. M., Singer, M. S., Izawa, M., Kannagi, R., Takeda, N., Rosen, S. D., and Muramatsu, T. (2004). N-acetylglucosamine 6-*O*-sulfotransferase-1 regulates expression of L-selectin ligands and lymphocyte homing. *J. Biol. Chem,* **279,** 35001–35008.

Uchimura, K., Muramatsu, H., Kadomatsu, K., Fan, Q. W., Kurosawa, N., Mitsuoka, C., Kannagi, R., Habuchi, O., and Muramatsu, T. (1998). Molecular cloning and characterization of an *N*-acetylglucosamine-6-*O*-sulfotransferase. *J. Biol. Chem.* **273,** 22577–22583.

Yada, T., Gotoh, M., Sato, T., Shionyu, M., Go, M., Kaseyama, H., Iwasaki, H., Kikuchi, N., Kwon, Y. D., Togayachi, A., Kudo, T., Watanabe, H., Narimatsu, H., and Kimata, K. (2003a). Chondroitin sulfate synthase-2. Molecular cloning and characterization of a novel human glycosyltransferase homologous to chondroitin sulfate glucuronyltransferase, which has dual enzymatic activities. *J. Biol. Chem.* **278,** 30235–30247.

Yada, T., Sato, T., Kaseyama, H., Gotoh, M., Iwasaki, H., Kikuchi, N., Kwon, Y. D., Togayachi, A., Kudo, T., Watanabe, H., Narimatsu, H., and Kimata, K. (2003b). Chondroitin sulfate synthase-3. Molecular cloning and characterization. *J. Biol. Chem.* **278,** 39711–39725.

Yamada, T., Ohtake, S., Sato, M., and Habuchi, O. (2004). Chondroitin 4-sulphotransferase-1 and chondroitin 6-sulphotransferase-1 are affected differently by uronic acid residues neighboring the acceptor GalNAc residues. *Biochem. J.* **384,** 567–575.

Yamauchi, S., Hirahara, Y., Usui, H., Takeda, Y., Hoshino, M., Fukuta, M., Kimura, J. H., and Habuchi, O. (1999). Purification and characterization of chondroitin 4-sulfotransferase from the culture medium of a rat chondrosarcoma cell line. *J. Biol. Chem.* **274,** 2456–2463.

Yamauchi, S., Mita, S., Matsubara, T., Fukuta, M., Habuchi, H., Kimata, K., and Habuchi, O. (2000). Molecular cloning and expression of chondroitin 4-sulfotransferase. *J. Biol. Chem.* **275,** 8975–8981.

# [15]  Measuring the Activities of the Sulfs: Two Novel Heparin/Heparan Sulfate Endosulfatases

*By* Kenji Uchimura, Megumi Morimoto-Tomita, and Steven D. Rosen

## Abstract

Sulfatases hydrolyze sulfate esters on a variety of molecules including glycosaminoglycans, sulfoglycolipids, and cytosolic steroids. These enzymes are found in a wide range of organisms with their basic enzymatic mechanisms broadly conserved. In mammals, many of the sulfatases localize in the lysosome and exhibit enzymatic activity on a small aryl substrate such as 4-methylumbelliferyl sulfate (4-MUS). They are known as *arylsulfatases.* Sulf-1 and Sulf-2 have been cloned and identified as sulfatases that release sulfate groups on the C-6 position of GlcNAc residue from an internal subdomain in intact heparin. Hence, these enzymes are endosulfatases. The Sulfs are secreted in an active form into conditioned medium of transfected Chinese hamster ovary (CHO) cells. In this chapter, arylsulfatase

METHODS IN ENZYMOLOGY, VOL. 416                                    0076-6879/06 $35.00
                      DOI: 10.1016/S0076-6879(06)16015-2

and endoglucosamine-6-sulfatase assays for the Sulfs are described. A solid-phase binding assay is also detailed, which allows investigation of the ability of the Sulfs to modulate the interaction of heparin-binding proteins with immobilized heparin. The example illustrated is vascular endothelial growth factor (VEGF). This assay is projected to be very useful in the investigation of the biological functions of the Sulfs.

Overview

Sulfatases catalyze the hydrolytic removal of sulfate esters from a diverse range of substrates. Sulfates on carbohydrates such as glycosaminoglycans, sulfoglycolipids, as well as that on steroid are hydrolyzed by sulfatases (Hanson et al., 2004). Sulfatase genes have been found in eukaryotes and prokaryotes (Hanson et al., 2004). In humans, some of them have been identified as causative genes for inherited diseases (Parenti et al., 1997). Mechanistic and structural features are highly conserved among the sulfatase members. A unique posttranslational modification generating an aldehyde group in the catalytic site (i.e., $\alpha$-formylglycine [FGly]) is shared by sulfatases (Cosma et al., 2003; Dierks et al., 2003; von Figura et al., 1998). Many of the mammalian sulfatases, with the notable exception of steroid sulfatase and a few others, are localized in lysosomes where they desulfate their substrates in an acidic environment. Many members of the sulfatase family are able to cleave a sulfate ester from a small aryl substance such as 4-MUS and are referred to as arylsulfatases (Parenti et al., 1997).

A highly novel member of the sulfatase family was identified in quail embryos (Dhoot et al., 2001) and named QSulf-1. Its rodent (Morimoto-Tomita et al., 2002; Ohto et al., 2002) and human (Morimoto-Tomita et al., 2002) orthologs and a closely related protein designated Sulf-2 have also been identified (Morimoto-Tomita et al., 2002). In contrast to previously known sulfatases, both HSulf-1 and HSulf-2, when expressed in CHO cells, are released into conditioned medium in active forms (Morimoto-Tomita et al., 2002). They exhibit activity in the 4-MUS assay with a neutral pH optimum (Morimoto-Tomita et al., 2002). Intriguingly, the HSulfs are active on intact heparin and 4-MUS. The Sulfs liberate sulfate on the C-6 position of glucosamine residue from internal trisulfated disaccharide units (i.e., IdoA[2S]-GlcNS[6S]) in heparin (Morimoto-Tomita et al., 2002). A similar activity of QSulf-1 has been demonstrated on heparan-sulfate chains (Ai et al., 2003; Viviano et al., 2004). Thus, the Sulfs are extracellular endosulfatases.

It is well established that various protein ligands such as growth factors, chemokines, and morphogens have the ability to bind to heparin/heparan sulfate (HS) proteoglycans on cell surfaces and in the extracellular matrix

(Esko and Selleck, 2002; Gallagher, 2001; Nakato and Kimata, 2002; Ornitz, 2000). To determine whether the Sulfs can modulate the interactions of these proteins with heparin/HS, we have developed an enzyme-linked immunosorbent assay (ELISA) that employs immobilized heparin and various protein ligands (Uchimura et al., 2006). We have found that binding of certain growth factors and chemokines to heparin is abolished or greatly reduced by pretreatment of the heparin with HSulf-2 (Uchimura et al., 2006).

HSulf-1 and HSulf-2 transcripts are up-regulated in human mammary carcinoma (Morimoto-Tomita et al., 2003, 2005). We further found that MCF-7 cell, a human breast cancer cell line, secretes an active form of HSulf-2 into conditioned medium (Morimoto-Tomita et al., 2005; Uchimura et al., 2006). Assessing whether HSulf-expressing cancer cells possess the enzymatic activities of Sulfs will be an important issue in investigating the potential pathogenic roles of these enzymes. Here, we describe assays for determining arylsulfatase and endoglucosamine-6-sulfatase activities in conditioned media (CM) of Sulf-expressing cells. We also provide an ELISA to determine the effect of HSulf-2 on the ability of immobilized heparin to support the binding of a heparin-binding factor VEGF.

## Preparation of HSulf-1 and HSulf-2

### Materials

F-12 Ham's medium (Cell Culture Facility at University of California, San Francisco, [CCF at UCSF]) supplemented 10% fetal bovine serum (FBS) (HyClone, Logan, UT) for CHO cells.

RPMI-1640 (CCF at UCSF) containing 10% FBS for MCF-7 cells.

OptiMEM I reduced serum medium (Invitrogen, Carlsbad, CA).

Phosphate-buffered saline (PBS) supplemented with 0.04% ethylenediaminetetraacetic acid (EDTA) (CCF at UCSF).

Solution of 0.05% trypsin and 0.02% versene in saline (CCF at UCSF).

Lipofectamine transfection reagent and PLUS reagent from Invitrogen.

Centriplus YM-30 and Centricon YM-30 Centrifugal Filter Units from Millipore (Billerica, MA).

Dialysis solution: 50 mM HEPES, pH 8.0.

### Method

1. CHO cells are cultured on 75 cm$^2$ flasks and passaged with PBS 0.04% EDTA until they reach 80% confluence. The cells are then

transiently transfected with a Myc-His–tagged HSulf-1 (pcDNA3.1/
Myc-His HSulf-1) or HSulf-2 (pcDNA3.1/Myc-His HSulf-2) expression
plasmid (Morimoto-Tomita et al., 2002) using Lipofectamine and PLUS
reagents according to the manufacturer's instructions. An empty vector is
used as a control ("mock" control). After 4-h incubation, the cells are
rinsed with warm PBS and OptiMEM and then incubated in 10 ml of
OptiMEM for 72 h.

2. A native form of HSulf-2 can be obtained from MCF-7 cells
(Uchimura et al., 2006). MCF-7 cells are maintained on 150 cm$^2$ flasks.
Confluent cells are passaged with a solution of trypsin/versene. Cells at
80% confluence in a 150-cm$^2$ flask are rinsed as above and incubated in
25 ml of OptiMEM for 72 h.

3. CM from transfected CHO cells or MCF-7 cells is collected in 50-ml
tubes and chilled on ice. Debris are removed by spinning the tubes 2500 rpm
at 4° for 10 min. The supernatant is transferred onto Centriplus YM-30 and
concentrated until its volume reaches 2–3 ml. The CM is further concentrated
with a Centricon YM-30 exchanging the buffer with 50 mM HEPES
pH 8.0. Aliquots (100 μl) of 100-fold concentrated CM are stored at −20°.

### Arylsulfatase Assay for HSulf-1 and HSulf-2 in CM

*Materials*

> 4-MUS (Sigma, St. Louis, MO) is dissolved at 50 mM in water. Note
> that a substrate solution (50 mM 4-MUS) is freshly dissolved as
> required in each assay.
> Lead acetate (Sigma) is dissolved at 400 mM in water and stored at
> room temperature.
> Reaction termination solution: 0.5 M Na$_2$CO$_3$/NaHCO$_3$, pH 10.7.
> A multiwell plate reader (CytoFluor II, Perseptive Biosystems,
> Framingham, MA).

*Method*

1. Concentrated and dialyzed CM derived from each CHO cell
transfectant (Fig. 1) or MCF-7 cell (Morimoto-Tomita et al., 2005) are
used as the enzyme sources. Note that the Myc-His–tagged proteins in CM
from CHO cell transfectants can be used as the source of enzyme as well.
The proteins can be bound to Ni-NTA resin (Qiagen) by rotating in 4°
overnight. A bead-bound recombinant protein or the material is eluted
from the resin with 250 mM imidazole/50 mM HEPES, pH 8.0, after
washing the resin with 50 mM HEPES, pH 8.0, three times.

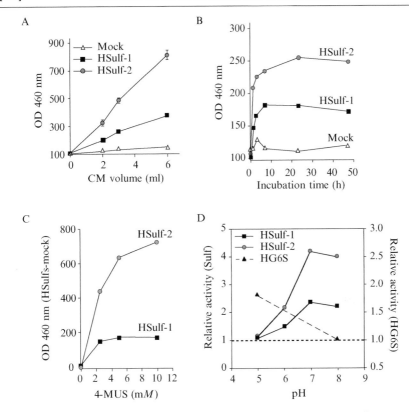

FIG. 1. Arylsulfatase activity in conditioned media (CM) derived from cells expressed HSulf-1 and HSulf-2. (A) The eluted material from the Ni-NTA resin was assayed for arylsulfatase activity at pH 8.0 for 2 h as a function of input volume of CM from pcDNA3.1/Myc-His HSulf-1–, pcDNA3.1/Myc-His HSulf-2–, or vector control (Mock)–transfected Chinese hamster ovary (CHO) cells (Morimoto-Tomita *et al.*, 2002). (B) The resin-bound material was examined for arylsulfatase activity as a function of time against 10 m$M$ 4-MUS substrate at pH 8.0. No activity was detected in the absence of added substrate (data not shown). (C) The concentrated CM was assayed for arylsulfatase activity at pH 8.0 for 2 h at different concentrations (1–10 m$M$) of 4-MUS. The activity in vector control material was subtracted from that of HSulf-transfected material. (D) Resin-bound HSulfs were tested for arylsulfatase activity at the indicated pH values for 1 h. The activity of each HSulf was measured relative to that of resins exposed to an equivalent volume of CM from vector control–transfected cells. The activity of human glucosamine-6-sulfatase (HG6S) (Kresse *et al.*, 1980) was determined (24 h incubation) relative to that of the buffer. Figure taken from Morimoto-Tomita *et al.*, 2002 with permission.

2. Two microliters of 100-fold concentrated CM are mixed with 10 m$M$ 4-MUS, 10 m$M$ lead acetate, and 50 m$M$ HEPES, pH 8.0 in a total volume of 30 $\mu$l. The reaction mixture is incubated at 37° for 2 h. Note that the enzymatic activity linearly increases up to 2 h incubation time when

recombinant material from CHO cell transfection is used (Fig. 1B) (Morimoto-Tomita et al., 2002). CM derived from mock transfectant of CHO cells is used to determine background level of arylsulfatase activity.

3. To terminate the reaction, an aliquot of 20 $\mu l$ of the reaction mix is transferred to a well of a 96-well plate containing 100 $\mu l$ of 0.5 $M$ $Na_2CO_3/NaHCO_3$.

4. Subsequently, the fluorescence of 4-MUF in the mixture is measured on a multiwell plate reader with an excitation wavelength of 360 nm and an emission wavelength of 460 nm. The OD 460 nm reports the arylsulfatase activity (Fig. 1).

## Endoglucosamine-6-Sulfatase Assay for HSulf-1 and HSulf-2

*Materials*

   Porcine intestinal heparin (Sigma) is dissolved at 10 $\mu g/\mu l$ in water and
      stored at $-20°$.
   Solutions, 0.5 $M$ HEPES, pH 7.5, and 0.5 $M$ $MgCl_2$, are stored at room
      temperature.
   Heparinase I (2 mU/$\mu l$) (EC 4.2.2.7), heparinase II (0.5 mU/$\mu l$), and
      heparinase III (0.2 mU/$\mu l$) (EC 4.2.2.8) from Sigma are stored in
      small aliquots at $-20°$.
   An Ultrafree-MC centrifugal filter unit (0.45 $\mu m$ durapore, Millipore).
   HPLC equipped with ultraviolet (UV) detector (Varian, Palo Alto, CA).
   A Partisil-10 SAX column (Whatman, Fairfield, NJ). Running buffer:
      0–600 m$M$ $KH_2PO_4$ in HPLC-grade water. The solution needs to be
      filtered and is prepared as required.
   Unsaturated disaccharide markers are from Sigma (i.e., ΔDiHS-0S,
      ΔDiHS-6S, ΔDiHS-NS, ΔDiHS-[N,6]diS, ΔDiHS-[N,2]diS, ΔDiHS-
      [N,2,6]triS).

*Method*

1. Endoglucosamine-6-sulfatase activity of recombinant HSulf-1 and HSulf-2, and CM of MCF-7 cells is determined using heparin as a substrate. Twenty $\mu l$ of 100-fold concentrated CM is mixed with 50 m$M$ HEPES, pH 7.8, 1 m$M$ $MgCl_2$ and 5 $\mu g$ of porcine intestinal heparin (Sigma) in a total volume of 100 $\mu l$. The mixture is incubated at 37°.

2. Immersing the reaction tubes in boiling water for 5 min stops the reaction.

3. The heparin is digested into disaccharides by incubating with a mix of 0.4 mU heparinase I, 0.1 mU heparinase II, and 0.04 mU heparinase III at 37° for 3 h overnight. The reaction is terminated by placing the tubes in boiling water for 5 min.

4. The mixture is filtered by centrifugation at 12,000$g$ for 2 min in an Ultrafree-MC filter. Filtered samples are kept on ice or at $-20°$ till injecting into a column.

5. The disaccharides of the digested heparin are then analyzed by HPLC on a Partisil-10 strong anion exchange column run at 41°. Disaccharides are eluted from the column by increasing ionic strength on the following schedule: 0–5 min, 12 m$M$ KH$_2$PO$_4$; 5–40 min, gradient from 12 to 600 mM; 40–45 min, 600 m$M$ (Fig. 2A).

6. Absorbance at 232 nm is monitored and components are identified by comparison with authentic unsaturated disaccharide markers. The enzymatic activity is defined by calculating the increase in the newly formed ΔDiHS-(N,2)diS in the digested heparin (Fig. 2A). The enzymatic activity proceeds linearly up to 8-h incubation when 20 $\mu$l of CM from MCF-7 cells is used (Fig. 1B). Note that 15 mU of chondroitinase ABC is employed to fragment the chondroitin sulfate into disaccharides after the incubation with CM in another assay using 5 $\mu$g of chondroitin 6-sulfate as a substrate in the standard reaction mix. Authentic markers from Seikagaku, Inc., are used (i.e., ΔDi-0S, ΔDi-6S, ΔDi-4S, ΔDi-diS$_D$, and ΔDi-diS$_E$) for a Partisil-10 HPLC.

## ELISA for Modulation of VEGF$_{165}$ Binding to Heparin by HSulf-2 in MCF-7 CM

### Materials

Albumin from bovine serum (BSA, further purified, pH 7, suitable for diluent in ELISA applications, Sigma). Heparin-BSA (heparin-BSA) (Sigma) is dissolved at 1 mg/ml in PBS+ (PBS with Ca$^{2+}$ and Mg$^{2+}$), 0.01% NaN$_3$, and stored at 4°. Blocking reagent: 3% BSA, 0.01% NaN$_3$ in PBS+ filtered by 0.45 $\mu$m filter, and stored at 4°. Wash buffer: 0.1% Tween-20 in PBS+.

A 96-well polystyrene ELISA plate (Immulon 2HB, DYNEX Technologies, Chantilly, VA).

The 165 amino acid form of human recombinant VEGF (VEGF$_{165}$) from R&D systems is dissolved at 0.02 mg/ml in 0.1% BSA/PBS and stored in small aliquots at $-20°$. The goat anti-human VEGF antibody (0.1 mg/ml) (R&D systems), biotinylated swine anti-goat IgG (H+L) (0.7 mg/ml) (Caltag, Burlingame, CA), and streptavidin conjugated with alkali phosphatase (1 mg/ml) (Caltag) are stored at 4°.

ImmunoPure PNPP tablets are from Pierce (Rockford, IL) and dissolved in an ice-cold solution of 10% diethanolamine, 0.5 m$M$ MgCl$_2$, and 0.02% NaN$_3$, pH 9.8, which is stored at 4°. Tablets of PNPP are freshly dissolved in each assay and kept on ice until use.

A microplate reader (Model 3550) from Bio-Rad (Hercules, CA).

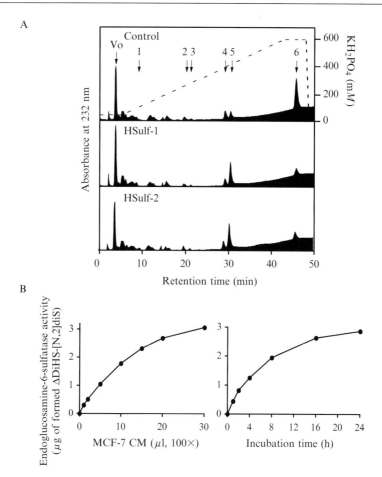

FIG. 2. Endoglucosamine-6-sulfatase activity in cultured media (CM) derived from cells expressed HSulf-1 and HSulf-2. (A) The CM (20 μl) of Chinese hamster ovary (CHO) cells transfected with the empty vector alone (Control), pcDNA3.1/Myc-His HSulf-1 (HSulf-1), or pcDNA3.1/Myc-His HSulf-2 (HSulf-2) was prepared. Heparin (5 μg) from the porcine intestine was incubated with CM and subsequently digested with a mix of bacterial heparinases. The resulting disaccharide fractions were analyzed by a Partisil-10 column on high-performance liquid chromatography (HPLC). The concentrations of $KH_2PO_4$ used for elution are indicated as dotted lines. The arrows denote the elution positions of unsaturated disaccharide markers: 1, ΔDiHS-0S; 2, ΔDiHS-6S; 3, ΔDiHS-NS; 4, ΔDiHS-(N,6)diS; 5, ΔDiHS-(N,2)diS; 6, ΔDiHS-(N,2,6)triS (Taken from Morimoto-Tomita *et al.*, with permission). (B) Endoglucosamine-6-sulfatase activity of MCF-7 CM on heparin as a function of CM volume (left, 4 h incubation) and reaction time (right, 10 μl CM) was determined. The activity was defined by calculating the amount of formed ΔDiHS-(N,2)diS in the standard assay (Taken from Uchimura *et al.*, 2006 with permission).

*Method*

1. The experimental design for the ELISA is schematically shown in Fig. 3A. To immobilize heparin, 100 ng/ml of heparin-BSA in PBS+ is added to the wells (100 μl/well) of a 96-well ELISA plate (Immulon 2HB). The plate is placed at 4° overnight.

2. The wells are washed three times with PBS+ containing 0.1% Tween-20 (PBS-T) and then blocked with the blocking reagent at room

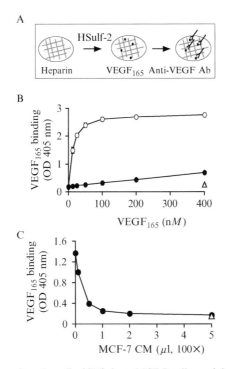

FIG. 3. HSulf-2 in cultured media (CM) from MCF-7 cells modulates the interaction of immobilized heparin with Vascular endothelial growth factor (VEGF)$_{165}$. (A) A solid-phase binding assay to measure an effect of HSulf-2 on the binding of VEGF$_{165}$ to heparin is schematically shown (Uchimura *et al.*, 2006). HSulf-2 was incubated with immobilized heparin before adding VEGF$_{165}$. A specific anti-VEGF antibody quantified the amount of VEGF$_{165}$ bound to immobilized heparin. (B) The binding of VEGF$_{165}$ to immobilized heparin–bovine serum albumin (BSA) as a function of concentration (0–400 n$M$) with or without MCF-7 CM (2 μl) treatment of heparin–BSA. The binding of VEGF$_{165}$ to untreated (○) or MCF-7 CM-treated (λ) heparin–BSA in the assay is shown. Δ denotes the binding to heparin–BSA treated with a mix of heparinases. (C) The effect of MCF-7 CM on VEGF$_{165}$ (25 n$M$) binding as a function of CM volume (0–5 μl) is shown (λ). Δ represents the binding of VEGF$_{165}$ to heparin–BSA treated with heparinases (Taken from Uchimura *et al.*, 2006 with permission).

temperature for 2 h. To determine the effects of HSulf-2 on the ligand-binding activities of heparin, the wells are washed as stated previously and then incubated with 100 $\mu$l of a reaction mixture containing 5 $\mu$mol of HEPES, pH 7.6, 1 $\mu$mol of $MgCl_2$, and enzyme (MCF-7 CM) at 37° overnight.

3. The wells are washed three times with PBS-T and then incubated with 25 $\mu$l of $VEGF_{165}$ (25 n$M$) in PBS at room temperature for 30 min. To obtain a dose–response curve, 0–200 n$M$ of $VEGF_{165}$ is added (Fig. 3B).

4. The wells are washed three times and incubated with 1 $\mu$g/ml of the goat anti-human VEGF antibody in 0.1% BSA/PBS+ (100 $\mu$l/well) at room temperature for 1 h.

5. The wells are washed as stated previously and incubated with 1.2 $\mu$g/ml of a biotinylated swine anti-goat IgG (H+L) in 0.1% BSA/PBS+ (100 $\mu$l/well) at room temperature for 30 min.

6. The wells are washed and incubated with 2 $\mu$g/ml of an alkali phosphatase-conjugated streptavidin in 0.1% BSA/PBS+ (100 $\mu$l/well) room temperature for 30 min. The wells are washed three times.

7. Finally, the wells are incubated with PNPP (Pierce) in diethanol-amine buffer, pH 9.8 (100 $\mu$l/well) at room temperature, for 5 min. $OD_{405}$ nm is read on a microplate reader (Fig. 3).

## Acknowledgments

Supported by grants from the Mizutani Foundation for Glycoscience and NIH (RO1-GM057411) to SDR and a UCSF Comprehensive Cancer Center Intramural Award from the Alexander and Margaret Stewart Trust to S. D. R. K. U. and M. M.-T. are Research Fellows of the Japan Society for the Promotion of Science.

## References

Ai, X., Do, A. T., Lozynska, O., Kusche-Gullberg, M., Lindahl, U., and Emerson, C. P., Jr. (2003). QSulf1 remodels the 6-$O$ sulfation states of cell surface heparan sulfate proteoglycans to promote Wnt signaling. *J. Cell. Biol.* **162,** 341–351.

Cosma, M. P., Pepe, S., Annunziata, I., Newbold, R. F., Grompe, M., Parenti, G., and Ballabio, A. (2003). The multiple sulfatase deficiency gene encodes an essential and limiting factor for the activity of sulfatases. *Cell* **113,** 445–456.

Dhoot, G. K., Gustafsson, M. K., Ai, X., Sun, W., Standiford, D. M., and Emerson, C. P., Jr. (2001). Regulation of Wnt signaling and embryo patterning by an extracellular sulfatase. *Science* **293,** 1663–1666.

Dierks, T., Schmidt, B., Borissenko, L. V., Peng, J., Preusser, A., Mariappan, M., and von Figura, K. (2003). Multiple sulfatase deficiency is caused by mutations in the gene encoding the human C(alpha)-formylglycine generating enzyme. *Cell* **113,** 435–444.

Esko, J. D., and Selleck, S. B. (2002). Order out of chaos: Assembly of ligand binding sites in heparan sulfate. *Annu. Rev. Biochem.* **71,** 435–471.

Gallagher, J. T. (2001). Heparan sulfate: Growth control with a restricted sequence menu. *J. Clin. Invest.* **108,** 357–361.

Hanson, S. R., Best, M. D., and Wong, C. H. (2004). Sulfatases: Structure, mechanism, biological activity, inhibition, and synthetic utility. *Angew. Chem. Int. Ed. Engl.* **43,** 5736–5763.

Kresse, H., Paschke, E., von Figura, K., Gilberg, W., and Fuchs, W. (1980). Sanfilippo disease type D: Deficiency of N-acetylglucosamine-6-sulfate sulfatase required for heparan sulfate degradation. *Proc. Natl. Acad. Sci. USA* **77,** 6822–6826.

Morimoto-Tomita, M., Uchimura, K., Bistrup, A., Lum, D. H., Egeblad, M., Boudreau, N., Werb, Z., and Rosen, S. D. (2005). Sulf-2, a proangiogenic heparan sulfate endosulfatase, is upregulated in breast cancer. *Neoplasia* **7,** 1001–1010.

Morimoto-Tomita, M., Uchimura, K., and Rosen, S. D. (2003). Novel extracellular sulfatase: Potential roles in cancer. *Trends Glycosci. Glycotechnol.* **15,** 159–164.

Morimoto-Tomita, M., Uchimura, K., Werb, Z., Hemmerich, S., and Rosen, S. D. (2002). Cloning and characterization of two extracellular heparin-degrading endosulfatases in mice and humans. *J. Biol. Chem.* **277,** 49175–49185.

Nakato, H., and Kimata, K. (2002). Heparan sulfate fine structure and specificity of proteoglycan functions. *Biochim. Biophys. Acta* **1573,** 312–318.

Ohto, T., Uchida, H., Yamazaki, H., Keino-Masu, K., Matsui, A., and Masu, M. (2002). Identification of a novel nonlysosomal sulphatase expressed in the floor plate, choroid plexus and cartilage. *Genes Cells* **7,** 173–185.

Ornitz, D. M. (2000). FGFs, heparan sulfate and FGFRs: Complex interactions essential for development. *Bioessays* **22,** 108–112.

Parenti, G., Meroni, G., and Ballabio, A. (1997). The sulfatase gene family. *Curr. Opin. Genet. Dev.* **7,** 386–391.

Uchimura, K., Morimoto-Tomita, M., Bistrup, A., Li, J., Lyon, M., Gallagher, J., Werb, Z., and Rosen, S. D. (2006). HSulf-2, an extracellular endoglucosamine-6-sulfatase, selectively mobilizes heparin-bound growth factors and chemokines: Effects on VEGF, FGF-1, and SDF-1. *BMC Biochem.* **7,** 2.

Viviano, B. L., Paine-Saunders, S., Gasiunas, N., Gallagher, J., and Saunders, S. (2004). Domain-specific modification of heparan sulfate by Qsulf1 modulates the binding of the bone morphogenetic protein antagonist Noggin. *J. Biol. Chem.* **279,** 5604–5611.

von Figura, K., Schmidt, B., Selmer, T., and Dierks, T. (1998). A novel protein modification generating an aldehyde group in sulfatases: Its role in catalysis and disease. *Bioessays* **20,** 5505–5510.

# [16]  Determination of Chemokine–Glycosaminoglycan Interaction Specificity

By Hiroto Kawashima

## Abstract

Chemokines are a large group of closely related cytokines that induce directed migration of leukocytes through interaction with a group of seven-transmembrane, G protein–coupled receptors. Biological properties of chemokines are influenced by their association with glycosaminoglycans (GAGs) on the cell surfaces and in the extracellular matrices. GAGs immobilize and enhance local concentration of chemokines, promoting their presentation to the receptors. This chapter describes a newly developed method to define oligosaccharide sequences in GAG chains that specifically interact with chemokines. A simple method to quantify the interaction between chemokines and GAGs based on surface plasmon resonance is also described. These methods will be useful in defining chemokine-GAG–binding specificities.

## Introduction

Chemokines are 8- to 10-kDa chemotactic proteins that mediate directional migration of leukocytes from the blood to the sites of inflammation. The ability of chemokines to bind to GAGs is crucial for their function. A soluble chemokine gradient is not stable, particularly under conditions of blood flow in the circulation, whereas binding of chemokines to GAGs, either at the cell surface of endothelial cells or in the extracellular matrix, serves to establish an immobilized chemokine gradient and "present" the molecules to leukocytes (Tanaka et al., 1993).

Proteoglycans are ubiquitous components of cell surface membranes, basement membranes, and extracellular matrices in various tissues. They belong to a family of macromolecules that consist of core proteins to which GAGs, sulfated polysaccharides, are attached. GAGs are linear polysaccharides made up of disaccharide units composed of hexosamine and hexuronic acid (or hexose). They are classified into chondroitin sulfate (CS), dermatan sulfate (DS), heparin, heparan sulfate (HS), keratan sulfate (KS), and hyaluronic acid (HA). Because of the high sulfate and carboxyl group content of their GAG moieties, proteoglycans have strong negative charges. This property allows them to interact with a wide range of proteins,

METHODS IN ENZYMOLOGY, VOL. 416                                    0076-6879/06 $35.00
DOI: 10.1016/S0076-6879(06)16016-4

including growth factors, enzymes, cytokines, lipoproteins, adhesion molecules, and chemokines (Bernfield *et al.*, 1999; Salmivirta *et al.*, 1996).

Our laboratory previously reported that oversulfated CS/DS chains containing GlcA$\beta$1/IdoA$\alpha$1–3GalNAc(4,6-*O*-disulfate) specifically interact with chemokines with moderate affinity (Kd 32.2 to 293 n*M*) and modulate their biological activities (Hirose *et al.*, 2001; Kawashima *et al.*, 2002). We also showed that a tetrasaccharide fragment composed of a tandem GlcA$\beta$1–3GalNAc(4,6-*O*-disulfate) unit directly interacts with chemokines (Kawashima *et al.*, 2002). These results indicated that a nonspecific electrostatic interaction is not the sole factor determining the interaction and that a specific carbohydrate structure is recognized by chemokines. In this chapter, a detailed protocol that has been used in these analyses is described. These methods will be of general use for the determination of chemokine-glycosaminoglycan–binding specificities.

## Materials

### Chemokines

1. Secondary lymphoid tissue chemokine (SLC) was purchased from DAKO Japan Co. Ltd. (Tokyo, Japan).
2. The C-terminally truncated form of secondary lymphoid tissue chemokine (SLC-T) was provided by Dr. Melissa Swope-Willis (Vertex Pharmaceutical Co., Cambridge, MA).

### GAGs

CS A (whale cartilage), CS C (shark cartilage), and CS E (squid cartilage) were purchased from Seikagaku Kogyo Co.

### Buffers

Buffer A: 0.05% Tween 20, 20 m*M* HEPES–NaOH, 0.15 *M* NaCl, 1 m*M* CaCl$_2$, 1 m*M* MgCl$_2$, pH 6.8

Buffer B: 20 m*M* HEPES–NaOH, 0.15 *M* NaCl, 1 m*M* CaCl$_2$, 1 m*M* MgCl$_2$, pH 6.8

## Methods

As an example, preparation (see the section "Preparation of Streptavidin-Conjugated Alkaline Phosphatase Coupled to Biotinylated Oligosaccharide Fragments from CS A, CS C, and CS E") and structural determination (see the section "Structural Determination of the Tetrasaccharide Fragments") of oligosaccharide fragments from CS A, CS C, and CS E, and an enzyme-linked immunosorbent assay (ELISA)–based binding assay (see the section "ELISA") are described here. Thereafter, a method to quantify the interaction between chemokines and glycosaminoglycans based on surface plasmon resonance is described (see the section "Kinetic Analysis Using BIAcore").

## Preparation of Streptavidin-Conjugated Alkaline Phosphatase Coupled to Biotinylated Oligosaccharide Fragments from CS A, CS C, and CS E

In the first step, oligosaccharide fragments are prepared from CS A, CS C, and CS E by digestion with sheep testicular hyaluronidase and fractionated by high-performance liquid chromatography (HPLC) on an amine-bound silica PA-03 column (Fig. 1A). The reducing terminus of each oligosaccharide is then labeled with biotin-LC hydrazide and coupled to streptavidin-alkaline phosphatase as described below.

1. CS E from squid cartilage (1 mg) is suspended in 450 $\mu$l of 50 m$M$ sodium acetate, 133 m$M$ NaCl, and 0.04% gelatin, pH 5.0, containing 0.6 mg (1800 units) sheep testicular hyaluronidase (Sigma, St. Louis, MO), and incubated at 37° for a total of 68.5 h.

2. An additional 2 mg (6000 units) of the enzyme is added at times 24 h and 45.5 h (see Note 1, later in this chapter).

3. CS A from whale cartilage and CS C from shark cartilage are suspended in 450 $\mu$l of 50 m$M$ sodium acetate, pH 5.0, containing 0.6 mg sheep testicular hyaluronidase, and incubated at 37° for 24 h.

4. A large portion of each digest (427.5 $\mu$l) is mixed with 400 $\mu$l $H_2O$, passed through a syringe filter (0.45-$\mu$m pore), and fractionated by HPLC on an amine-bound silica YMC-Pack PA-03 column (YMC, Inc., Wilmington, NC, see Note 2, later in this chapter) with a linear gradient from 16 m$M$ to 1 $M$ NaH$_2$PO$_4$ over 60 min. The eluates are monitored by absorbance at 210 nm.

5. Each fraction is applied to a Sephadex G-25 column (1 × 30 cm, Amersham-Pharmacia) equilibrated with distilled water, and the eluates are monitored by absorbance at 210 nm. The concentration of oligosaccharides in each fraction is calculated by comparing its absorbance with that of GlcNAc with a defined concentration. The fractions are divided into several tubes (each tube should contain 20–80 $\mu$g oligosaccharides) and evaporated to dryness.

6. To each tube containing 20–80 $\mu$g oligosaccharides, 125 m$M$ biotin-LC-hydrazide (Pierce, Rockford, IL) and 1 $M$ NaCNBH$_3$ in dimethylsulfoxide/acetic acid (7:3) is added, and the reaction mixture is incubated for 3 h at 65° and then for 12.5–18.5 h at 37°.

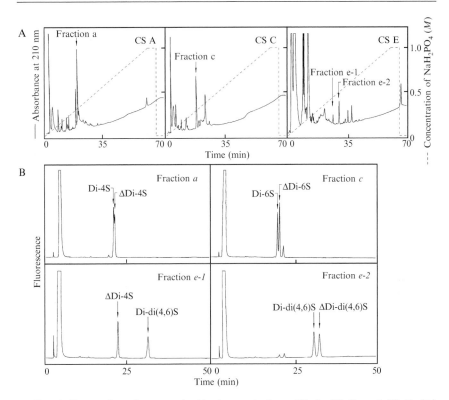

Fig. 1. Preparation of tetrasaccharide fragments from CS A, CS C, and CS E. (A) Fractionation of oligosaccharides obtained by sheep testicular hyaluronidase digestion of CS A, CS C, and CS E by high-performance liquid chromatography (HPLC) on an amine-bound silica PA-03 column using a linear salt gradient (dashed line). (B) Fractions *a*, *c*, *e-1*, and *e-2* were digested with chondroitinase ACII (0.3 U/ml) at 37° for 1 h, derivatized with 2-AB, and analyzed by HPLC on an amine-bound silica PA-03 column with a linear gradient elution from 16 to 606 m*M* NaH$_2$PO$_4$ over 45 min. The elution positions of 2-AB–derivatized standard disaccharides are shown (arrows).

7. To remove unreacted biotin-LC-hydrazide, each reaction mixture is applied to a Sephadex G-25 column as described in Step 5 and the fractions containing biotinylated oligosaccharides are collected and evaporated to dryness (see Note 3, later in this chapter).

8. A portion of the fraction containing 16 pmol of each biotinylated oligosaccharide is dissolved in 0.5 ml buffer A and incubated overnight with 1 μl (2 pmol) of streptavidin-alkaline phosphatase (Promega) at 4°.

9. The reaction mixture is diluted three times with buffer A and applied to the wells of 96-well flat-bottomed microtiter plates (Costar

EIA/RIA plate No. 3690, Corning, Inc., Corning, NY) coated with chemokines as described in the section "ELISA," Steps 1 and 2.

## Structural Determination of the Tetrasaccharide Fragments

The carbohydrate structure of each oligosaccharide is determined by enzyme treatment and 2-AB derivatization, followed by HPLC analyses as described below.

1. A portion of non-biotinylated tetrasaccharide fragments (see the section "Preparation of Streptavidin-Conjugated Alkaline Phosphatase Coupled to Biotinylated Oligosaccharide Fragments from CS A, CS C, and CS E," Step 5) is incubated with chondroitinase ACII (0.3 U/ml) in 3% acetic acid adjusted to pH 7.0 with triethylamine at 37° for 1 h.

2. After evaporation, the derivatization of the disaccharide products with 2-AB (2-aminobenzamide) is performed according to the method of Kinoshita and Sugahara (1999) (see Note 4, later in this chapter).

3. Oligosaccharides derivatized with 2-AB are purified by partition between chloroform and distilled water (1:1). This should be repeated five times.

4. The aqueous phase, containing the derivatized disaccharides, is analyzed by HPLC, as described in Step 4 in the section "Preparation of Streptavidin-Conjugated Alkaline Phosphatase Coupled to Biotinylated Oligosaccharide Fragments from CS A, CS C, and CS E," except that the elution is performed with a linear gradient from 16 to 606 m$M$ NaH$_2$PO$_4$ over 45 min and that the eluates are monitored using a fluorescence detector with excitation and emission wavelengths of 330 and 420 nm, respectively. The structure of each disaccharide fragment is determined by comparing its retention time with those of standard disaccharides (see Note 5, later in this chapter). An example of structural determination is shown in Fig. 1B (see Note 6, later in this chapter).

## Elisa

The binding of biotinylated oligosaccharides coupled with streptavidin-alkaline phosphatase to chemokines coated on an ELISA plate can be determined with high sensitivity as described below.

1. Wells of 96-well flat-bottomed microtiter plates (Costar EIA/RIA plate No. 3690, Corning, Inc., see Note 7, later in this chapter) are incubated overnight with 25 $\mu$l/well of chemokines (5 $\mu$g/ml) in phosphate-buffered saline (PBS) at 4°.

2. The wells are blocked with 100 μl/well of 3% bovine serum albumin (BSA) in PBS for 2 h.

3. The wells are incubated for 1 h with 25 μl/well of streptavidin-alkaline phosphatase coupled with biotinylated oligosaccharides prepared in Step 9 in the section "Preparation of Streptavidin-Conjugated Alkaline Phosphatase Coupled to Biotinylated Oligosaccharide Fragments from CS A, CS C, and CS E."

4. After washing the wells three times with buffer A, Blue Phos substrate (Kirkegaard & Perry Laboratories, Gaithersburg, MD) is added, and the plate is incubated for 30 min to 2 h. The optical density is read at 620 nm in a microtiter plate reader (InterMed Co., Tokyo, Japan). A typical result obtained by this method is shown in Fig. 2B (see Note 8, later in this chapter).

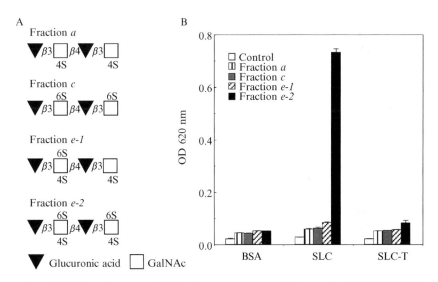

Fig. 2. Binding of a tetrasaccharide fragment composed of repeating GlcAβ1–3GalNAc (4,6-O-disulfate) units to secondary lymphoid tissue chemokine (SLC). (A) Structure of the oligosaccharides in fractions a, c, e-1, and e-2. Closed triangle, GlcA; open square, GalNAc; 4S, 4-O-sulfation; 6S, 6-O-sulfation; β3, β1–3 linkage; β4, β1–4 linkage. (B) Binding of streptavidin-conjugated alkaline phosphatase coupled to or without biotinylated fractions a, c, e-1, or e-2 to wells coated with 5 μg/ml bovine serum albumin (BSA), SLC, or the C-terminally truncated form of secondary lymphoid tissue chemokine (SLC-T). Binding was detected by enzyme-linked immunosorbent assay (ELISA) with the Blue Phos substrate. Each bar represents the mean ± standard deviation of quadruplicate determinations.

*Kinetic Analysis Using BIAcore*

Herein a method to determine the affinity kinetics of the interactions between GAGs and chemokines by surface plasmon resonance (SPR) is described.

1. This experiment is performed on a BIAcore biosensor (BIAcore AB, Uppsala, Sweden). All experiments are performed at 25°. The running buffer, which is used for the washing and dissociation phase, is buffer B.

2. About 1.8–2.0 kilo resonance units (kRU) (1 kRU = 1 ng/mm²) streptavidin is covalently immobilized on the B1 sensor chip via its primary amine groups, using an amine coupling kit (Amersham Biosciences) according to the instructions provided by the manufacturer.

3. The remaining activated groups are blocked with 150 μl of 1 *M* ethanolamine–HCl, pH 8.5.

4. Each GAG biotinylated at the reducing end with biotin-LC-hydrazide (Pierce) according to the method of Amara *et al.* (1999) (see Note 9, later in this chapter) is then injected over the sensor chip surface to obtain an immobilization level of about 150 RU (see Note 10, later in this chapter).

5. Different concentrations of chemokines are injected at 30 μl/min from time 0 to 90 s; after this, running buffer is injected and the response in resonance units is recorded as a function of time.

6. The sensor chip surface is regenerated with 300 μl of 1 *M* NaCl (see Note 11, later in this chapter).

7. Affinity kinetic parameters are determined with the BIAevaluation 3.0 software (BIAcore AB) using a single-site binding model according to the instructions provided by the manufacturer. Typical sensorgrams obtained by this method are shown in Fig. 3 (see Note 12, later in this chapter).

*Notes*

1. Compared to CS A and CS C, CS E is relatively resistant to the sheep testicular hyaluronidase. However, it can be fragmented under these conditions.

2. After repeated use of the YMC-Pack PA-03 column, the retention time of each oligosaccharide is shortened. Standard oligosaccharides should be applied to the column at the beginning and end of the analysis. Typically, a column can be used more than 100 times.

3. The efficiency of biotinylation should be examined at this point. If 100% of the oligosaccharides are biotinylated, non-reducing terminus-derived disaccharides but not reducing terminus-derived unsaturated

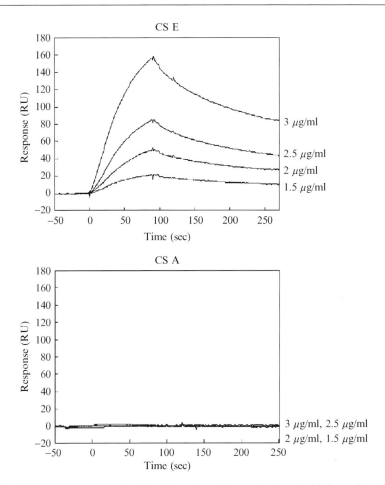

FIG. 3. Sensorgrams recording the interactions of secondary lymphoid tissue chemokine (SLC) with immobilized glycosaminoglycans (GAGs) in the BIAcore. Various concentrations of SLC were continuously injected over the sensor chip surface coupled to CS E or CS A at a flow rate of 30 μl/min from time 0 to 90 s; after this, running buffer was injected and the response in resonance units was recorded as a function of time.

disaccharides are detected after the chondroitinase ACII digestion of a portion of the biotinylated oligosaccharide fractions, followed by 2-AB derivatization (see the section "Structural Determination of the Tetrasaccharide Fragments," Steps 1 through 4, and Note 4) and HPLC analysis. Typically, about 100% biotinylation is attained under the condition described in the method.

4. Derivatization of disaccharides with 2-AB is performed as follows: (a) 2-AB (9.5 mg) is dissolved in 200 $\mu$l of DMSO:acetic acid (7:3, v/v). (b) 6.2 mg of NaCNBH$_3$ (Aldrich Chemicals Co.) is dissolved in 100 $\mu$l of the solution prepared in (a). (c) Five microliters of the solution prepared in (b) is added to the evaporated disaccharide sample and incubated at 65° for 2 h.

5. Standard oligosaccharides are prepared as follows: Each disaccharide in the unsaturated chondro-disaccharide kit (C kit; Seikagaku Kogyo Co.) is separately labeled with 2-AB as described in the section "Structural Determination of the Tetrasaccharide Fragments," Steps 2 and 3, and Note 4.

6. After fraction e-1 is treated with chondroitinase ACII and derivatized with 2-AB, Di-di(4,6)S and ΔDi-4S are detected at a molar ratio of approximately 1:1 on HPLC, indicating that fraction e-1 represents GlcA β1–3GalNAc (4, 6-O-disulfate)β1–4GlcAβ1–3GalNAc(4-O-sulfate) (Figs. 1B and 2A). Similarly, the structures of fraction a, c, and e-2 are determined to represent GlcAβ1–3GalNAc(4-O-sulfate)β1–4GlcAβ1–3GalNAc(4-O-sulfate), GlcAβ1–3GalNAc(6-O-sulfate)β1–4GlcAβ1–3GalNAc(6-O-sulfate), and GlcAβ1–3GalNAc(4, 6-O-disulfate)β1–4GlcAβ1–3GalNAc(4, 6-O-disulfate), respectively.

7. The surface area of a well of this 96-well flat-bottomed microtiter plate is half of that of conventional 96-well plates. The amount of the reagents needed for the assay can be reduced by using this plate.

8. In an example shown in Fig. 2B, only fraction e-2 binds to SLC. This fraction does not interact with SLC-T that lacks the C-terminal 32 amino acids containing basic amino acid clusters. These results indicate that the tetrasaccharide composed of a tandem GlcAβ1–3GalNAc(4,6-O-disulfate) unit specifically interacts with the C-terminal region of SLC.

9. Biotinylation of GAGs is performed as follows: Each GAG suspended in PBS at 1 m$M$ is incubated for 24 h at room temperature with 10 m$M$ biotin-LC-hydrazide. After the reaction, 1/10 volume of 1 $M$ Tris–HCl (pH 8.0) is added to the mixture and incubated for 1 h at room temperature to terminate the reaction. The mixture is then extensively dialyzed against water to remove unreacted biotin-LC-hydrazide.

10. The immobilization level should be about 150 RU. If too much GAGs are immobilized on a sensor chip, k$_{off}$ value cannot be appropriately determined because of the re-binding of the dissociated chemokines to the sensor chip surface.

11. No significant change in the baseline should be observed after surface regeneration. If it is, a new sensor chip should be prepared.

12. In an example shown in Fig. 3, SLC binds dose dependently to CS E immobilized on the sensor chip surface. In contrast, SLC does not

interact with CS A. Evaluation of the affinity kinetic parameters indicates that SLC interacts with CS E with moderate affinity (Kd 85.8 n$M$). The $k_{on}$ and $k_{off}$ values are $4.15 \times 10^4 \ M^{-1} \ s^{-1}$ and $3.56 \times 10^{-3} \ s^{-1}$, respectively.

## Acknowledgment

I would like to thank Professor Minoru Fukuda for providing me the opportunity to write this chapter.

## References

Amara, A., Lorthioir, O., Valenzuela, A., Magerus, A., Thelen, M., Montes, M., Virelizier, J.-L., Delepierre, M., Baleux, F., Lortat-Jacob, H., and Arenzana-Seisdedos, F. (1999). Stromal cell-derived factor-1$\alpha$ associates with heparan sulfates through the first $\beta$-strand of the chemokines. *J. Biol. Chem.* **274,** 23916–23925.

Bernfield, M., Gotte, M., Park, P. W., Reizes, O., Fitzgerald, M. L., Lincecum, J., and Zako, M. (1999). Functions of cell surface heparan sulfate proteoglycans. *Annu. Rev. Biochem.* **68,** 729–777.

Hirose, J., Kawashima, H., Yoshie, O., Tashiro, K., and Miyasaka, M. (2001). Versican interacts with chemokines and modulates cellular responses. *J. Biol. Chem.* **276,** 5228–5234.

Kawashima, H., Atarashi, K., Hirose, M., Hirose, J., Yamada, S., Sugahara, K., and Miyasaka, M. (2002). Oversulfated chondroitin/dermatan sulfates containing GlcA$\beta$1/IdoA$\alpha$1–3GalNAc(4,6-*O*-disulfate) interact with L- and P-selectin and chemokines. *J. Biol. Chem.* **277,** 12921–12930.

Kinoshita, A., and Sugahara, K. (1999). Microanalysis of glycosaminoglycan-derived oligosaccharides labeled with a fluorophore 2-aminobenzamide by high-performance liquid chromatography: Application to disaccharide composition analysis and exosequencing of oligosaccharides. *Anal. Biochem.* **269,** 367–378.

Salmivirta, M., Lidholt, K., and Lindahl, U. (1996). Heparan sulfate: A piece of information. *FASEB J.* **10,** 1270–1279.

Tanaka, Y., Adams, D. H., and Shaw, S. (1993). Proteoglycans on endothelial cells present adhesion-inducing cytokines to leukocytes. *Immunol. Today* **14,** 111–115.

# [17]  Identification of Proteoglycan-Binding Proteins

*By* Takashi Muramatsu,
Hisako Muramatsu, and Tetsuhito Kojima

## Abstract

There are many proteins that bind to proteoglycans; they include proteins in extracellular matrices, growth factors or cytokines, plasma proteins, transmembrane proteins, and cytoplasmic proteins as listed in this chapter. Proteins that bind to a proteoglycan have been searched by using a proteoglycan

METHODS IN ENZYMOLOGY, VOL. 416
0076-6879/06 $35.00
DOI: 10.1016/S0076-6879(06)16017-6

as a ligand. Alternatively, a ligand protein has been used to find a proteoglycan as a binding partner. When the glycosaminoglycan (GAG) portion of a proteoglycan is responsible for the binding, a native proteoglycan is necessary for the analysis of binding. When the protein portion is responsible for the binding, a recombinant core protein without GAG chains may be used for analysis. This chapter describes experimental procedures dealing with two native proteoglycans, versican (PG-M) and syndecan-4 (ryudocan). Versican has been identified as a proteoglycan with binding capability to a growth factor, midkine. Purified syndecan-4 has been used to identify proteins that bind to the proteoglycan.

Overview

A large number of proteins bind to proteoglycans; they include proteins in extracellular matrices such as collagens, growth factors or cytokines, plasma proteins, transmembrane proteins, and cytoplasmic proteins (Table I). Furthermore, a proteoglycan usually recognizes more than a class of proteins. This divergent pattern of binding properties allows a proteoglycan as a hub to make up a molecular complex.

Although many proteins belong to proteoglycans (Varki et al., 1999), we have dealt with only typical examples of proteoglycans, namely perlecan, syndecans, glypicans, aggrecan, versican, neurocan, decorin, biglycan, and phosphacan. Perlecan is a heparan sulfate proteoglycan located in basement membrane, with a large core protein (400 kDa). Syndecan-1, syndecan-2 (fibroglycan), syndecan-3 (N-syndecan), and syndecan-4 (ryudocan) form a family of transmembrane heparan sulfate proteoglycans with a short core protein (31–45 kDa). Glypicans are phosphatidylinositol-anchored heparan sulfate proteoglycans with about 60 kDa core proteins. Aggrecan, versican (PG-M), and neurocan form a family of chondroitin sulfate proteoglycans of high molecular weights (220–145 kDa), with a lectin domain at the C-terminus. Decorin and biglycan consist of a family of secreted chondroitin sulfate proteoglycans with short core proteins (36–38 kDa). Phosphacan carries chondroitin sulfate chains as ligand-binding domains and is an extracellular form of a transmembrane tyrosine phosphatase.

A new binding protein for a proteoglycan can be found after affinity chromatography using the proteoglycan as a ligand. Affinity chromatography on a column of a binding protein can also reveal a proteoglycan as a specific binding partner. The strategy to study the binding proteins depends on the nature of the binding sites. Growth factors and cytokines often bind more strongly to GAG portion of proteoglycans, whereas there are many proteins that bind strongly to the protein portion. In the case of binding to the GAG portion, proteoglycans to be studied should be

TABLE I

EXAMPLES OF PROTEOGLYCAN-BINDING PROTEINS

| Ligand proteins | Proteoglycans |
|---|---|
| **Proteins in extracellular matrices** | |
| Collagen | |
| Type I | Syndecan-1 (San Antonio et al., 1994), decorin (Hedbom and Heinegard, 1993; Schonherr et al., 1995), biglycan (Schonherr et al., 1995) |
| Type II | Syndecan-1 (San Antonio et al., 1994), decorin (Hedbom and Heinegard, 1993) |
| Type III | Syndecan-1 (San Antonio et al., 1994) |
| Type IV | Syndecan-1 (San Antonio et al., 1994) |
| Type V | Syndecan-1 (San Antonio et al., 1994), syndecan-3 (Erdman et al., 2002) |
| Type VI | Syndecan-1 (San Antonio et al., 1994), decorin and biglycan (Wiberg et al., 2001) |
| Type IX | Biglycan (Chen et al., 2006) |
| Type XIII | Perlecan (Tu et al., 2002) |
| Fibronectin | Syndecan-1 (Elenius et al., 1990), decorin (Schmidt et al., 1991) |
| Laminin | Syndecan-1 (Hoffman et al., 1998), syndecan-2 and syndecan-4 (Utani et al., 2001) |
| Tenascins | Syndecan-1 (Salmivirta et al., 1991), aggrecan (Day et al., 2004), versican (Aspberg et al., 1995), neurocan (Milev et al., 1997, 1998a), phosphacan (Milev et al., 1997, 1998a) |
| Link protein | Aggrecan (Shi et al., 2004), versican (Matsumoto et al., 2003) |
| Fibulins | Perlecan (Brown et al., 1997), aggrecan and versican (Aspberg et al., 1999; Day et al., 2004; Olin et al., 2001) |
| Nidogens | Perlecan (Brown et al., 1997; Kohfeldt et al., 1998) |
| β-Amyloid protein | Perlecan, decorin, and biglycan (Snow et al., 1995) |
| Matrilin-1 | Decorin (Wiberg et al., 2003) |

(continued)

TABLE I (*continued*)

| Ligand proteins | Proteoglycans |
| --- | --- |
| **Growth factors and cytokines** | |
| Basic fibroblast growth factor | Perlecan (Knox et al., 2002), syndecan-3 (Chernousov and Carey, 1993), syndecan-4 (Kojima et al., 1996), glypican-3 (Song et al., 1997), glypican-4 (Hagihara et al., 2000), phosphacan (Milev et al., 1998b) |
| Hepatocyte growth factor | Syndecan-1 (Derksen et al., 2002) |
| Transforming growth factor-$\beta$ | Decorin and biglycan (Fukushima et al., 1993; Hildebrand et al., 1994) |
| Tumor necrosis factor-$\alpha$ | Decorin and biglycan (Tufvesson and Westergren-Thorsson, 2002) |
| Interleukin-8 | Syndecan-2 (Halden et al., 2004) |
| Midkine | Syndecan-1 (Mitsiadis et al., 1995), syndecan-4 (Kojima et al., 1996), glypican-2 (Kurosawa et al., 2001), versican (Zou et al., 2000), phosphacan (Maeda et al., 1999) |
| Pleiotrophin | Syndecan-3 (Raulo et al., 1994), phosphacan (Maeda et al., 1996) |
| **Plasma proteins** | |
| Apolipoprotein E | Biglycan (Klezovitch et al., 2000) |
| Low-density lipoprotein | Decorin (Pentikainen et al., 1997) |
| C1q | Decorin and biglycan (Groeneveld et al., 2005; Krumdieck et al., 1992) |
| Mannose-binding lectin | Decorin and biglycan (Groeneveld et al., 2005) |
| Fibrinogen | Decorin (Dugan et al., 2003) |
| Tissue factor pathway inhibitor | Syndecan-4 (Kojima et al., 1996) |

**Transmembrane proteins**

| | |
|---|---|
| Epidermal growth factor receptor | Decorin (Iozzo et al., 1999) |
| Fibroblast growth factor receptor | Perlecan (Konx et al., 2002) |
| Insulin-like growth factor-I receptor | Decorin (Schonherr et al., 2005) |
| N-CAM | Neurocan (Friedlander et al., 1994), phosphacan (Milev et al., 1994, 1996) |
| L1 | Neurocan (Friedlander et al., 1994), phosphacan (Milev et al., 1994) |
| TAG-1 | Neurocan and phosphacan (Milev et al., 1996; Oleszewski et al., 2000) |
| Scavenger receptor | Biglycan and decorin (Santiago-Garcia et al., 2003) |
| Selectins | Versican (Kawashima et al., 2000) |
| Sodium channel | Phosphacan (Ratcliffe et al., 2000) |
| Amyloid precursor protein | Glypican (Williamson et al., 1996) |
| α-Dystroglycan | Biglycan (Bowe et al., 2000) |

**Cytoplasmic proteins**

| | |
|---|---|
| Syntenin | Syndecans (Grootjans et al., 1997) |
| CASK | Syndecan-2 (Cohen et al., 1998; Hsueh et al., 1998) |
| Ezrin | Syndecan-2 (Granes et al., 2000) |
| Protein kinase Cα | Syndecan-4 (Lim et al., 2003) |
| α-Actinin | Syndecan-4 (Greene et al., 2003) |

isolated from the source in which GAGs with high affinity binding sites are produced. In the case of binding to the protein portion, the protein may be produced as a recombinant one. For high-molecular-weight proteoglycans, an active domain instead of the full-length protein may be used as a ligand to search binding proteins (Aspberg et al., 1995, 1999; Day et al., 2004; Matsumoto et al., 2003; Olin et al., 2001; Shi et al., 2004). Recombinant proteins have also been employed in a yeast two-hybrid system to find a binding partner (Cohen et al., 1998; Grootjans et al., 1997; Hsueh et al., 1998).

After finding a new binding protein, binding affinities are usually established by Scatchard plot analysis or surface plasmon resonance. Atomic energy microscopy is also used (Chen et al., 2006). To study the biological significance of the binding, the required next step is to verify the co-localization of the two interacting proteins in the tissue or on the cells; this point can be demonstrated by immunochemical methods (Wiberg et al., 2003). Finally, genetic manipulation will establish the functional significance of the binding.

Because proteoglycans with GAG chains capable of ligand binding activity are of significant interest in glycobiology, the examples in the following sections deal with versican and syndecan-4 with such characteristics.

## Versican as a Binding Protein to Midkine in the Midgestation Mouse Embryos

Midkine is a multifunctional factor promoting growth, survival, and migration of various cells (Muramatsu, 2002; Muramatsu et al., 2003). The midkine receptor is composed of several molecules including $\alpha 4\beta 1$- or $\alpha 6\beta 1$-integrin, low-density lipoprotein (LDL) receptor–related protein, and proteoglycans (Muramatsu et al., 2004). Midkine binds strongly to over-sulfated structures in chondroitin sulfates (E units) and heparan sulfate (trisulfated units) (Muramatsu et al., 2003; Zou et al., 2003). In midgestation mouse embryos, in which midkine is expected to exert many functions, the predominant midkine-binding proteoglycan is versican as revealed by protein sequencing after deglycosylation (Zou et al., 2000). The strong binding of midkine to versican has been confirmed after isolation of versican by immunoaffinity chromatography using affinity-purified antibodies to it (Zou et al., 2000). Because versican is not a transmembrane protein, versican is considered to modulate the midkine activity, as binding to midkine can enhance midkine action by concentrating it to the cell periphery or inhibit the action by competing with the binding to a component of the receptor (Zou et al., 2000).

*Materials*

    CNBr-activated Sepharose 4B (Amersham Biosciences)
    Q-Sepharose fast-flow (Amersham Biosciences)
    HisTrap HP (Amersham Biosciences)
    Glutathione–Sepharose 4 Fast Flow (Amersham Biosciences)
    Expression vector; pET-16b (Novagen), pGEX-2T (Amersham Biosciences)
    Midkine–Sepharose prepared by coupling 50 mg of recombinant human midkine to 10 ml of CNBr-activated Sepharose 4B (Muramatsu *et al.*, 2003)

*Solutions*

    Buffer A: 10 m$M$ Tris, pH 7.5, 0.5 m$M$ phenylmethylsulfonylfluoride (PMSF) and protease inhibitor cocktail tablets (Complete, Roche; 1 tablet/50 ml)
    Buffer B: Buffer A containing 0.5 $M$ sucrose, 0.1 $M$ KCl, 10 m$M$ MgCl$_2$, 2 m$M$ CaCl$_2$
    Buffer C: Buffer A containing 0.15 $M$ NaCl, 0.1 $M$ sucrose, 8 m$M$ CHAPS
    Buffer D: Buffer A containing 0.15 $M$ NaCl, 8 m$M$ CHAPS
    Buffer E: Buffer A containing 2 $M$ NaCl, 5 m$M$ ethylenediamine-tetraacetic acid (EDTA), 8 m$M$ CHAPS
    Buffer F: Buffer A containing 4 m$M$ CHAPS
    Extraction buffer: 20 m$M$ Tris, pH 8.0, 0.15 $M$ NaCl, 1% Triton X-100
    Coupling buffer: 0.1 $M$ NaHCO$_3$, pH 8.3, 0.5 $M$ NaCl

*Isolation of Midkine-Binding Proteoglycans from Mouse Embryos*

    1. Thirteen-day mouse embryos ($\sim$65 g) from 30 pregnant ICR mice are homogenized in 400 ml of ice-cold buffer A.

    2. The homogenates are mixed with an equal volume of buffer B. The mixture is centrifuged at 800$g$ for 10 min and then 8000$g$ for 10 min to remove the nuclear and mitochondria fractions and the supernatant is ultracentrifuged at 110,000$g$ for 1.5 h at 4°. The pellet is extracted with 150 ml of buffer C at 4° for 2 h. The extract is again ultracentrifuged at 110,000$g$ for 1 h.

    3. The supernatant is mixed with 3 ml of midkine–Sepharose and rotated at 4° overnight. The resin is then packed into a column, and washed with 20-column volume of buffer D and eluted with 6-column volume of buffer E.

4. The midkine-binding fraction is diluted 10-fold with buffer F and then applied to a 4-ml Q-Sepharose fast-flow column equilibrated with buffer F containing 0.2 $M$ NaCl. The column is washed with 10 column volume of buffer F containing 0.5 $M$ NaCl and the proteoglycan fraction is eluted with 5 column volume of buffer F containing 1.5 $M$ NaCl.

5. The aforementioned procedure is repeated eight times, and proteoglycans obtained from about 520 g of the embryos of 240 pregnant mice are combined.

*Preparation of Affinity-Purified Antibodies to Versican*

1. A portion of the mouse versican/PG-M cDNA (bases 5510–6328; Ito *et al.*, 1995; Zimmermann *et al.*, 1994) is cloned into expression vector pET-16b for histidine-tagged fusion protein preparation or pGEX-2T for GST fusion protein preparation. The constructs are transformed into BL21 *Escherichia coli* cells. Transformed cells are grown in Luria-Bertani medium containing 50 $\mu$g/ml ampicillin. Expression of the fusion proteins is induced by the addition of isopropylthio-$\beta$-D-galactoside to a final concentration of 1 m$M$. After 3 h induction, the bacterial cells are spun down at 3000$g$ for 10 min.

2. For histidine-tagged fusion proteins, the pellet of bacteria is extracted with 6 $M$ guanidine, pH 8.0, by sonicating for 3 min at 4° and standing at room temperature for 1 h. The solution is cleared of bacterial debris by centrifugation at 12,000$g$ for 20 min followed by filtration of the supernatant with a 0.45-$\mu$m syringe driven filter unit. The fusion protein is purified by using a 1-ml His-Trap HP column according to the manufacturer's instructions. The histidine-tagged fusion proteins are separated on 12% gel by sodium dodecylsulfate (SDS)–polyacrylamide gel electrophoresis (PAGE), and a 43-kDa major band is excised and purified by electroelution. The purified fusion protein is emulsified in incomplete Freund's adjuvant and used to immunize a New Zealand white rabbit as described previously (Muramatsu *et al.*, 1993).

3. For GST fusion proteins, the bacterial pellet is extracted with extraction buffer by sonicating for 3 min at 4°. After centrifuging at 10,000$g$ for 20 min, the supernatant is directly applied to a 5-ml glutathione–Sepharose 4 Fast Flow column. After a wash with Extraction buffer, the GST–versican fusion protein is eluted with 10 m$M$ glutathione and dialyzed against Coupling buffer. The versican–Sepharose is prepared by coupling 50 mg of the purified GST–versican fusion protein to 10 ml of CNBr-activated Sepharose.

4. The resulting antiserum (10 ml) is applied to a column of versican–Sepharose (4 ml) equilibrated with 20 m$M$ sodium phosphate buffer,

pH 7.2. The column is washed with 100 ml of the same buffer, and the antibodies are eluted with 0.1 $M$ glycine buffer, pH 2.7, and dialyzed against Coupling buffer.

### Isolation of Versican from the Midkine-Binding Proteoglycans by Immunoaffinity Chromatography

1. Affinity-purified anti-versican antibodies (35 mg) are coupled to CNBr-activated Sepharose (7 ml).

2. Midkine-binding proteoglycans isolated from embryos, as mentioned earlier, are diluted six-fold with buffer F and then applied to the antibody column. The column is washed with 10-column volume of 20 m$M$ sodium phosphate, pH 7.2, 0.2 $M$ NaCl. Versican is eluted with 6-column volume of 0.1 $M$ sodium phosphate, pH 6.6, 3.5 $M$ potassium thiocyanate.

3. The eluate is dialyzed against 20 m$M$ sodium phosphate, pH 7.2, 0.15 $M$ NaCl, and then concentrated with Centricon YM-30 (Millipore). The amount of versican obtained is about 1.5 mg as a protein.

### Comments

The binding of $^{125}$I-labeled versican to midkine has been measured by solid-phase binding assay (Zou et al., 2000). The purified versican strongly binds to midkine with a $K$d of 1.0 n$M$. Digestion with chondroitinase ABC, AC-1, or B decreases the binding with midkine to 6, 8, and 10%, respectively, compared to the undigested control. The finding that all the chondroitinases virtually abolish the binding indicates that midkine binds to hybrid chondroitin sulfate chains with a domain of dermatan sulfate structure.

### Preparation of Syndecan-4 and Identification of Its Binding Proteins

Syndecan-4 (ryudocan) has been purified from culture medium of EAhy926 cells (Kojima et al., 1992a,b, 1996) by combination of immunoaffinity chromatography and ion exchange chromatography. Solid-phase binding assay using the purified syndecan-4 has revealed proteins which bind to syndecan-4 with high affinity (Kojima et al., 1996).

### Materials

DEAE Sephacel (Amersham Biosciences)
CNBr-activated Sepharose (Amersham Biosciences)
PD-10 column (Amersham Biosciences)
A synthetic peptide of human syndecan-4 ectodomain (residues 27–39):
NH$_2$-YCDEDVVGPGQESDD-COOH

Basic fibroblast growth factor (bFGF) (Seikagaku Co., Tokyo, Japan)
Chemically synthesized human midkine (Peptide Institute, Suita, Japan)
Recombinant human tissue factor pathway inhibitor (TFPI) (Chemo-
Sero-Therapeutic Research Institute, Kumamoto, Japan).

*Solutions*

Phosphate-buffered saline (PBS): 8 m$M$ $Na_2HPO_4$, 1.5 m$M$ $KH_2PO_4$,
pH 7.4, 0.137 m$M$ NaCl, 2.7 m$M$ KCl
PBS containing 0.1% bovine serum albumin (0.1% BSA-PBS).

*Preparation of Monoclonal Anti-human Syndecan-4 Antibody and*
  *Its Conjugation to Sepharose*

The synthetic peptide, $NH_2$-YCDEDVVGPGQESDD-COOH is conjugated to ovalbumin with *meta*-maleimidobenzoyl-N-hydroxysuccinimide
ester (Harlow and Lane, 1988). The hybridoma cells producing mouse anti-
human syndecan-4 monoclonal antibody has been obtained by fusion of
spleen cells from immunized BALB/c mice with the ovalbumin-peptide
and P3U1 parent cells. Four hundred million of mouse spleen cells is
mixed with one hundred million of P3U1 cells, and fused in the presence
of 50% PEG-4000. Hybridomas are grown in HAT medium and selected
by a radioimmunoassay. The monoclonal antibody (9 mg) is coupled to
CNBr-activated Sepharose (2 ml).

*Preparation of Human Syndecan-4 from Conditioned Media of*
  *EAhy926 Cells*

The EAhy926 cell line (Edgell *et al.*, 1983), a hybrid of human umbilical
vein endothelial cell and human lung carcinoma cell, is a generous gift from
Dr. Cora-Jean S. Edgell (University of North Carolina, Chapel Hill).
EAhy926 cells are cultured in Dulbecco Modified Eagle Medium (DMEM)
supplemented with 10% fetal bovine serum (FBS), 100 units/ml of penicillin,
and 100 $\mu$g/ml of streptomycin sulfate. The flasks are gassed with 5% $CO_2$ at
37°. The media is exchanged every 2 days and the conditioned media are
harvested when the cultures attain a stationary density for 7 days.

1. Conditioned media of EAhy926 cells (1500 ml) is adjusted to
contain 1 m$M$ EDTA and 1 m$M$ PMSF, and applied to a DEAE Sephacel
column (2.5 × 10 cm) at a flow rate of 20 ml/h.
2. After washing the column with five column volumes of 50 m$M$ Na
acetate, pH 6.0, 0.4 $M$ NaCl, 1 m$M$ EDTA, followed by five column volumes

of 50 m$M$ Tris, pH 7.4, 0.3 $M$ NaCl, 1 m$M$ EDTA, bound proteoglycans are eluted with 50 m$M$ Tris, pH 7.4, 1 $M$ NaCl, 1 m$M$ EDTA.

3. After dialysis against 50 m$M$ Tris, pH 7.4, 0.15 $M$ NaCl, 1 m$M$ EDTA, CHAPS is added to a concentration of 0.6%, and the proteoglycans eluted from the DEAE Sephacel column are charged to the anti-human syndecan-4 column (1 ml).

4. The column is washed with five column volumes of 50 m$M$ Tris, pH 7.4, 0.15 $M$ NaCl, 1 m$M$ EDTA, 0.6% CHAPS. Proteoglycans enriched in syndecan-4 are eluted with three column volumes of 50 m$M$ glycine, pH 2.5, 0.15 $M$ NaCl, followed by immediate neutralization with one-fifth the volume of 1 $M$ Tris, pH 7.8.

5. The eluant is loaded onto a second DEAE Sephacel column (0.5 ml). The column is washed with five column volumes of 50 m$M$ Na acetate, pH 5.0, 0.4 $M$ NaCl, followed by five column volumes of 10 m$M$ Tris, pH 7.4, 0.3 $M$ NaCl, 1 m$M$ EDTA. Syndecan-4 is eluted with 10 m$M$ Tris, pH 7.4, 1 $M$ NaCl, 1 m$M$ EDTA. The overall yields of human syndecan-4 from conditioned media of EAhy926 cells average 2.5 ng of protein/$10^6$ cells for the entire purification procedure. These results are similar to purification yields of rat syndecan-4 from rat fat pad microvascular endothelial cells as previously described (Kojima *et al.*, 1992a).

*Solid-Phase Binding Assay*

Purified syndecan-4 is radiolabeled with Na$^{125}$I by the chloramine-T method, as described previously (Kojima *et al.*, 1992a). The radiolabeled protein is separated from free Na$^{125}$I by gel filtration chromatography through a PD-10 desalting column and purified on a DEAE Sephacel column (0.3 ml) again as outlined earlier. A solid-phase ligand binding assay is performed as described by Chernousov and Carey (1993), with a minor modification.

1. Various proteins (0.25–5.0 $\mu$g/well) in 100 $\mu$l of 50 m$M$ NaHCO$_3$, pH 9.2, are coated onto microtiter plate wells (Falcon 3911) by overnight incubation at 4°.

2. After three washes with 100 $\mu$l of 0.1% BSA-PBS, the wells are incubated with 1% BSA in PBS for 4 h at room temperature to block nonspecific binding sites.

3. The wells are rinsed three times with 0.1% BSA-PBS and incubated with $^{125}$I-radiolabeled syndecan-4 (3000–7000 cpm/well) in 100 $\mu$l of 0.1% BSA-PBS at 4° overnight.

4. After three rinses with 0.1% BSA-PBS, to remove any unbound material, the wells are dried and individually separated. The amount

of bound radiolabeled syndecan-4 in each well is quantified with a gamma-counter.

### Concentration-Dependent Syndecan-4 Binding to bFGF, Midkine, and TFPI

Microtiter wells are coated with bFGF (0.25 $\mu$g/well), midkine (0.25 $\mu$g/well), or TFPI (2.5 $\mu$g/well). The wells are then incubated with various concentrations of $^{125}$I-labeled syndecan-4 (specific activity = 4.2 $\times$ 10$^5$ cpm/$\mu$g). After washing, the amount of bound radioactivity is counted. The Scatchard analyses of $^{125}$I-syndecan-4 binding to immobilized bFGF, midkine, and TFPI are performed individually.

### Comments

A solid-phase ligand binding assay indicates that bFGF, midkine, and TFPI exhibit significant $^{125}$I-syndecan-4 binding (Kojima *et al.*, 1996). In contrast, epidermal growth factor, insulin-like growth factor, granulocyte colony stimulating factor (CSF), monocyte-CSF, granulocyte–monocyte CSF, interleukin-3, and antithrombin III show no significant affinity for syndecan-4, even when the concentrations of these proteins are increased to 5 $\mu$g/well. Heparitinase treatment destroys the ability of $^{125}$I-syndecan-4 binding to bFGF, midkine, and TFPI, whereas chondroitinase ABC treatment does not, indicating that heparan sulfate chains are responsible for the binding.

The concentration-dependent and saturable bindings of syndecan-4 to bFGF, midkine, and TFPI are observed, and Scatchard analyses showed that the apparent Kds are 0.50 n$M$ for bFGF, 0.30 n$M$ for midkine, and 0.74 n$M$ for TFPI.

### References

Aspberg, A., Adam, S., Kostka, G., Timpl, R., and Heinegard, D. (1999). Fibulin-1 is a ligand for the C-type lectin domains of aggrecan and versican. *J. Biol. Chem.* **274**, 20444–20449.

Aspberg, A., Binkert, C., and Ruoslahti, E. (1995). The versican C-type lectin domain recognizes the adhesion protein tenascin-R. *Proc. Natl. Acad. Sci. USA* **92**, 10590–10594.

Brown, J. C., Sasaki, T., Gohring, W., Yamada, Y., and Timpl, R. (1997). The C-terminal domain V of perlecan promotes beta1 integrin-mediated cell adhesion, binds heparin, nidogen and fibulin-2 and can be modified by glycosaminoglycans. *Eur. J. Biochem.* **250**, 39–46.

Chen, C. H., Yeh, M. L., Geyer, M., Wang, G. J., Huang, M. H., Heggeness, M. H., Hook, M., and Luo, Z. P. (2006). Interactions between collagen IX and biglycan measured by atomic force microscopy. *Biochem. Biophys. Res. Commun.* **339**, 204–208.

Chernousov, M. A., and Carey, D. J. (1993). N-syndecan (syndecan 3) from neonatal rat brain binds basic fibroblast growth factor. *J. Biol. Chem.* **268**, 16810–16814.

Cohen, A. R., Woods, D. F., Marfatia, S. M., Walther, Z., Chishti, A. H., and Anderson, J. M. (1998). Human CASK/LIN-2 binds syndecan-2 and protein 4.1 and localizes to the basolateral membrane of epithelial cells. *J. Cell Biol.* **142**, 129–138.

Day, J. M., Olin, A. I., Murdoch, A. D., Canfield, A., Sasaki, T., Timpl, R., Hardingham, T. E., and Aspberg, A. (2004). Alternative splicing in the aggrecan G3 domain influences binding interactions with tenascin-C and other extracellular matrix proteins. *J. Biol. Chem.* **279**, 12511–12518.

Derksen, P. W., Keehnen, R. M., Evers, L. M., van Oers, M. H., Spaargaren, M., and Pals, S. T. (2002). Cell surface proteoglycan syndecan-1 mediates hepatocyte growth factor binding and promotes Met signaling in multiple myeloma. *Blood* **99**, 1405–1410.

Dugan, T. A., Yang, V. W., McQuillan, D. J., and Hook, M. (2003). Decorin binds fibrinogen in a Zn2+-dependent interaction. *J. Biol. Chem.* **278**, 13655–13662.

Edgell, C-J. S., McDonald, C. C., and Graham, J. B. (1983). Permanent cell line expressing human factor VIII-related antigen established by hybridization. *Proc. Natl. Acad. Sci. USA* **80**, 3734–3737.

Elenius, K., Salmivirta, M., Inki, P., Mali, M., and Jalkanen, M. (1990). Binding of human syndecan to extracellular matrix proteins. *J. Biol. Chem.* **265**, 17837–17843.

Erdman, R., Stahl, R. C., Rothblum, K., Chernousov, M. A., and Carey, D. J. (2002). Schwann cell adhesion to a novel heparan sulfate binding site in the N-terminal domain of α4 type V collagen is mediated by syndecan-3. *J. Biol. Chem.* **277**, 7619–7625.

Friedlander, D. R., Milev, P., Karthikeyan, L., Margolis, R. K., Margolis, R. U., and Grumet, M. (1994). The neuronal chondroitin sulfate proteoglycan neurocan binds to the neural cell adhesion molecules Ng-CAM/L1/NILE and N-CAM, and inhibits neuronal adhesion and neurite outgrowth. *J. Cell Biol.* **125**, 669–680.

Fukushima, D., Butzow, R., Hildebrand, A., and Ruoslahti, E. (1993). Localization of transforming growth factor β binding site in betaglycan. Comparison with small extracellular matrix proteoglycans. *J. Biol. Chem.* **268**, 22710–22715.

Granes, F., Urena, J. M., Rocamora, N., and Vilaro, S. (2000). Ezrin links syndecan-2 to the cytoskeleton. *J. Cell Sci.* **113**, 1267–1276.

Greene, D. K., Tumova, S., Couchman, J. R., and Woods, A. (2003). Syndecan-4 associates with α-actinin. *J. Biol. Chem.* **278**, 7617–7623.

Groeneveld, T. W., Oroszlan, M., Owens, R. T., Faber-Krol, M. C., Bakker, A. C., Arlaud, G. J., McQuillan, D. J., Kishore, U., Daha, M. R., and Roos, A. (2005). Interactions of the extracellular matrix proteoglycans decorin and biglycan with C1q and collectins. *J. Immunol.* **175**, 4715–4723.

Grootjans, J. J., Zimmermann, P., Reekmans, G., Smets, A., Degeest, G., Durr, J., and David, G. (1997). Syntenin, a PDZ protein that binds syndecan cytoplasmic domains. *Proc. Natl. Acad. Sci. USA* **94**, 13683–13688.

Hagihara, K., Watanabe, K., Chun, J., and Yamaguchi, Y. (2000). Glypican-4 is an FGF2-binding heparan sulfate proteoglycan expressed in neural precursor cells. *Dev. Dyn.* **219**, 353–367.

Halden, Y., Rek, A., Atzenhofer, W., Szilak, L., Wabnig, A., and Kungl, A. J. (2004). Interleukin-8 binds to syndecan-2 on human endothelial cells. *Biochem. J.* **377**, 533–538.

Harlow, E, and Lane, D. (1988). "Antibodies: A Laboratory Manual." Cold Spring Harbor Laboratory, Cold Spring Harbor, NY.

Hedbom, E., and Heinegard, D. (1993). Binding of fibromodulin and decorin to separate sites on fibrillar collagens. *J. Biol. Chem.* **268**, 27307–27312.

Hildebrand, A., Romaris, M., Rasmussen, L. M., Heinegard, D., Twardzik, D. R., Border, W. A., and Ruoslahti, E. (1994). Interaction of the small interstitial proteoglycans biglycan, decorin and fibromodulin with transforming growth factor β. *Biochem. J.* **302**, 527–534.

Hoffman, M. P., Nomizu, M., Roque, E., Lee, S., Jung, D. W., Yamada, Y., and Kleinman, H. K. (1998). Laminin-1 and laminin-2 G-domain synthetic peptides bind syndecan-1 and are involved in acinar formation of a human submandibular gland cell line. *J. Biol. Chem.* **273**, 28633–28641.

Hsueh, Y. P., Yang, F. C., Kharazia, V., Naisbitt, S., Cohen, A. R., Weinberg, R. J., and Sheng, M. (1998). Direct interaction of CASK/LIN-2 and syndecan heparan sulfate proteoglycan and their overlapping distribution in neuronal synapses. *J. Cell Biol.* **142**, 139–151.

Iozzo, R. V., Moscatello, D. K., McQuillan, D. J., and Eichstetter, I. (1999). Decorin is a biological ligand for the epidermal growth factor receptor. *J. Biol. Chem.* **274**, 4489–4492.

Ito, K., Shinomura, T., Zako, M., Ujita, M., and Kimata, K. (1995). Multiple forms of mouse PG-M, a large chondroitin sulfate proteoglycan generated by alternative splicing. *J. Biol. Chem.* **270**, 958–965.

Kawashima, H., Hirose, M., Hirose, J., Nagakubo, D., Plaas, A. H., and Miyasaka, M. (2000). Binding of a large chondroitin sulfate/dermatan sulfate proteoglycan, versican, to L-selectin, P-selectin, and CD44. *J. Biol. Chem.* **275**, 35448–35456.

Klezovitch, O., Formato, M., Cherchi, G. M., Weisgraber, K. H., and Scanu, A. M. (2000). Structural determinants in the C-terminal domain of apolipoprotein E mediating binding to the protein core of human aortic biglycan. *J. Biol. Chem.* **275**, 18913–18918.

Knox, S., Merry, C., Stringe,r S., Melrose, J., and Whitelock, J. (2002). Not all perlecans are created equal: Interactions with fibroblast growth factor (FGF) 2 and FGF receptors. *J. Biol. Chem.* **277**, 14657–14665.

Kohfeldt, E., Sasaki, T., Gohring, W., and Timpl, R. (1998). Nidogen-2: A new basement membrane protein with diverse binding properties. *J. Mol. Biol.* **282**, 99–109.

Kojima, T., Katsumi, A., Yamazaki, T., Muramatsu, T., Nagasaka, T., Ohsumi, K., and Saito, H. (1996). Human ryudocan from endothelial-like cells binds basic fibroblast growth factor, midkine, and tissue factor pathway inhibitor. *J. Biol. Chem.* **271**, 5914–5920.

Kojima, T., Leone, C. W., Marchildon, G. A., Marcum, J. A., and Rosenberg, R. D. (1992a). Isolation and characterization of heparan sulfate proteoglycans produced by cloned rat microvascular endothelial cells. *J. Biol. Chem.* **267**, 4859–4869.

Kojima, T., Shworak, N. S., and Rosenberg, R. D. (1992b). Molecular cloning and expression of two distinct cDNA encoding heparan sulfate proteoglycan core proteins from a rat endothelial cell line. *J. Biol. Chem.* **267**, 4870–4877.

Krumdieck, R., Hook, M., Rosenberg, L. C., and Volanakis, J. E. (1992). The proteoglycan decorin binds C1q and inhibits the activity of the C1 complex. *J. Immunol.* **149**, 3695–3701.

Kurosawa, N., Chen, G. Y., Kadomatsu, K., Ikematsu, S., Sakuma, S., and Muramatsu, T. (2001). Glypican-2 binds to midkine: The role of glypican-2 in neuronal cell adhesion and neurite outgrowth. *Glycoconj. J.* **18**, 499–507.

Lim, S. T., Longley, R. L., Couchman, J. R., and Woods, A. (2003). Direct binding of syndecan-4 cytoplasmic domain to the catalytic domain of protein kinase C$\alpha$ (PKC$\alpha$) increases focal adhesion localization of PKC$\alpha$. *J. Biol. Chem.* **278**, 13795–13802.

Maeda, N., Ichihara-Tanaka, K., Kimura, T., Kadomatsu, K., Muramatsu, T., and Noda, M. A. (1999). Receptor-like protein-tyrosine phosphatase PTP$\zeta$/RPTP$\beta$ binds a heparin-binding growth factor midkine. Involvement of arginine 78 of midkine in the high affinity binding to PTP?. *J. Biol. Chem.* **274**, 12474–12479.

Maeda, N., Nishiwaki, T., Shintani, T., Hamanaka, H., and Noda, M. (1996). 6B4 proteoglycan/phosphacan, an extracellular variant of receptor-like protein-tyrosine phosphatase$\zeta$/RPTP$\beta$, binds pleiotrophin/heparin-binding growth-associated molecule (HB-GAM). *J. Biol. Chem.* **271**, 21446–21452.

Matsumoto, K., Shionyu, M., Go, M., Shimizu, K., Shinomura, T., Kimata, K., and Watanabe, H. (2003). Distinct interaction of versican/PG-M with hyaluronan and link protein. *J. Biol. Chem.* **278**, 41205–41212.

Milev, P., Chiba, A., Haring, M., Rauvala, H., Schachner, M., Ranscht, B., Margolis, R. K., and Margolis, R. U. (1998a). High affinity binding and overlapping localization of neurocan and phosphacan/protein-tyrosine phosphatase-$\zeta/\beta$ with tenascin-R, amphoterin, and the heparin-binding growth-associated molecule. *J. Biol. Chem.* **273,** 6998–7005.

Milev, P., Fischer, D., Haring, M., Schulthess, T., Margolis, R. K., Chiquet-Ehrismann, R., and Margolis, R. U. (1997). The fibrinogen-like globe of tenascin-C mediates its interactions with neurocan and phosphacan/protein-tyrosine phosphatase-$\zeta/\beta$. *J. Biol. Chem.* **272,** 15501–15509.

Milev, P., Friedlander, D. R., Sakurai, T., Karthikeyan, L., Flad, M., Margolis, R. K., Grumet, M., and Margolis, R. U. (1994). Interactions of the chondroitin sulfate proteoglycan phosphacan, the extracellular domain of a receptor-type protein tyrosine phosphatase, with neurons, glia, and neural cell adhesion molecules. *J. Cell Biol.* **127,** 1703–1715.

Milev, P., Maurel, P., Haring, M., Margolis, R. K., and Margolis, R. U. (1996). TAG-1/axonin-1 is a high-affinity ligand of neurocan, phosphacan/protein-tyrosine phosphatase-$\zeta/\beta$ and N-CAM. *J. Biol. Chem.* **271,** 15716–15723.

Milev, P., Monnerie, H., Popp, S., Margolis, R. K., and Margolis, R. U. (1998b). The core protein of the chondroitin sulfate proteoglycan phosphacan is a high-affinity ligand of fibroblast growth factor-2 and potentiates its mitogenic activity. *J. Biol. Chem.* **273,** 21439–21442.

Mitsiadis, T. A., Salmivirta, M., Muramatsu, T., Muramatsu, H., Rauvala, H., Lehtonen, E., Jalkanen, M., and Thesleff, I. (1995). Expression of the heparin-binding cytokines, midkine (MK) and HB-GAM (pleiotrophin) is associated with epithelial-mesenchymal interactions during fetal development and organogenesis. *Development* **121,** 37–51.

Muramatsu, H., Shirahama, H., Yonezawa, S., Murata, H., and Muramatsu, T. (1993). Midkine (MK), a retinoic acid-inducible growth/differentiation factor: Immunochemical evidence for the function and distribution. *Dev. Biol.* **159,** 392–402.

Muramatsu, H., Zou, P., Suzuki, H., Oda, Y., Chen, G. Y., Sakaguchi, N., Sakuma, S., Maeda, N., Noda, M., and Muramatsu, T. (2004). $\alpha4\beta1$- and $\alpha6\beta1$-integrins are functional receptors for midkine, a heparin-binding growth factor. *J. Cell Sci.* **117,** 5405–5415.

Muramatsu, T. (2002). Midkine and pleiotrophin: Two related proteins involved in development, survival, inflammation and tumorigenesis. *J. Biochem.* **132,** 359–371.

Muramatsu, T., Muramatsu, H., Kaneda, N., and Sugahara, K. (2003). Recognition of Glycosaminoglycans by midkine. *Methods Enzymol* **363,** 365–376.

Oleszewski, M., Gutwein, P., von der Lieth, W., Rauch, U., and Altevogt, P. (2000). Characterization of the L1-neurocan-binding site. Implications for L1-L1 homophilic binding. *J. Biol. Chem.* **275,** 34478–34485.

Olin, A. I., Morgelin, M., Sasaki, T., Timpl, R., Heinegard, D., and Aspberg, A. (2001). The proteoglycans aggrecan and versican form networks with fibulin-2 through their lectin domain binding. *J. Biol. Chem.* **276,** 1253–1261.

Pentikainen, M. O., Oorni, K., Lassila, R., and Kovanen, P. T. (1997). The proteoglycan decorin links low density lipoproteins with collagen type I. *J. Biol. Chem.* **272,** 7633–7638.

Ratcliffe, C. F., Qu, Y., McCormick, K. A., Tibbs, V. C., Dixon, J. E., Scheuer, T., and Catterall, W. A. (2000). A sodium channel signaling complex: Modulation by associated receptor protein tyrosine phosphatase $\beta$. *Nat. Neurosci.* **3,** 437–444.

Raulo, E., Chernousov, M. A., Carey, D. J., Nolo, R., and Rauvala, H. (1994). Isolation of a neuronal cell surface receptor of heparin binding growth-associated molecule (HB-GAM). Identification as N-syndecan (syndecan-3). *J. Biol. Chem.* **269,** 12999–13004.

Salmivirta, M., Elenius, K., Vainio, S., Hofer, U., Chiquet-Ehrismann, R., Thesleff, I., and Jalkanen, M. (1991). Syndecan from embryonic tooth mesenchyme binds tenascin. *J. Biol Chem.* **266,** 7733–7739.

San Antonio, J. D., Karnovsky, M. J., Gay, S., Sanderson, R. D., and Lander, A. D. (1994). Interactions of syndecan-1 and heparin with human collagens. *Glycobiology* **4,** 327–332.

Santiago-Garcia, J., Kodama, T., and Pitas, R. E. (2003). The class A scavenger receptor binds to proteoglycans and mediates adhesion of macrophages to the extracellular matrix. *J. Biol. Chem.* **278,** 6942–6946.

Schmidt, G., Hausser, H., and Kresse, H. (1991). Interaction of the small proteoglycan decorin with fibronectin. Involvement of the sequence NKISK of the core protein. *Biochem. J.* **280,** 411–414.

Schonherr, E., Sunderkotter, C., Iozzo, R. V., and Schaefer, L. (2005). Decorin, a novel player in the insulin-like growth factor system. *J. Biol. Chem.* **280,** 15767–15772.

Schonherr, E., Witsch-Prehm, P., Harrach, B., Robenek, H., Rauterberg, J., and Kresse, H. (1995). Interaction of biglycan with type I collagen. *J. Biol. Chem.* **270,** 2776–2783.

Shi, S., Grothe, S., Zhang, Y., O'Connor-McCourt, M. D., Poole, A. R., Roughley, P. J., and Mort, J. S. (2004). Link protein has greater affinity for versican than aggrecan. *J. Biol. Chem.* **279,** 12060–12066.

Snow, A. D., Kinsella, M. G., Parks, E., Sekiguchi, R.T, Miller, J. D., Kimata, K., and Wight, T. N. (1995). Differential binding of vascular cell-derived proteoglycans (perlecan, biglycan, decorin, and versican) to the $\beta$-amyloid protein of Alzheimer's disease. *Arch. Biochem. Biophys.* **320,** 84–95.

Song, H. H., Shi, W., and Filmus, J. (1997). OCI-5/rat glypican-3 binds to fibroblast growth factor-2 but not to insulin-like growth factor-2. *J. Biol. Chem.* **272,** 7574–7577.

Tu, H., Sasaki, T., Snellman, A., Gohring, W., Pirila, P., Timpl, R., and Pihlajaniemi, T. (2002). The type XIII collagen ectodomain is a 150-nm rod and capable of binding to fibronectin, nidogen-2, perlecan, and heparin. *J. Biol. Chem.* **277,** 23092–23099.

Tufvesson, E., and Westergren-Thorsson, G. (2002). Tumour necrosis factor-$\alpha$ interacts with biglycan and decorin. *FEBS Lett.* **530,** 124–128.

Utani, A., Nomizu, M., Matsuura, H., Kato, K., Kobayashi, T., Takeda, U., Aota, S., Nielsen, P. K., and Shinkai, H. (2001). A unique sequence of the laminin $\alpha$3 G domain binds to heparin and promotes cell adhesion through syndecan-2 and -4. *J. Biol. Chem.* **276,** 28779–28788.

Varki, A., Cummings, R., Esko, J., Freeze, H., Hart, G., and Marth, J. (eds.) (1999). "Essentials of Glycobiology." Cold Spring Harbor Press, Cold Spring Harbor, NY.

Wiberg, C., Hedbom, E., Khairullina, A., Lamande, S. R., Oldberg, A., Timpl, R., Morgelin, M., and Heinegard, D. (2001). Biglycan and decorin bind close to the *N*-terminal region of the collagen VI triple helix. *J. Biol. Chem.* **276,** 18947–18952.

Wiberg, C., Klatt, A. R., Wagener, R., Paulsson, M., Bateman, J. F., Heinegard, D., and Morgelin, D. (2003). Complexes of matrilin-1 and biglycan or decorin connect collagen VI microfibrils to both collagen II and aggrecan. *J. Biol. Chem.* **278,** 37698–37704.

Williamson, T. G., Mok, S. S., Henry, A., Cappai, R., Lander, A. D., Nurcombe, V., Beyreuther, K., Masters, C. L., and Small, D. H. (1996). Secreted glypican binds to the amyloid precursor protein of Alzheimer's disease (APP) and inhibits APP-induced neurite outgrowth. *J. Biol. Chem.* **271,** 31215–31221.

Zimmermann, D. R., Dours-Zimmermann, M. T., Schubert, M., and Bruckner-Tuderman, L. (1994). Versican is expressed in the proliferating zone in the epidermis and in association with the elastic network of the dermis. *J. Cell Biol.* **124,** 817–825.

Zou, K., Muramatsu, H., Ikematsu, S., Sakuma, S., Salama, R. H. M., Shinomura, T., Kimata, K., and Muramatsu, T. (2000). A heparin-binding growth factor, midkine, binds to a chondroitin sulfate proteoglycan, PG-M/versican. *Eur. J. Biochem.* **267,** 4046–4053.

Zou, P., Zou, K., Muramatsu, H., Ichihara-Tanaka, K., Habuchi, O., Ohtake, S., Ikematsu, S., Sakuma, S., and Muramatsu, T. (2003). Glycosaminoglycan structures required for strong binding to midkine, a heparin-binding growth factor. *Glycobiology* **13,** 35–42.

[18]   Functions of Glycans Revealed by Gene
       Inactivation of L-Selectin Ligand
       Sulfotransferases in Mice

*By* HIROTO KAWASHIMA

## Abstract

Lymphocyte homing is mediated by a specific interaction between L-selectin and its sulfated glycoprotein ligands expressed on high endothelial venules (HEVs) in lymph nodes. To examine the significance of sulfation of L-selectin ligands, gene targeting mice deficient in both *N*-acetylglucosamine-6-*O*-sulfotransferase (GlcNAc6ST)-1 and GlcNAc6ST-2 (HEC-GlcNAc6ST/LSST) have been generated. In the double-knockout mice, binding of MECA-79 antibody to lymph node HEV was completely abolished, indicating that extended core 1 *O*-glycans containing GlcNAc-6-*O*-sulfate is completely diminished in those mice. Furthermore, the mutant mice showed approximately 75% less lymphocyte homing to the peripheral lymph nodes (PLNs) and significantly less contact hypersensitivity response than wild-type mice, demonstrating that GlcNAc6ST-1 and GlcNAc6ST-2 play a major role in L-selectin ligand biosynthesis in HEVs. In this chapter, the detailed protocols that have been used for the functional assays of these sulfotransferase double-knockout mice are described.

## Introduction

Lymphocyte migration from the bloodstream into the secondary lymphoid organs is initiated by adhesive interactions between lymphocyte homing receptors and their respective ligands on HEVs. L-selectin was initially identified as a lymphocyte homing receptor and shown to be important in lymphocyte migration into the PLNs. Subsequently, L-selectin was also found to be expressed on other types of leukocytes and involved in rolling interactions of leukocytes in the postcapillary venules of inflamed tissues. The study in L-selectin–deficient mice also confirmed the important role of L-selectin in lymphocyte homing to PLNs and leukocyte migration to the sites of inflammation.

As with other members of the selectin family, the extracellular domain of L-selectin consists of an N-terminal calcium-dependent lectin domain followed by an epidermal growth factor-like domain and then short consensus repeats that are similar to those found in complement regulatory proteins. Studies of the carbohydrate-based ligands for L-selectin

METHODS IN ENZYMOLOGY, VOL. 416                                    0076-6879/06 $35.00

expressed on HEV of lymph nodes have identified four sialomucins, Gly-CAM-1, CD34, podocalyxin-like protein and Sgp200. The ability of these ligands to function entirely depends on their decoration with specific carbohydrate structures, including 6-sulfo sialyl Lewis X (Sialic acidα2–3Galβ1–4[Fucα1–3(sulfo-6)]GlcNAcβ1-R), which contains fucose, sialic acid, and sulfate. The 6-sulfo sialyl Lewis X structure is present either in the core 2 or extended core 1 branch or in both branches of L-selectin ligand O-glycans (Yeh et al., 2001) (Fig. 1). The same study revealed that the MECA-79 antibody, which is widely used to detect HEVs in lymph nodes or HEV-like vessels at the sites of chronic inflammation, recognizes O-glycans containing 6-sulfo N-acetylglucosamine in the extended core 1 structure.

To determine sulfation requirement of L-selectin ligand O-glycans, gene mutant mice deficient in the HEV-restricted sulfotransferase, N-acetyl-glucosamine-6-O-sulfotransferase-2 (GlcNAc6ST-2, HEC-GlcNAc6ST/LSST), were generated (Hemmerich et al., 2001; Hiraoka et al., 2004). In GlcNAc6ST-2 knockout mice, binding of MECA-79 antibody to lymph node HEV was almost abolished, except for the binding observed in the abluminal lining of HEVs, suggesting that GlcNAc-6-O-sulfation in extended core 1 branch is mediated mainly by GlcNAc6ST-2.

In mice, four members of GlcNAc-6-O-sulfotransferases have been reported. One of the family members, GlcNAc6ST-1, is widely expressed in various tissues including lymph node HEVs. Kawashima et al. (2005) and Uchimura et al. (2005) crossbred GlcNAc6ST-2 knockout mice with GlcNAc6ST-1 knockout mice (Uchimura et al., 2004) to determine the cooperative role of GlcNAc6ST-1 and GlcNAc6ST-2 in L-selectin ligand biosynthesis. Immunofluorescence studies revealed that binding of MECA-79 antibody to lymph node HEVs of the double-knockout mice was completely abolished, indicating that GlcNAc-6-O-sulfation in the extended core 1 branch of O-glycans in HEVs is absent in the double-knockout mice. Whereas GlcNAc6ST-1 and GlcNAc6ST-2 single-knockout mice showed approximately 20% and 50% reduction in lymphocyte homing, respectively, GlcNAc6ST-1 and GlcNAc6ST-2 double-knockout mice showed approximately 75% reduction in lymphocyte homing. It was also shown that contact hypersensitivity (CHS) responses were significantly diminished in the double-knockout mice because of the reduction in lymphocyte trafficking to the draining lymph nodes. These results demonstrate the essential role of GlcNAc6ST-1 and GlcNAc6ST-2 in L-selectin ligand biosynthesis in HEVs and their importance in immune surveillance. Herein, the detailed protocols for genotyping of mutant mice, as well as lymphocyte homing assay, L-selectin–immunoglobulin M (IgM) staining and the CHS response assay that have been performed in these studies are described.

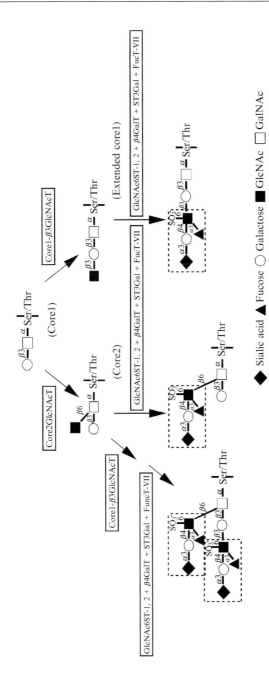

Fig. 1. Biosynthesis of L-selectin ligand oligosaccharides. Core 2 branched O-glycan containing both core 2 branch and extended core 1 structure (*right*), extended core 1 structure (*middle*), and biantennary O-glycan containing both core 2 branch and extended core 1 structure (*left*) modified with 6-sulfo sialyl Lewis X (indicated by the dot line–boxed area) function as L-selectin ligand oligosaccharides in HEVs. Extended core 1 structure modified with GlcNAc-6-O-sulfate is recognized by the MECA-79 antibody. Each roman character and number represents a carbohydrate linkage. Core2 GlcNAcT, Core2 $\beta$1,6-N-acetylglucosaminyltransferase; Core1-$\beta$3G6ST, Core1 extension $\beta$1,3-N-acetylglucosaminyltransferase; GlcNAc6ST, GlcNAc-6-O-sulfotransferase; $\beta$4GalT, $\beta$1,4-galactosyltransferase; ST3Gal, $\alpha$2,3-sialyltransferase; FucT-VII, fucosyltransferase-VII.

## Materials

### Mice

GlcNAc6ST-1 and GlcNAc6ST-2 double-knockout mice are generated by breeding GlcNAc6ST-1 (Uchimura *et al.*, 2004) and GlcNAc6ST-2 (Hiraoka *et al.*, 2004) knockout mice. Mice should be treated in accordance with guidelines of the National Institutes of Health (NIH) and the U.S. Department of Agriculture (USDA).

### Buffer

Phosphate-buffered saline (PBS). For 1 liter of solution, use the following:
0.2 g $KH_2PO_4$
2.9 g $Na_2HPO_4 \cdot 12H_2O$
0.2 g KCl
8.0 g NaCl

## Methods

### Genotyping

#### Preparation of Genomic DNA

1. Cut the mouse tail at 5 mm from the tip. Put the tail piece into a 1.5-ml Eppendorf tube.
2. Add 0.2 ml Lysis buffer containing 0.2 mg/ml proteinase K (Roche Diagnostics Co., Indianapolis, IN), 1% sodium dodecylsulfate (SDS), 100 m$M$ ethylenediaminetetraacetic acid (EDTA), 100 m$M$ NaCl, and 50 m$M$ Tris–HCl, pH 8.0, and incubate overnight at 45° (see Note 1, later in this chapter).
3. Vortex the mixture vigorously, and centrifuge at 5000 rpm briefly (see Note 2).
4. Add 0.1 ml 10 $M$ ammonium acetate, and vortex vigorously.
5. Centrifuge at 12,000 rpm for 5 min.
6. Transfer 0.25 ml of the supernatant to a new 1.5-ml Eppendorf tube (see Note 3), add 0.75 ml of ethanol, and mix well by inverting the tube several times.
7. Place the tube at room temperature for 10 min (see Note 4).
8. Centrifuge at 12,000 rpm for 5 min.
9. Discard the supernatant.

10. Add 0.5 ml 80% ethanol into the tube, and centrifuge at 12,000 rpm for 1 min.

11. Remove the supernatant carefully and air-dry the sample for 15 min.

12. Dissolve the sample in 0.1 ml 10 m*M* Tris–HCl, pH 8.0.

*Polymerase Chain Reaction*

1. For simultaneous detection of wild-type and mutant alleles of GlcNAc6ST-1, primers 5′-TCTATGAGCCTGTGTGGCACGT-3′ (G6W5), 5′-GCATACCACCTTGTTAGTGGC-3′ (G6W3), and 5′-TGACAACGTCGAGCACAGCTG-3′ (NeoP5) are used. Mix the following (see Note 5):

    AccuPrime Taq DNA SuperMix II (Invitrogen Co., Carlsbad, CA) 5 μl (see Note 6)

    H₂O 3.2 μl

    G6W5 (20 pmol/μl) 0.4 μl

    G6W3 (20 pmol/μl) 0.1 μl

    NeoP5 (20 pmol/μl) 0.3 μl

    Genomic DNA (1:10 diluted) 1 μl

2. For simultaneous detection of wild-type and mutant alleles of Glc-NAc6ST-2, primers 5′-AAGAAAGGGAGGCTGCTGATGTTC-3′ (F2W), 5′-TCCACCATATCAAAGGGCTGCTGA-3′ (R2W), and 5′-AACTTCAGGGTCAGCTTGCCGT-3′ (R1M) are used. Mix the following:

    AccuPrime Taq DNA SuperMix II 5 μl

    H₂O 2.8 μl

    F2W (10 pmol/μl) 0.4 μl

    R2W (10 pmol/μl) 0.4 μl

    R1M (20 pmol/μl) 0.4 μl

    Genomic DNA (1:10 diluted) 1 μl

3. Incubate the reaction tubes in a thermal cycler at 94° for 2 min to denature the template and activate the enzyme.

4. Perform 40 cycles of polymerase chain reaction (PCR) amplification as follows:

    30 s at 94° (denaturing)

    30 s at 60° (annealing)

    1 min at 68° (extension).

5. Incubate the reaction tubes at 68° for 5 min.

6. Maintain the reaction at 4°. The samples can be stored at −20° until use.

7. Analyze the PCR products by 2% agarose gel electrophoresis and visualize by ethidium bromide staining. A typical result is shown in Fig. 2B.

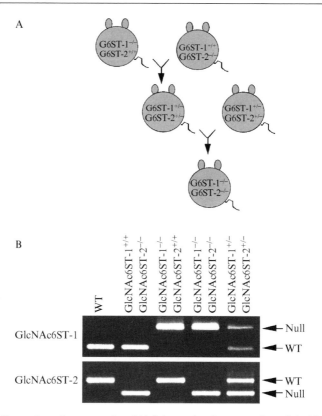

FIG. 2. Generation of mutant mice. (A) Scheme for the generation of double-knockout mice. G6ST-1, GlcNAc6ST-1; G6ST-2, GlcNAc6ST-2. (B) Polymerase chain reaction (PCR) analysis of genomic DNA) derived from wild-type (WT) and mutant mice. Genomic DNA from WT and mutant mice were subjected to genomic PCR as described in the "Methods" section. Upper panel, GlcNAc6ST-1 genotyping; lower panel, GlcNAc6ST-2 genotyping. Arrows indicate PCR products derived from WT or null alleles.

## Lymphocyte Homing Assay

### Preparation of CMFDA-Labeled Lymphocytes

1. Collect mesenteric lymph nodes (MLNs) and spleens from five C57BL/6 female mice (8–10 weeks old) into 5 ml RPMI 1640 medium containing 10% fetal calf serum (FCS), 10 m$M$ HEPES, 100 U/ml penicillin, 100 $\mu$g/ml streptomycin (RPMI-Med) (see Note 7).

2. Lymphocytes are squeezed out from the MLNs and spleens separately by frosted glass slides.

3. Add 5 ml RPMI-Med to the cell suspension from MLNs, mix well, and pass it through a cell strainer (70 $\mu$m). Collect the passed-through fraction into a new tube, and centrifuge at 1500 rpm for 5 min (see Note 8).

4. Add 5 ml RPMI-Med and place the MLN cell suspension on ice until use.

5. Add 5 ml RPMI-Med to the spleen cell suspension, mix well, and stand the tube on ice for 5 min.

6. Collect 9 ml of the spleen cell suspension from the top, pass it through a cell strainer (70 $\mu$m), and centrifuge at 1500 rpm for 5 min.

7. Remove supernatant, add 3 ml ACK lysing buffer (150 m$M$ NH$_4$Cl, 1 m$M$ KHCO$_3$, 0.1 m$M$ EDTA, pH 7.2–7.3), and incubate at room temperature for 5 min (see Note 9).

8. Add 12 ml RPMI-Med, mix well, and centrifuge at 1500 rpm for 5 min.

9. Remove supernatant, add 10 ml RPMI-Med, and pass the sample through a cell strainer (70 $\mu$m). The passed-through fraction should be collected into a new tube.

10. Combine the cell suspension from Steps 4 and 9, and centrifuge at 1500 rpm for 5 min.

11. Remove supernatant, add 5 ml RPMI 1640 containing 10 m$M$ HEPES, 100 U/ml penicillin, 100 $\mu$g/ml streptomycin, and 5 $\mu M$ 5-chloromethyl fluorescence diacetate (CMFDA), and incubate at 37° for 30 min.

12. Add 5 ml RPMI-Med, pass the cell suspension through a cell strainer (70 $\mu$m), and centrifuge at 1500 rpm for 5 min.

13. Remove supernatant.

14. Add 10 ml RPMI 1640 containing 5% FCS, 10 m$M$ HEPES, 100 U/ml penicillin, and 100 $\mu$g/ml streptomycin, and centrifuge at 1500 rpm for 5 min.

15. Remove supernatant, and suspend the cells in sterile PBS at the cell density of $2.5 \times 10^7$ cells/300 $\mu$l. Place the cell suspension on ice until use.

### Determination of Lymphocyte Homing

1. Inject 300 $\mu$l of the CMFDA-labeled cells into the tail vein of each mouse (7 to 8 weeks old) (see Note 10).

2. After 1 h, sacrifice the animals and collect PLNs, MLNs, and Peyer's patches (PPs) into 2 ml ice-cold PBS (see Note 11).

3. Lymphocytes are squeezed out from the lymphoid organs and the cell suspension is passed through a cell strainer (70 $\mu$m). The passed-through cell suspension should be collected into new tubes. Keep the cell suspensions on ice.

4. The cells are subjected to fluorescence-activated cell sorter (FACS) analysis to determine the fractional content of fluorescent cells (see Note 12). The data should be summarized as shown in Fig. 3.

### L-Selectin–IgM Staining

#### Preparation of L-Selectin–IgM

1. The cDNA encoding the Fc region of human IgM is amplified with oligonucleotides, 5'-CGGGATCCTGTGATTGCTGAGCTGCCTC CCA-3', 5'-GCTCTAGATCAGTAGCAGGTGCCAGCTGTGT-3', and subcloned into the *BamH*I/XbaI site of pcDNA1.1, using pcDNAI/E-selectin–IgM (Maly *et al.*, 1996) as a template. The resultant plasmid is designated pcDNA1.1/IgM. To construct pcDNA1.1/L-selectin–IgM, L-selectin cDNA is excised from pCDM8/L-selectin–IgG (Sueyoshi *et al.*, 1994) by *EcoR*I (blunted) and *BamH*I, and subcloned into *Hind*III (blunted) and *BamH*I sites of pcDNA1.1/IgM.

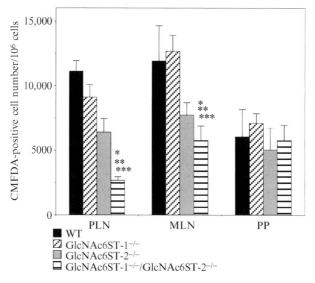

FIG. 3. Lymphocyte trafficking to different secondary lymphoid organs. CMFDA-labeled lymphocytes ($2.5 \times 10^7$ cells) were injected into tail veins of different mouse lines. One hour later, fluorescent lymphocytes in lymphocyte suspensions from lymphoid organs were quantified by flow cytometry. At least four recipient mice were tested in each experiment. For PLNs, $*p < .01$ vs. wild-type mice, $**p < .01$ vs. GlcNAc6ST-1–deficient mice, and $***p < .01$ vs. GlcNAc6ST-2–deficient mice. For MLNs, $*p < .01$ vs. wild-type mice, $**p < .01$ vs. GlcNAc6ST-1–deficient mice, and $***p < .1$ vs. GlcNAc6ST-2–deficient mice.

2. Transfect Cos-1 cells with pcDNA1.1/L-selectin–IgM expression vector using Lipofectamine reagent (Invitrogen) according to the manufacturer's instruction.

3. Forty eight hours after transfection, collect the culture supernatant.

4. Centrifuge the supernatant 10,000 rpm for 5 min.

5. Collect the supernatant and store it at $-80°$ until use.

### Detection of L-Selectin–IgM Binding to HEV

1. Prepare frozen sections (7 $\mu$m in thickness) of PLNs, MLNs, and PPs from wild-type and mutant mice.

2. Air-dry the sections for 20 min.

3. Fix the sections with acetone for 5 min.

4. Incubate the sections with 3% bovine serum albumin (BSA) to block nonspecific binding sites.

5. Incubate the tissue sections overnight at $4°$ with culture supernatant of Cos-1 cells transfected with a pcDNA1.1/L-selectin–IgM expression vector (see Note 14).

6. After washing three times with Dulbecco Modified Eagle Medium (DMEM) containing 10% FCS, the sections are incubated with Alexa Fluor 594–conjugated goat anti-human IgM (Invitrogen) diluted 1:1000 at $4°$ for 1 h (see Note 15).

7. After washing three times, mount the sections with Vecta Shield mounting medium (Vector).

8. Analyze the sections by fluorescence microscopy.

### Determination of Contact Hypersensitivity Response

1. Shave the dorsal skin of mice on Day 0 (see Note 16).

2. Apply 25 $\mu$l of 0.5% 2,4-dinitrofluorobenzene (DNFB, Sigma) in acetone:olive oil (4:1, v/v) onto the shaved skin of mice on Days 0 and 1 (see Note 17).

3. On day 5, treat the right ear with 20 $\mu$l of 0.2% DNFB (10 $\mu$l/side of the pinna), and the left ear with the vehicle (see Note 18).

4. Measure ear swelling using a thickness gauge before and 24 h after treatment (see Note 19). A typical result is shown in Fig. 4.

### Notes

1. Occasionally, mix gently for efficient tissue destruction.

2. Vigorous vortexing is necessary to precipitate proteins in Steps 4 and 5 efficiently.

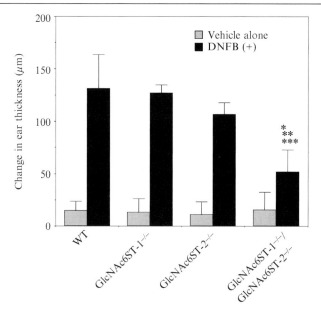

Fig. 4. Ear swelling 24 h after challenge with DNFB (black bars) and vehicle alone (gray bars) in wild-type (WT), GlcNAc6ST-1–deficient, GlcNAc6ST-2–deficient, and double–deficient mice (n = 5). *$p < .01$ vs. WT mice, **$p < .01$ vs. GlcNAc6ST-1–deficient mice, ***$p < .01$ vs. GlcNAc6ST-2–deficient mice.

3. Care should be taken not to take the precipitates.

4. This step should be done at room temperature to avoid precipitation of SDS. In addition, incubation time should not be prolonged, because unnecessary precipitates may be formed.

5. Shown here is the composition of reagents in one reaction tube. It is recommended to prepare the premixture without genomic DNA before adding it to the DNA samples.

6. This mixture includes buffer, anti-*Taq* DNA polymerase antibody, thermostable AccuPrime protein, $Mg^{2+}$, deoxyribonucleotide triphosphates, and recombinant *Taq* polymerase. The thermostable AccuPrime protein enhances specific primer-template hybridization during every cycle of PCR. Reaction reagents from other commercial sources can also be used, although optimization of PCR cycles may be needed.

7. Surgical tissue must be kept moist and used immediately. Delays in cell preparation will decrease cell viability.

8. For minimum cell loss due to adherence, polypropylene tubes are recommended over polystyrene tubes. As indicated in the method, addition of FCS to the RPMI-Med is also needed to minimize cell loss.

9. This step is needed to burst red blood cells. Longer incubation will damage other cell types including lymphocytes.

10. Intravenous injection is a difficult procedure that requires patience and experience. Inexperienced investigators should practice this skill before the actual experiments. It is recommended to warm the tail in warm water (55°) to dilate vessels before injection.

11. The major PLNs are axillary, brachial, cervical, and inguinal lymph nodes. These lymph nodes are usually quite small in normal unimmunized mice housed in specific pathogen-free facilities. Lymph nodes are yellow-white small tissues with a round shape. To collect lymph nodes, grasp them with forceps gently and pull them free of attached tissue. Adjacent fat should be dissected.

12. It is critical to know the forward and side scatters, and fluorescence parameters of the cells being analyzed so as not to collect false data. It is necessary to appropriately adjust parameter settings of the flow cytometer using unlabeled and CMFDA-labeled lymphocytes before sample analysis.

13. Dispense the supernatant into small aliquots before storage to avoid repeated freezing and thawing.

14. Moisture chamber should be used for the incubation.

15. Keep sections wet during the washing steps.

16. A 50-ml polypropylene tube with a square cut (1.5 × 7 cm) from the top and several small holes at the bottom can be used as a restrainer of mice. When placing a mouse in the tube, make sure that it is facing to the small holes at the bottom to secure its respiration, and then push the bottom of the mouse toward the bottom of the tube with five to six pieces of Kim wipe (Kimberly-Clark Worldwide, Inc.) to hold it. Shave the dorsal skin of the mouse through the square cut using a pair of electric hair clippers.

17. Wear plastic gloves and safety glasses when handling DNFB. Note that it is a strong hapten causing contact hypersensitivity.

18. It is better to perform this step with somebody else who helps hold the animal during the step.

19. A person who has no information about the genotypes of the animals should measure ear thickness to obtain results without any prejudice.

## Acknowledgment

I would like to thank Professor Minoru Fukuda for providing me the opportunity to write this chapter.

## References

Hemmerich, S., Bistrup, A., Singer, M. S., van Zante, A., Lee, J. K., Tsay, D., Peters, M., Carminati, J. L., Brennan, T. J., Carver-Moore, K., Leviten, M., Fuentes, M. E., Ruddle, N. H., and Rosen, S. D. (2001). Sulfation of L-selectin ligands by an HEV-restricted sulfotransferase regulates lymphocyte homing to lymph nodes. *Immunity* **15**, 237–247.

Hiraoka, N., Kawashima, H., Petryniak, B., Nakayama, J., Mitoma, J., Marth, J. D., Lowe, J. B., and Fukuda, M. (2004). Core 2 branching $\beta$1,6-$N$-acetylglucosaminyltransferase and high endothelial venule-restricted sulfotransferase collaboratively control lymphocyte homing. *J. Biol. Chem.* **279**, 3058–3067.

Kawashima, H., Petryniak, B., Hiraoka, N., Mitoma, J., Huckaby, V., Nakayama, J., Uchimura, K., Kadomatsu, K., Muramatsu, T., Lowe, J. B., and Fukuda, M. (2005). $N$-Acetylglucosamine-6-$O$-sulfotransferases 1 and 2 cooperatively control lymphocyte homing through L-selectin ligand biosynthesis in high endothelial venules. *Nat. Immunol.* **6**, 1096–1104.

Maly, P., Thall, A., Petryniak, B., Rogers, C. E., Smith, P. L., Marks, R. M., Kelly, R. J., Gersten, K. M., Cheng, G., Saunders, T. L., Camper, S. A., Camphausen, R. T., Sullivan, F. X., Isogai, Y., Hindsgaul, O., von Andrian, U. H., and Lowe, J. B. (1996). The $\alpha$(1,3) fucosyltransferase Fuc-TVII controls leukocyte trafficking through an essential role in L-, E-, and P-selectin ligand biosynthesis. *Cell* **86**, 643–653.

Sueyoshi, S., Tsuboi, S., Sawada-Hirai, R., Dang, U. N., Lowe, J. B., and Fukuda, M. (1994). Expression of distinct fucosylated oligosaccharides and carbohydrate-mediated adhesion efficiency directed by two different $\alpha$-1,3-fucosyltransferase. Comparison of E- and L-selection-mediated adhesion. . *J. Biol. Chem.* **269**, 32342–32350.

Uchimura, K., Gauguet, J.-M., Singer, M. S., Tsay, D., Kannagi, R., Muramatsu, T., von Andrian, U. H., and Rosen, S. D. (2005). A major class of L-selectin ligands is eliminated in mice deficient in two sulfotransferases expressed in high endothelial venules. *Nat. Immunol.* **6**, 1105–1113.

Uchimura, K., Kadomatsu, K., El-Fasakhany, F. M., Singer, M. S., Izawa, M., Kannagi, R., Takeda, N., Rosen, S. D., and Muramatsu, T. (2004). $N$-acetylglucosamine 6-$O$-sulfotransferase-1 regulates expression of L-selectin ligands and lymphocyte homing. *J. Biol. Chem.* **279**, 35001–35008.

Yeh, J. C., Hiraoka, N., Petryniak, B., Nakayama, J., Ellies, L. G., Rabuka, D., Hindsgaul, O., Marth, J. D., Lowe, J. B., and Fukuda, M. (2001). Novel sulfated lymphocyte homing receptors and their control by a Core1 extension $\beta$1,3-$N$-acetylglucosaminyltransferase. *Cell* **105**, 957–969.

# Section VI

# Functions Revealed by Forced Expression or Gene Knockout

[19]   Expression of Specific Carbohydrates by
       Transfection with Carbohydrate
       Modifying Enzymes

By JUNYA MITOMA and MINORU FUKUDA

Abstract

The identification of cDNAs encoding glycosyltransferases and carbohydrate-modifying enzymes such as sulfotransferases has allowed expression of a given enzyme in cells that lack the enzyme or express it at very low levels. By comparing the function and/or structure of carbohydrates expressed in cells before and after transfection, we can determine the function of the ectopically expressed enzyme. This assay is less time consuming than assaying function by obtaining cells deficient in a given enzyme. Moreover, it is a more definitive method for establishing the function of the enzyme because the result is derived from an enzyme introduced by transfection. Using this method, an enormous amount of knowledge relevant to the structure and function of glycoenzymes has been derived from such studies. In this chapter, we describe methods used to obtain mammalian cells that have acquired new carbohydrate structures and function following transfection of mammalian expression vectors harboring glycoenzymes.

Overview

A variety of cell surface oligosaccharides are synthesized in cells by sequential activities of glycosyltransferases and carbohydrate-modifying enzymes such as sulfotransferases. In the past 2 decades, more than a hundred cDNAs encoding specific carbohydrate-modifying enzymes have been cloned (Fukuda et al., 1996, 2001). These cDNAs have been identified using three strategies: (1) protein purification and subsequent amino acid sequence determination followed by conventional cDNA cloning based on that sequence (Masri et al., 1988; Narimatsu et al., 1986; Yamamoto et al., 1990); (2) complementation of defective gene(s) by introducing a cDNA or genomic DNA library (Kumar and Stanley, 1989, 1990); or (3) expression cloning and selection of a cDNA directing expression of a specific carbohydrate in transfected versus un-transfected cells (Bierhuizen and Fukuda, 1992; Fukuda et al., 1996; Larsen et al., 1989; Nakayama et al., 1995; Ong et al., 1998).

METHODS IN ENZYMOLOGY, VOL. 416                                    0076-6879/06 $35.00
Copyright 2006, Elsevier Inc. All rights reserved.                 DOI: 10.1016/S0076-6879(06)16019-X

Extending the third and potentially the second approach, one can express in cells specific carbohydrates and test their function by introducing cDNA(s) encoding a specific enzyme into parental cells that lack the enzyme or express it at low levels. Several factors should be considered in designing such an experiment. First, one must choose cells or cell lines that lack one or possibly more of the carbohydrate-modifying enzymes required to synthesize that structure. Second, one must be able to detect a functional or structural consequence of expression of the newly introduced cDNA. Third, it is preferable to have an idea as to the function of the enzyme before testing it. Fourth, the cDNA should be cloned into a mammalian expression vector harboring a powerful promoter such as the Cytomegalovirus promoter in order to achieve high levels of expression.

Mammalian cells expressing SV40 large T and polyoma large T are particularly useful for this purpose. A transfected expression vector harboring an SV40 or polyomavirus origin of replication can be amplified many-fold in mammalian cells, because SV40 large T or polyoma large T bind to those respective origins, enabling transfected plasmids to replicate episomally. Indeed, it was critical to express polyoma large T gene together with a cDNA library constructed in pcDNA1, which contains a polyoma replication origin, to clone cDNA encoding Core2$\beta$1,6-$N$-acetylglucosaminyltransferase-1 (Core2GlcNAcT-1) in Chinese hamster ovary (CHO) cells. CHO cells were particularly well suited to this protocol because polyoma large T functions well in rodent cells (Bierhuizen and Fukuda, 1992). COS-1 cells, on the other hand, express SV40 large T because they are stably transformed with a portion of the SV40 virus. Vectors such as pcDNA1, which also contain an SV40 origin of replication, can be amplified in COS-1 cells, resulting in high expression of encoded proteins. When recipient cells do not express SV40 large T antigen, they can be transfected with a vector encoding it, making them capable of amplifying an appropriate vector (de Chaseval and de Villartay, 1991). These procedures are particularly useful when transient expression of a mammalian expression vector results in a low yield of a desired protein.

As one of the first examples of determining the function of glycosyltransferases, Lowe *et al.* (1990) expressed $\alpha$1,3-fucosyltransferases in CHO cells and tested for E-selectin ligand activity. Their studies showed that E-selectin binds to transfected cells when sialyl Lewis X is expressed on the cell surface. These studies are a classic example of how expression of a specific carbohydrate can be correlated with function, in this case, binding of E-selectin (Lowe *et al.*, 1991).

CHO cells are among the most suitable cells to analyze mucin-type $O$-glycan structures, because CHO cells synthesize minimal amounts of core 2 $O$-glycan structure and synthesize only core 1 (Gal$\beta$1-3GalNAc-$O$-Ser/Thr)

(Sasaki *et al.*, 1987). Because CHO cells lack almost all fucosyltransferases, without introduction of exogenous cDNA, this core 1 oligosaccharide can be modified with sialic acid by sialyltransferases, but not with *N*-acetyllactosamine or fucose, which are often needed to make complex and physiologically very important oligosaccharides such as blood group H, sialyl Lewis X and sialyl Lewis A.

Many mutant CHO cells defective in key enzymes are available (see Stanley and Ioffe, 1995; Chapter 11 in this volume by Patnaik and Stanley, 2006). For example, Lec1 CHO cells lack *N*-acetylglucosaminyltransferase-I, which is essential for synthesis of complex-type *N*-glycans (Kumar *et al.*, 1989, 1990). Using Lec1 cells, we can build up functional carbohydrate-capping structures only in mucin-type *O*-glycans if, for example, Core2GlcNAcT-1 is coexpressed. If we use Lec2 cells, which lack Golgi-sialylation (Deutscher *et al.*, 1984), extensive modification of *N*-acetyllactosamine can be achieved. This cell line was, thus, used to analyze HNK-1 glycan biosynthesis, because $\beta$1, 3-glucuronyltransferase can form the HNK-1 glycan precursor without competing with a sialyltransferase (Ong *et al.*, 1998). In the following paragraphs, we provide a protocol for a typical experiment using CHO cells.

## Plasmid Preparation and Cell Culture

All cDNAs used in this section encode human proteins, although cDNAs encoding mouse glycoenzymes can be used. To transfect CHO cells with exogenous glycosyltransferases, plasmids containing cDNA are purified by conventional methods such as alkaline/sodium dodecylsulfate (SDS) method followed by PEG 6000 precipitation, cesium chloride density gradient ultracentrifugation (Sambrook *et al.*, 1989), or plasmid purification kits such as Qiagen Plasmid Maxi Kit. Plasmid vectors we have used include pcDNA1, pcDNA3, pcDNA3.1/Hyg, pcDNA3.1/Zeo, and pCMV/Bsd (Invitrogen). The first three vectors use the Cytomegalovirus promoter to drive expression of foreign genes. pCMV/Bsd (Blasticidin S) (Invitrogen), for example, can be co-transfected with an expression vector lacking an antibiotic-resistance gene appropriate for eukaryotic cells, such as pcDNA1. At least 3 h before transfection, CHO cells are treated with trypsin/ethylenediaminetetraacetic acid (EDTA) or enzyme-free Cell Dissociation solution (Specialty Media, Phillipsburg, NJ) and seeded onto cell culture dishes at 10–80% confluency in alpha-MEM with 10% fetal bovine serum (alpha-MEM/FBS). Lipofectamine and PLUS reagents are used for liposome-mediated transfection. If cells are approximately 50% confluent, we use the lipid/DNA ratio recommended by the manufacturer. This ratio should be adjusted according to cell confluency.

Transient Expression

*Materials*

CHO-K1 cells
Lipofectamine reagent (Invitrogen)
PLUS reagent (Invitrogen)
Opti-MEM (Invitrogen)
Alpha-MEM
FBS
1 mg/ml plasmid harboring glycosyltransferase cDNA

*Methods*

When transfecting CHO cells in a 10-cm dish, 4 $\mu$g of plasmid is diluted in 0.75 ml of Opti-MEM, and 20 $\mu$l of PLUS reagent is added. When transfecting two or more plasmids, plasmids are combined in an equal amount, with a total amount of 8 $\mu$g or less. Thirty microliters of Lipofect-amine reagent is diluted separately in 0.75 ml of the same medium. Fifteen minutes after dilution, both solutions are combined and incubated another 15 min. During the incubation, the culture medium is replaced with 5 ml of Opti-MEM. The liposome-plasmid mixture is then applied to cells in 5 ml of serum-free medium. Two to six hours later, the medium is changed to alpha-MEM/FBS. Twenty-four to forty-eight hours later, cells are analyzed for expression of target antigen. Alpha-MEM without FBS and antibiotics can substitute for Opti-MEM. Figures 1 and 2 show the results of transient transfection.

Stable Expression

*Additional Materials for Stable Expression*

Geneticin (Invitrogen)
Hygromycin B (Calbiochem)
Zeocin (Invitrogen)
Blasticidin S (Invitrogen)

*Methods*

Stable transfection uses methods similar to transient transfection. If the transfected plasmid does not contain an appropriate antibiotic-resistance gene, such as the *neo* gene for Geneticin or the Hygromycin B–resistance gene, it should be co-transfected with one-fifth to one-fifteenth the amount

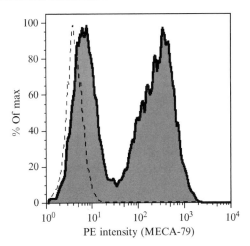

FIG. 1. Flow cytometric analysis with MECA-79 antibody. Chinese hamster ovary (CHO) cells stably transfected with CD34 and FucT-VII are transiently transfected with Core1-$\beta$3GlcNAcT and GlcNAc6ST-2. Transient transfection resulted in more than 50% of cells becoming MECA-79 positive (bold solid line). Dashed line represents CHO/CD34/FucT-VII without primary antibody.

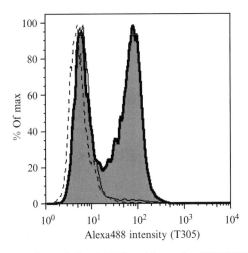

FIG. 2. Flow cytometric analysis of T305-positive cells. CHO/PSGL-1/FucT-VII cells became T305 positive after transient co-transfection with CD43 and Core2GlcNAcT-1 (solid bold line). Transfection of CD43 alone did not produce antibody-positive cells (solid thin line). Dashed thin line represents a control without the primary antibody.

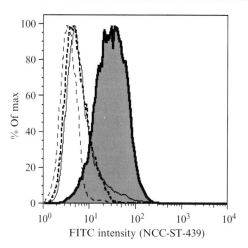

FIG. 3. Flow cytometric analysis of various stably transfected Chinese hamster ovary (CHO) lines using the NCC-ST-439 antibody. This figure shows CHO/PSGL-1/Core2GlcNAcT-1/FucT-VII without primary antibody (dashed thin line), CHO/PSGL-1/Core2 GlcNAcT-1 (solid thin line), CHO/PSGL-1/FucT-VII (dashed bold line), and CHO/PSGL-1/Core2GlcNAcT-1/FucT-VII (solid bold line).

of a plasmid harboring such a gene. Following transfection, 24 h after changing the medium to alpha-MEM/FBS, either 1 mg/ml G418, 0.4 mg/ml Hygromycin B (for pcDNA3.1/Hyg), 0.2 mg/ml Zeocin (for pcDNA3.1/Zeo), or 10 μg/ml Blasticidin S (for pCMV/Bsd) is added. Medium should be changed to fresh medium containing appropriate antibiotics every 3–4 days. Well-separated colonies should appear approximately 2 wk after transfection, and a single colony can be picked. Alternatively, cells can be trypsinized, seeded again at 20 cells/10-cm dish and colonies should form in approximately 5 days. Cells can also be sorted with a cell sorter to a single cell if desired. Figure 3 represents the result of stable transfection.

## Stable Expression of Multiple Genes

Two methods are available to obtain cells expressing multiple genes, which is necessary if a desired antigen requires the activity of multiple enzymes. One is simultaneous transfection of multiple plasmids. For example, MECA-79 antigen requires the function of two enzymes, $\beta$1,3-$N$-acetylglucosaminyl transferase-3 (Yeh et al., 2001) and $N$-acetylglucosamine-6-sulfotransferase-2 (Hiraoka et al., 1999). To obtain MECA-79–positive CHO cells, equal amounts of expression vectors encoding both enzymes are transfected with one-tenth the amount of plasmid coding the antibiotic-resistance gene, in case the plasmids do not harbor one. The advantage of this method is that multiple genes can be screened with one antibiotic. The other method is to transfect multiple genes sequentially. Once a stable cell line that expresses the desired gene has been obtained, the second transfection using a different antibiotic-resistance gene is carried out on that cell line. This is the method of choice if one is working with a line of cells expressing one gene and desires to transfect another gene.

Flow Cytometry

Detection of oligosaccharides is often done using a specific monoclonal antibody such as CSLEX-1 for nonsulfated sialyl Lewis X, HECA-452 for both sulfated and nonsulfated sialyl Lewis X (Mitsuoka *et al.*, 1997), and MECA-79 for peripheral node addressin (Streeter *et al.*, 1988). Typical antibodies and their known structural requirements are listed in Table I. There are several methods, such as flow cytometry and immunofluorescence microscopy, to evaluate stained cells. Here, we describe flow cytometry, a fast and quantitative method.

*Materials*

Phosphate-buffered saline (PBS) containing 0.1 mg/ml bovine serum albumin (BSA) (PBS/BSA)
Cell Dissociation solution (Specialty Media)

TABLE I
TYPICAL MONOCLONAL ANTIBODIES FOR CELL SURFACE CARBOHYDRATES

| Antibody | Structure | cDNA |
|---|---|---|
| CSLEX-1 | Sialyl Lewis X (nonsulfated) | FucT-VII |
| 2H5 | Sialyl Lewis X | FucT-VII |
| HECA-452 | Sialyl Lewis X (6-sulfated and nonsulfated) | FucT-VII |
| MECA-79 | 6-sulfo-*N*-acetyllactosamine on core 1 extended *O*-glycan | Core1-$\beta$3GlcNAcT-1, GlcNAc6ST-1/2 |
| T305 | CD43 modified with core 2 branch | CD43, Core2GlcNAcT-1 |
| NCC-ST-439 | Sialyl Lewis X on core 2 branch | FucT-VII, Core2GlcNAcT-1 |
| HNK-1 | 3-sulfoglucuronyl glycans | GlcAT-S/P, HNK ST |

*Note:* Antibodies and required structures are listed, as are cDNAs required for expression of epitopes in CHO cells. FucT-TVII, fucosyltransferase VII; Core1-$\beta$3GlcNAcT, core1$\beta$1,3- *N*-acetylglucosaminyltransferase; GlcNAc6ST-1/2, *N*-acetylglucosamine-6-sulfotransferase-1 and -2; CD43, leukosialin/sialophorin; Core2GlcNAcT-1, core2$\beta$1,6-*N*-acetylglucosaminyltransferase-1 GlcAT-S/P, $\beta$1,3-glucuronyltransferase-S and -P; HNK-1 ST, HNK-1 sulfotransferase.

Primary antibodies such as MECA-79, T305, and NCC-ST-439
Secondary antibodies labeled with FITC, Alexa Fluor 488, or Cy2
Secondary antibodies labeled with phycoerythrin (PE)
Falcon 2054 tubes

*Methods*

Cells in a 10-cm dish are treated with 1 ml of Cell Dissociation solution for 10 min at room temperature. Nine milliliters of PBS/BSA is added and cells are well dissociated by repeated pipetting. One milliliter of cell suspension is added to round-bottomed polystyrene tubes such as Falcon 2054 and centrifuged at 1000 rpm for 3 min. To the cell pellet, 20–100 $\mu$l of antibody solution is added, gently mixed by tapping the tube, and incubated for 5–30 min at 4°. The proper antibody dilution should be determined for each lot. Usually, a culture supernatant of hybridoma cells can be used directly or at a 1:10 dilution in PBS/BSA, ascites can be diluted approximately 1:1000, and purified antibody can be used at a concentration of 1 to approximately 10 $\mu$g/ml. The cells are washed two times with 0.5 ml of ice-cold PBS/BSA, and FITC- or PE-labeled secondary antibody is added to the tube. After 5 min at 4°, cells are washed once again and resuspended in 0.5 ml PBS/BSA. FACScan, FACSort, FACSCalibur and FACSCanto (BD) can be used to detect stained cells.

*Examples*

MECA-79 ANTIGEN. The MECA-79 antigen, also known as peripheral node addressin (PNAd), is an important marker of high endothelial venules. This antigen is recognized by the monoclonal antibody MECA-79, established by Streeter *et al.* (1988). The structure responsible for antibody recognition is Gal$\beta$1-4(6-sulfo)GlcNAc$\beta$1-3Gal$\beta$1-3GalNAc$\alpha$-*O*-R (Yeh *et al.*, 2001). A critical step to synthesize this structure is the addition of GlcNAc by core 1 extension enzyme (Core1-$\beta$3GlcNAcT, also called Core1-$\beta$1,3-*N*-acetylglucosaminyltransferase, or $\beta$1,3-*N*-acetylglucosaminyltransferase-3). The modification of GlcNAc residues with a sulfate group is also essential for formation of this epitope (Hemmerich *et al.*, 1994). Sulfation is done by *N*-acetylglucosamine-6-sulfotransferases-1 and -2 (GlcNAc6ST-1 and -2, respectively) (Hiraoka *et al.*, 1999; Uchimura *et al.*, 2004). Because CHO cells do not express these enzymes but do express galactosyltransferases, introduction of cDNA for Core1-$\beta$3GlcNAcT and GlcNAc6ST-1 or -2 is required to express the MECA-79 antigen. Figure 1 shows flow cytometric analysis of CHO cells stably expressing CD34, fucosyltransferase-VII, and GlcNAc6ST-2 after transient transfection of Core1-$\beta$3GlcNAcT. CD34 expression facilitates efficient scaffolding of *O*-glycans and fucosyltransferase-VII is required for selectin binding activity.

Introduction of Core1-$\beta$3GlcNAcT results in establishment of MECA-79–positive cells. In this experiment, transfection efficiency is more than 50%. Cells expressing GlcNAc6ST-2 can be detected by this procedure, because cells expressing both GlcNAc6ST-2 and Core1-$\beta$3GlcNAcT are MECA-79 positive. By a similar procedure, we established MECA-79–positive CHO cells stably expressing CD34, FucT-VII, Core1-$\beta$3GlcNAcT, and GlcNAc6ST-2. Lymphocyte rolling was supported over this cell line, demonstrating L-selectin ligand activity of MECA-79–positive 6-sulfo sialyl Lewis X (Yeh *et al.*, 2001).

T305 ANTIGEN. Leukosialin, also known as sialophorin or CD43, is a mucin expressed in lymphocytes that functions as a counter-receptor for the macrophage adhesion receptor sialoadhesin (van den Berg TK *et al.*, 2001) and the E-selectin ligand (Fuhlbrigge *et al.*, 2006). Modification with a core 2 branch attenuates the function of CD43 (Tsuboi and Fukuda, 1997; Fukuda and Tsuboi, 1999). T305, a monoclonal antibody for CD43, recognizes only CD43 modified with a core 2 branch (Bierhuizen *et al.*, 1994; Piller *et al.*, 1991; Sportsman *et al.*, 1985). Figure 2 shows a typical flow cytometric analysis using T305. CHO cells stably expressing PSGL-1 and FucT-VII are negative for this antigen (not shown). In this experiment, transient transfection of CD43 alone did not alter the T305 staining profile (thin solid line in Fig. 2). However, transient co-transfection of both CD43 and Core2 GlcNAcT-1 resulted in more than 50% T305-positive cells. This assay is also useful to detect CHO cells expressing Core2GlcNAcT-1, because those cells are positive for T305 after transient expression of CD43.

NCC-ST-439 ANTIGEN. Core 2 *O*-glycan, Gal$\beta$1→4(GlcNAc$\beta$1→6)Gal$\beta$1→3GalNAc→*O*-Ser/Thr, is a well-known, physiologically significant major core structure of mucin-type *O*-glycans. P-selectin glycoprotein ligand-1 (PSGL-1) must be modified with this core structure with sialyl Lewis X for recognition by P-selectin (Kumar *et al.*, 1996). There is no specific antibody that recognizes core 2 branch itself, but the monoclonal antibody NCC-ST-439 preferentially recognizes sialyl Lewis X on core 2 *O*-glycans (Kobayashi *et al.*, 2004; Kumamoto *et al.*, 1998). Figure 3 shows staining of CHO cells stably expressing various glycosyltransferases. Stable transfection of FucT-VII (Fig. 3, dotted bold line) or Core2GlcNAcT-1 (Fig. 3, thin solid line) alone in CHO cells expressing PSGL-1 as a scaffold protein does not make cells positive for this antibody. If we stably transfect both Core2GlcNAcT-1 and FucT-VII in these cells, they become NCC-ST-439-positive (Fig. 3, bold solid line).

## Acknowledgments

The authors thank Aleli Morse for organizing the manuscript. The work in our laboratory was supported by the National Institutes of Health (NIH) grants CA33000, CA33895, CA48737, and CA71932.

## References

Bierhuizen, M. F., and Fukuda, M. (1992). Expression cloning of a cDNA encoding UDP-GlcNAc:Gal β1-3-GalNAc-R (GlcNAc to GalNAc) β1-6GlcNAc transferase by gene transfer into CHO cells expressing polyoma large tumor antigen. *Proc. Natl. Acad. Sci. USA* **89,** 9326–9330.

Bierhuizen, M. F., Maemura, K., and Fukuda, M. (1994). Expression of a differentiation antigen and poly-*N*-acetyllactosaminyl *O*-glycans directed by a cloned core 2 beta-1,6-*N*-acetylglucosaminyltransferase. *J. Biol. Chem.* **269,** 4473–4479.

de Chaseval, R., and de Villartay, J. P. (1991). High level transient gene expression in human lymphoid cells by SV40 large T antigen boost. *Nucleic Acids Res.* **20,** 245–250.

Deutscher, S. L., Nuwayhid, N., Stanley, P., Briles, E. I., and Hirschberg, C. B. (1984). Translocation across Golgi vesicle membranes: A CHO glycosylation mutant deficient in CMP-sialic acid transport. *Cell* **39,** 295–299.

Fuhlbrigge, R. C., King, S. L., Sackstein, R., and Kupper, T. S. (2006). CD43 is a ligand for E-selectin on CLA[+] human T cells. *Blood* **107,** 1421–1426.

Fukuda, M., and Tsuboi, S. (1999). Mucin-type *O*-glycans and leukosialin. *Biochim. Biophys. Acta* **1455,** 205–217.

Fukuda, M., Bierhuizen, M. F., and Nakayama, J. (1996). Expression cloning of glycosyltransferases. *Glycobiology* **6,** 683–689.

Fukuda, M., Hiraoka, N., Akama, T. O., and Fukuda, M. N. (2001). Carbohydrate-modifying sulfotransferases: Structure, function, and pathophysiology. *J. Biol. Chem.* **27,** 47747–47750.

Hemmerich, S., Butcher, E. C., and Rosen, S. D. (1994). Sulfation-dependent recognition of high endothelial venules (HEV)-ligands by L-selectin and MECA 79, and adhesion-blocking monoclonal antibody. *J. Exp. Med.* **180,** 2219–2226.

Hiraoka, N., Petryniak, B., Nakayama, J., Tsuboi, S., Suzuki, M., Yeh, J. C., Izawa, D., Tanaka, T., Miyasaka, M., Lowe, J. B., and Fukuda, M. (1999). A novel, high endothelial venule-specific sulfotransferase expresses 6-sulfo sialyl Lewis[X], an L-selectin ligand displayed by CD34. *Immunity* **11,** 79–89.

Kobayashi, M., Mitoma, J., Nakamura, N., Katsuyama, T., Nakayama, J., and Fukuda, M. (2004). Induction of peripheral lymph node addressin in human gastric mucosa infected by Helicobacter pylori. *Proc. Natl. Acad. Sci. USA* **101,** 17807–17812.

Kumamoto, K., Mitsuoka, C., Izawa, M., Kimura, N., Otsubo, N., Ishida, H., Kiso, M., Yamada, T., Hirohashi, S., and Kannagi, R. (1998). Specific detection of sialyl Lewis X determinant carried on the mucin GlcNAcβ1→6GalNAcalpha core structure as a tumor-associated antigen. *Biochem. Biophys. Res. Commun.* **247,** 514–517.

Kumar, R., and Stanley, P. (1989). Transfection of a human gene that corrects the Lec1 glycosylation defect: Evidence for transfer of the structural gene for *N*-acetylgluco-saminyltransferase I [published erratum in *Mol. Cell. Biol.* **10,** 3857. *Mol. Cell. Biol.* **9,** 5713–5717.

Kumar, R., Yang, J., Larsen, R. D., and Stanley, P. (1990). Cloning and expression of *N*-acetylglucosaminyltransferase I, the medial Golgi transferase that initiates complex *N*-linked carbohydrate formation. *Proc. Natl. Acad. Sci. USA* **87,** 9948–9952.

Kumar, R., Camphausen, R. T., Sullivan, F. X., and Cumming, D. A. (1996). Core2 β-1, 6-*N*-acetylglucosaminyltransferase enzyme activity is critical for P-selectin glycoprotein ligand-1 binding to P-selectin. *Blood* **88,** 3872–3879.

Larsen, R. D., Rajan, V. P., Ruff, M. M., Kukowska-Latallo, J., Cummings, R. D., and Lowe, J. B. (1989). Isolation of a cDNA encoding a murine UDPgalactose:β-D-galactosyl-1, 4-*N*-acetyl-D-glucosaminide α-1,3-galactosyltransferase: Expression cloning by gene transfer. *Proc. Natl. Acad. Sci. USA* **86,** 8227–8231.

Lowe, J. B., Stoolman, L. M., Nair, R. P., Larsen, R. D., Berhend, T. L., and Marks, R. M. (1990). ELAM-1-dependent cell adhesion to vascular endothelium determined by a transfected human fucosyltransferase cDNA. *Cell* **63**, 475–484.

Lowe, J. B., Kukowska-Latallo, J. F., Nair, R. P., Larsen, R. D., Marks, R. M., Macher, B. A., Kelly, R. J., and Ernst, L. K. (1991). Molecular cloning of a human fucosyltransferase gene that determines expression of the Lewis x and VIM-2 epitopes but not ELAM-1-dependent cell adhesion. *J. Biol. Chem.* **266**, 17467–17477.

Masri, K. A., Appert, H. E., and Fukuda, M. N. (1988). Identification of the full-length coding sequence for human galactosyltransferase ($\beta$-$N$-acetylglucosaminide: $\beta$1,4-galactosyltransferase). *Biochem. Biophys. Res. Commun.* **157**, 657–663.

Mitsuoka, C., Kawakami-Kimura, N., Kasugai-Sawada, M., Hiraiwa, N., Toda, K., Ishida, H., Kiso, M., Hasegawa, A., and Kannagi, R. (1997). Sulfated sialyl Lewis X, the putative L-selectin ligand, detected on endothelial cells of high endothelial venules by a distinct set of anti-sialyl Lewis X antibodies [published erratum in *Biochem. Biophys. Res. Commun.* **233**, 576]. *Biochem. Biophys. Res. Commun.* **230**, 546–551.

Nakayama, J., Fukuda, M. N., Fredette, B., Ranscht, B., and Fukuda, M. (1995). Expression cloning of a human polysialyltransferase that forms the polysialylated neural cell adhesion molecule present in embryonic brain. *Proc. Natl. Acad. Sci. USA* **92**, 7031–7035.

Narimatsu, H., Sinha, S., Brew, K., Okayama, H., and Qasba, P. K. (1986). Cloning and sequencing of cDNA of bovine $N$-acetylglucosamine ($\beta$1–4)galactosyltransferase. *Proc. Natl. Acad. Sci. USA* **83**, 4720–4724.

Ong, E., Yeh, J. C., Ding, Y., Hindsgaul, O., and Fukuda, M. (1998). Expression cloning of a human sulfotransferase that directs the synthesis of the HNK-1 glycan on the neural cell adhesion molecule and glycolipids. *J. Biol. Chem.* **273**, 5190–5195.

Patnaik, S. K., and Stanley, P. (2006). Lectin-resistant CHO glycosylation mutants. *Methods Enzymol.* **416**, 159–182.

Piller, F., Le Deist, F., Weinberg, K. I., Parkman, R., and Fukuda, M. (1991). Altered $O$-glycan synthesis in lymphocytes from patients with Wiskott-Aldrich syndrome. *J. Exp. Med.* **173**, 1501–1510.

Sambrook, J., Fritsch, E. F., and Maniatis, T. (1989). "Molecular Cloning, A Laboratory Manual," 2nd ed., pp. l.21–l.52. Cold Spring Harbor Laboratory Press, Plainview, NY, USA.

Sasaki, H., Bothner, B., Dell, A., and Fukuda, M. (1987). Carbohydrate structure of erythropoietin expressed in Chinese hamster ovary cells by a human erythropoietin cDNA. *J. Biol. Chem.* **262**, 12059–12076.

Sportsman, J. R., Park, M. M., Cheresh, D. A., Fukuda, M., Elder, J. H., and Fox, R. I. (1985). Characterization of a membrane surface glycoprotein associated with T-cell activation. *J. Immunol.* **135**, 158–164.

Stanley, P., and Ioffe, E. (1995). Glycosyltransferase mutants: Key to new insights in glycobiology. *FASEB J.* **9**, 1436–1444.

Streeter, P. R., Rouse, B. T., and Butcher, E. C. (1988). Immunohistologic and functional characterization of a vascular addressin involved in lymphocyte homing into peripheral lymph nodes. *J. Cell. Biol.* **107**, 1853–1862.

Tsuboi, S., and Fukuda, M. (1997). Branched $O$-linked oligosaccharides ectopically expressed in transgenic mice reduce primary T-cell immune responses. *EMBO J.* **16**, 6364–6373.

Uchimura, K., Kadomatsu, K., El-Fasakhany, F. M., Singer, M. S., Izawa, M., Kannagi, R., Takeda, N., Rosen, S. D., and Muramatsu, T. (2004). $N$-acetylglucosamine 6-$O$-sulfotransferase-1 regulates expression of L-selectin ligands and lymphocyte homing. *J. Biol. Chem.* **279**, 35001–35008.

van den Berg, T. K., Nath, D., Ziltener, H. J., Vestweber, D., Fukuda, M., van Die, I., and Crocker, P. R. (2001). Cutting edge: CD43 functions as a T cell counterreceptor for the macrophage adhesion receptor sialoadhesin (Siglec-1). *J. Immunol.* **166,** 3637–3640.

Yamamoto, F., Marken, J., Tsuji, T., White, T., Clausen, H., and Hakomori, S. (1990). Cloning andcharacterization of DNA complementary to human UDP-GalNAc: Fucα1→2Galα1→3GalNAc transferase (histo-blood group A transferase) mRNA. *J. Biol. Chem.* **265,** 1146–1151.

Yeh, J. C., Hiraoka, N., Petryniak, B., Nakayama, J., Ellies, L. G., Rabuka, D., Hindsgaul, O., Marth, J. D., Lowe, J. B., and Fukuda, M. (2001). Novel sulfated lymphocyte homing receptors and their control by a Core1 extension β1,3-N-acetylglucosaminyltransferase. *Cell* **105,** 957–969.

# [20]  N-Glycan Structure Analysis Using Lectins and an α-Mannosidase Activity Assay

*By* Tomoya O. Akama and Michiko N. Fukuda

## Abstract

α-Mannosidase IIx (MX) and α-mannosidase II (MII) are homologous enzymes whose critical roles in N-glycan processing were established in large part by analysis of the MII/MX double-knockout mouse. To analyze the structures of N-glycans synthesized in the mutant mice, we employed lectin blot and lectin histochemistry in addition to mass spectrometry analysis and two-dimensional high-performance liquid chromatography (HPLC) mapping. We also produced soluble MII and MX by transfecting mammalian cells with expression vectors and determined substrate specificity of MX. This chapter describes methods using lectins to analyze N-glycans in knockout mice and provides a protocol to assay α-mannosidase activity using soluble MX.

## Overview

N-glycosylation is the major form of post-translational modification of newly synthesized proteins through the sorting pathway. The major biosynthetic steps for N-glycans in vertebrates have been established (Kornfeld and Kornfeld, 1985; Schachter, 1991). A key conversion of high mannose

METHODS IN ENZYMOLOGY, VOL. 416
0076-6879/06 $35.00
DOI: 10.1016/S0076-6879(06)16020-6

to complex-type oligosaccharides occurs in the medial Golgi, where GlcNAc-transferase I (GlcNAc-TI) adds GlcNAc to form a hybrid-type N-glycan, GlcNAc$_1$Man$_5$GlcNAc$_2$ (Schachter et al., 1983). The Golgi enzyme MII then removes two mannosyl residues to form GlcNAc$_1$Man$_3$GlcNAc$_2$ (Moremen et al., 1994; Tulsiani et al., 1982), which is further modified by GlcNAc-transferase II (GlcNAc-TII) (Harpaz and Schachter, 1980) to form GlcNAc$_2$Man$_3$GlcNAc$_2$, the precursor of complex-type N-glycans.

Targeted mutation in mouse is a powerful means to define gene function. Knockout mice often reveal unanticipated biological roles of the targeted gene product. For example, when the gene encoding GlcNAc-TI was disrupted in the mouse, GlcNAc-TI nulls died at embryonic day 10 (E10) because of multisystemic defects in various morphogenic processes, including neural tube formation (Ioffe and Stanley, 1994; Metzler et al., 1994). Because GlcNAc-TI nulls can synthesize only high mannose-type N-glycans, the knockout phenotype demonstrates that in the mouse high mannose-type N-glycans alone cannot support embryonic development beyond E10. On the other hand, mice lacking GlcNAc-TII develop embryonically and are born (Wang et al., 2002). Because GlcNAc-TII nulls synthesize hybrid-type N-glycans, these findings suggest that hybrid-type N-glycans can support embryogenesis in the mouse. However, newborn GlcNAc-TII-null pups show severe gastrointestinal, hematopoietic, osteogenic, and neuronal abnormalities, phenotypes resembling human congenital disorders of glycosylation IIa (CDGIIa) (Schachter and Jaeken, 1999; Tan et al., 1996). These findings indicate that hybrid-type N-glycans are not sufficient to maintain normal postnatal development in the mouse and in humans (Wang et al., 2001).

Although MII catalyzes the step in N-glycan biosynthesis after GlcNAc-TI and before GlcNAc-TII (Kornfeld and Kornfeld, 1985; Schachter, 1991), MII-null mice were born and were apparently normal but exhibited mild dyserythropoiesis, a phenotype resembling the human genetic disease congenital dyserythropoietic anemia type II (CDAII) or *HEMPAS* (Chui et al., 1997). Furthermore, MII nulls synthesized complex-type N-glycans, despite the complete absence of MII activity in tissues. These findings led to a proposal of an alternative pathway for MII activity (Chui et al., 1997), a pathway that included two candidate enzymes, $\alpha$-mannosidase III, a cobalt-dependent broad specificity $\alpha$-mannosidase expressed in rat liver microsomes (Bonay and Hughes, 1991; Bonay et al., 1992), and $\alpha$-mannosidase IIx (MX).

Human MX (Misago et al., 1995) is the product of the *MAN2A2* gene, which is homologous to *MAN2A1* for human MII. Previous studies suggest that human MX is enzymatically active and functions in N-glycan biosynthesis (Misago et al., 1995; Oh-Eda et al., 2001). When *Man2a2*, the mouse ortholog of *MAN2A2*, was disrupted, MX nulls were apparently normal except that mutant males were subfertile (Akama *et al.*, 2002). Although MX-null mice produced complex-type *N*-glycans, *N*-linked carbohydrate structures in the MX-null mouse testis differed from those expressed in wild-type mice, suggesting that MX also processes *N*-glycans. To determine the role of MX on *N*-glycan biosynthesis *in vivo*, we generated *Man2a1* and *Man2a2* double gene knockouts (DKOs) and found that almost all double-null embryos died shortly after birth because of respiratory failure, which represents a much more severe phenotype than that seen in single nulls. Structural analysis of *N*-glycans showed that double nulls completely lacked complex-type *N*-glycans, demonstrating an essential role of both enzymes in *N*-glycan processing (Akama *et al.*, 2006).

In this chapter, we describe methods used to analyze *N*-glycans by lectin blots and lectin histochemistry. Detailed *N*-glycan structure analysis by mass spectrometry and two-dimensional HPLC analyses are described elsewhere in this book. This chapter also provides protocols for MII and MX activity assays using recombinant enzymes.

Lectin Blot Analysis of Glycoproteins Produced by MII/MX
DKO Proteins

Lectins are convenient probes for analyzing alterations in *N*-glycan carbohydrate structure. Biotinylated lectins are commercially available. Lectin binding to a specific set of glycoproteins can be visualized by peroxidase-conjugated avidin on a blot. Avidin conjugated with other enzymes, such as alkaline phosphatase, can be also used.

*Materials*

PVDF filter protein blot (made using common methods for immunoblot analysis)

Blocking solution: 5% bovine serum albumin (BSA) dissolved in phosphate-buffered saline (PBS)

Avidin blocking solution: 20 μg/ml avidin in blocking solution

Washing buffer: PBS with 0.05% Tween 20

Lectin solution: 2.5 μg/ml biotinylated lectin and 10 μM biotin in blocking solution

Horseradish peroxidase (HRP)–avidin solution: 1 μg/ml HRP-conjugated avidin (or streptavidin) and 0.3% BSA in PBS

Chemiluminescent reagent (e.g., SuperSignal West Pico Kit, Pierce).

*Procedures*

1. Block a blotted filter with Avidin blocking solution for at least 1 h at room temperature (or overnight at 4°). Free avidin will mask endogenous

biotin-binding proteins that produce extra bands and high background (Vaitaitis *et al.*, 1999). Because BSA does not have any carbohydrate modification on its protein backbone, this blocking solution can be universally used for any lectin. Another blocking solution such as 10% skim milk in PBS may also be used; however, the blocking solution must not include carbohydrate structures that will be recognized by the lectin of interest. The blocking reagent can be reused by adding $NaN_3$ to 0.05% and stored at 4°.

2. Wash the filter three times for 5 min each with washing buffer.

3. Incubate the filter with Lectin solution for 1 h at room temperature. Free biotin will neutralize the biotin-binding site of avidin, which binds to endogenous biotin-binding proteins.

4. Wash the filter three times for 5 min each with washing buffer.

5. Incubate the filter with HRP–avidin solution for 30 min at room temperature. Although HRP–avidin can be diluted by PBS without BSA, addition of 0.3% BSA may decrease background. $NaN_3$ must not be included in the solution because it is a strong inhibitor of HRP.

6. Wash the filter three times for 5 min each with washing buffer.

7. React the filter with a chemiluminescent reagent according to manufacturer's instructions and expose it to an x-ray film (Fig. 1).

The concentration of biotinylated lectin and HRP-labeled avidin should be titrated and adjusted for each experiment. Although lectin blot analysis is easy to perform, the method may not identify precise carbohydrate structures because of the broad specificity of lectins for a group of carbohydrate structures. Lectin blot analysis combined with glycosidase treatment should confirm the presence of a specific carbohydrate structure on proteins.

## Lectin Histochemistry of MII/MX DKO Embryos

Lectin histochemistry detects the presence of a specific carbohydrate structure in specific tissue types or subcellular locations. Figure 2 shows tissue sections from paraffin-embedded embryos stained with biotinylated lectins. Fluorescent dye–conjugated avidin can also be used to detect lectin binding to tissues and cultured cells.

### Materials

Paraformaldehyde-fixed paraffin sections of mouse embryos
Hemo-D (or toluene)
100% ethanol
95% ethanol
0.3% $H_2O_2$ in methanol: dilute 30% $H_2O_2$ with 100 volumes of 100% methanol

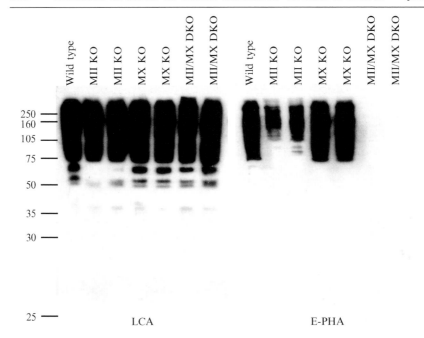

FIG. 1. Lectin blot analysis of MII/MX double-knockout mouse embryos. Membrane proteins (20 μg on each lane) prepared from E15 embryos were resolved by sodium dodecylsulfate (SDS)–polyacrylamide gel electrophoresis (PAGE), followed by electrotransferring to PVDF membranes. A blotted membrane was probed with LCA lectin, which binds to high-mannose type N-glycans (left), or with E-PHA lectin, which binds to complex type N-glycans with bisecting GlcNAc (right). Note that no signal was detected in the MII/MX double knockout by E-PHA lectin, suggesting that DKO mice synthesize no complex-type N-glycans.

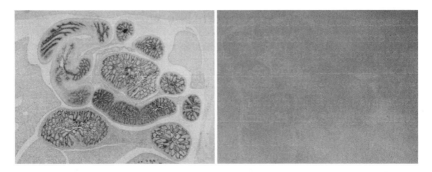

FIG. 2. Lectin histochemistry of the MII/MX double-knockout mouse. Sections of tissues from newborn wild-type (left) or MII/MX DKO (right) mouse were probed using LPHA lectin, which binds to complex-type N-glycans with a GlcNAcβ1–6Man branch. Strong signals were detected in intestinal epithelial cells in wild-type mice, and signals were not seen in the mutant.

PBS

Blocking solution: 5% BSA dissolved in PBS.

Pap-pen

Avidin blocking solution: 20 $\mu$g/ml avidin in blocking solution

Lectin solution: 2.5 $\mu$g/ml biotinylated lectin and 10 $\mu M$ biotin in blocking solution

HRP–avidin solution: 1 $\mu$g/ml HRP-conjugated avidin (or streptavidin) and 0.3% BSA in PBS DAB substrate kit (Vector labs).

*Procedures*

1. Place paraformaldehyde-fixed paraffin sections of mouse embryos on glass slides.
2. De-paraffinize sections by Hemo-D (or toluene), followed by immersion in 100% ethanol.
3. Incubate sections in 0.3% $H_2O_2$ in 100% methanol at room temperature for at least 1 h (or 4° for overnight) to inactivate endogenous peroxidase.
4. Wash sections in 95% ethanol once and then wash three times with PBS.
5. Circle tissue sections using the Pap-pen.
6. Apply avidin blocking solution to sections and incubate at room temperature for 1 h.
7. Wash sections three times with PBS.
8. Overlay lectin solution on sections and incubate for 1 h at room temperature. Wash three times with PBS.
9. Overlay diluted HRP–avidin solution and incubate sections at room temperature for 30 min.
10. Wash sections three times with PBS.
11. React sections with DAB peroxidase substrate solution (Vector). Other peroxidase substrates such as AEC (Zymed) can be used instead of DAB.

Preparation of Recombinant Soluble MII and MX

MII activity in tissues can be measured using immunoprecipitated MII (Chui *et al.*, 1997). However, this method cannot be applied to endogenous MX because no MX antibody suitable for immunoprecipitation is available. Production of recombinant MX enzyme is the only way to detect MX activity *in vitro*. Because both MII and MX are type II membrane proteins with a membrane spanning domain near the N-terminal and a catalytic domain in the lumen of the Golgi apparatus, soluble enzyme can be generated by replacing the membrane-spanning domain with a cleavable signal sequence

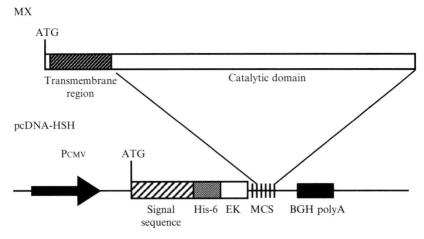

FIG. 3. Construction of pcDNA-HSH-MX expression vector. pcDNA-HSH has a cleavable human GC-SF signal sequence followed by a 6xHis peptide (His-6) and an enterokinase recognition sequence (EK). When the vector is transfected into cultured mammalian cells, the peptide encoded by sequence inserted at the multicloning site (MCS) is expressed and secreted into the medium. Secreted protein can be detected by an anti-Xpress antibody (Invitrogen) and purified by a nickel affinity column.

(Angata *et al.*, 2001). Using the pcDNA-HSH expression vector and polymerase chain reaction (PCR) cloning, an expression vector encoding soluble MX protein can be constructed (Fig. 3). Soluble MX is then recovered from the medium of cultured cells transfected with that expression vector. Concentrated medium containing soluble MX is used as a source of enzymatic activity. Soluble MX protein can also be purified by a nickel affinity column using a 6xHis tag sequence fused to sequences encoding the soluble protein. Soluble MII can be similarly prepared.

*Procedures*

1. Culture Cos-1 cells in two 10-cm dishes until 50% confluent.
2. Transfect expression vector into Cos-1 cells by Lipofectamine reagent (Invitrogen) according to manufacturer's instruction.
3. One day after transfection, replace culture medium with serum-free OptiMEM and continue to culture transfected cells for 2 more days.
4. Recover culture medium into the appropriate tube and centrifuge to remove dead cells and debris.

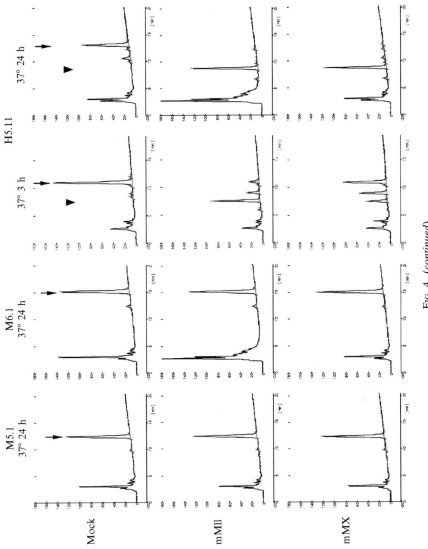

Fig. 4. (*continued*)

5. Concentrate the supernatant into less than 1 ml using a Centricon-30 filter (Millipore) according to the manufacturer's instruction.
6. Concentrate the medium further into less than 200 $\mu$l using a Microcon-30 filter (Millipore).
7. Recover the concentrated medium into a new microtube, add an equal volume of glycerol, and store at $-20°$ until use.

## MII and MX Activity Assays

Both MII and MX hydrolyze a hybrid type $N$-glycan structure, GlcNAc$_1$-Man$_5$-GlcNAc$_2$ (Akama et al., 2006). This carbohydrate substrate tagged with pyridilamine (PA) can be prepared from tissues of the MII-deficient mouse, because the mouse exhibits reduced $N$-glycan processing efficiency (for preparation of PA-labeled $N$-glycan, see volume 415, Chapter 6). A PA-labeled substrate of the carbohydrate structure is used as a substrate for MX activity and the product is analyzed by two-dimensional HPLC analysis (for preparation of PA-labeled $N$-glycans and HPLC analysis, see volume 415, Chapter 6).

### Materials

Concentrated medium as an enzyme source 10X Reaction buffer: 1 $M$ NaOAc, pH 5.8
PA-labeled substrate.

### Procedures

1. Incubate 15 pmol substrate with 3 $\mu$l of concentrated medium containing soluble MX or MII in 50 $\mu$l of 0.1 $M$ NaOAc at 37°.
2. Stop enzymatic reaction by placing sample tube into boiling water for 5 min.
3. Analyze digestion products by two-dimensional HPLC (Fig. 4).

---

FIG. 4. Mannosidase activity detected by amide column high-performance liquid chromatography (HPLC). Each PA-tagged oligosaccharide, M5.1 (Man$_5$GlcNAc$_2$), M6.1(Man$_6$GlcNAc$_2$), and H5.11 (GlcNAc$_1$Man$_5$GlcNAc$_2$) was incubated with concentrated medium of Cos-1 cells transfected with pcDNA-HSH (mock), pcDNA-HSH-MII (MII), and pcDNA-HSH-MX (MX). The digest was analyzed by amide-column HPLC. Elution positions of the initial substrate (arrow) and final product (arrowhead) are shown.

# References

Akama, T. O., Nakagawa, H., Sugihara, K., Narisawa, S., Ohyama, C., Nishimura, S., O'Brien, A. D., Moremen, K. W., Millan, J. L., and Fukuda, M. N. (2002). Germ cell survival through carbohydrate-mediated interaction with Sertoli cells. *Science* **295,** 124–127.

Akama, T. O., Nakagawa, H., Wong, N. K., Sutton-Smith, M., Dell, A., Morris, H. R., Nakayama, J., Nishimura, S., Pai, A., Moremen, K. W., Marth, J. D., Fukuda, M. N., (2006). Essential and mutually compensatory roles of α-mannosidase II and α-mannosidase II$_X$ in N-glycan processing *in vivo* in mice. *Proc. Natl. Acad. Sci. USA* **103**(24), 8983–8988.

Angata, K., Yen, T.-Y., El-Battari, A., Macher, B. A., and Fukuda, M. (2001). Unique disulfide bond structures found in ST8SiaIV polysialyl-transferase are required for activity. *J. Biol. Chem.* **276,** 15369–15377.

Bonay, P., and Hughes, R. C. (1991). Purification and characterization of a novel broad-specificity (alpha 1–2, alpha 1–3 and alpha 1–6) mannosidase from rat liver. *Eur. J. Biochem.* **197,** 229–238.

Bonay, P., Roth, J., and Hughes, R. C. (1992). Subcellular distribution in rat liver of a novel broad-specificity (alpha 1–2, alpha 1–3 and alpha 1–6) mannosidase active on oligomannose glycans. *Eur. J. Biochem.* **205,** 399–407.

Chui, D., Oh-eda, M., Liao, Y.-F., Penneerselvan, K., Lal, A., Marek, K. W., Freeze, H., Moremen, K. W., Fukuda, M. N., and Marth, J. D. (1997). Alpha-mannosidase-II deficiency results in dyserythropoiesis and unveils an alternate pathway in oligosaccharide biosynthesis. *Cell* **90,** 157–167.

Harpaz, N., and Schachter, H. (1980). Control of glycoprotein synthesis. Bovine colostrum UDP-N-acetylglucosamine:alpha-D-mannoside beta 2-N-acetylglucosaminyltransferase I. Separation from UDP-N-acetylglucosamine:Alpha-D-mannoside beta 2-N-acetylglucosaminyltransferase II, partial purification, and substrate specificity. *J. Biol. Chem.* **255,** 4885–4893.

Ioffe, E., and Stanley, P. (1994). Mice lacking N-acetylglucosaminyltransferase I activity die at midgestation, revealing an essential role for complex or hybrid N-linked carbohydrates. *Proc. Natl. Acad. Sci. USA* **91,** 728–732.

Kornfeld, R., and Kornfeld, S. (1985). Assembly of asparagine-linked oligosaccharides. *Annu. Rev. Biochem.* **54,** 631–664.

Metzler, M., Gertz, A., Sarker, M., Schachter, H., Scrader, J. M., and Marth, J. D. (1994). Complex asparagine-linked oligosaccharides required for morphogenic events during post-implantation development. *EMBO J.* **13,** 2056–2065.

Misago, M., Liao, Y.-F., Kudo, S., Eto, S., Mattei, M.-G., Moremen, K. W., and Fukuda, M. N. (1995). Molecular cloning and expression of cDNAs encoding human α-mannosidase II and a novel α-mannosidase IIx isozyme. *Proc. Natl. Acad. Sci. USA* **92,** 11766–11770.

Moremen, K. W., Trimble, R. B., and Herscovics, A. (1994). Glycosidases of asparagine-linked oligosaccharide processing pathway. *Glycobiology* **4,** 113–125.

Oh-Eda, M., Nakagawa, H., Akama, T. O., Lowitz, K., Misago, M., Moremen, K. W., and Fukuda, M. N. (2001). Overexpression of the Golgi-localized enzyme alphamannosidase IIx in Chinese hamster ovary cells results in the conversion of hexamannosyl-N-acetylchitobiose to tetramannosyl-N-acetylchitobiose in the *N*-glycan-processing pathway. *Eur. J. Biochem.* **268,** 1280–1288.

Schachter, H. (1991). The "yellow brick road" to branched complex *N*-glycans. *Glycobiology* **1,** 453–461.

Schachter, H., and Jaeken, J. (1999). Carbohydrate-deficient glycoprotein syndrome type II. *Biochim. Biophys. Acta* **1455,** 179–192.

Schachter, H., Narashimhan, S., Gleeson, P., and Vella, G. (1983). Control of branching during the biosynthesis of asparagine-linked oligosaccharides. *Can. J. Biochem.* **61,** 1049–1066.

Tan, J., Jaeken, J., and Schachter, H. (1996). Mutations in the MGAT2 gene controlling complex *N*-glycan synthesis cause carbohydrate-deficient glycoprotein syndrome type II, an autosomal recessive disease with defective brain development. *Am. J. Hum. Genet.* **59,** 810–817.

Tulsiani, D. R., Hubbard, S. C., Robbins, P. W., and Touster, O. (1982). Alpha-D-mannosidases of rat liver Golgi membranes. Mannosidase II is the GlcNAcMAN5cleaving enzyme in glycoprotein biosynthesis and mannosidases Ia and Ib are the enzymes converting Man9 precursors to Man5 intermediates. *J. Biol. Chem.* **257,** 3660–3668.

Vaitaitis, G. M., Sanderson, R. J., Kimble, E. J., Elkins, N. D., and Flores, S. C. (1999). Modification of enzyme-conjugated streptavidin-biotin Western blot technique to avoid detection of endogenous biotin-containing proteins. *BioTechniques* **26,** 854–858.

Wang, Y., Schachter, H., and Marth, J. D. (2002). Mice with a homozygous deletion of the Mgat2 gene encoding UDP-N-acetylglucosamine:alpha-6-D-mannoside beta1,2-N-acetyl-glucosaminyltransferase II: A model for congenital disorder of glycosylation type IIa. *Biochim. Biophys. Acta* **1573,** 301–311.

Wang, Y., Tan, J., Sutton-Smith, M., Ditto, D., Panico, M., Campbell, R. M., Varki, N. M., Long, J. M., Jaeken, J., Levinson, S. R., Wynshaw-Boris, A., Morris, H. R., Le, D., Dell, A., Schachter, H., and Marth, J. D. (2001). Modeling human congenital disorder of glycosylation type IIa in the mouse: Conservation of asparagine-linked glycan-dependent functions in mammalian physiology and insights into disease pathogenesis. *Glycobiology* **11,** 1051–1070.

# [21]   Targeted Disruption of the Gene Encoding Core 1 β1-3-Galactosyltransferase (T-Synthase) Causes Embryonic Lethality and Defective Angiogenesis in Mice

*By* Lijun Xia and Rodger P. McEver

## Abstract

The biosynthesis of the core 1 *O*-glycan (Galβ1-3GalNAcα1-Ser/Thr, T antigen) is controlled by core 1 β1–3-galactosyltransferase (T-synthase), which catalyzes the addition of Gal to GalNAcα1-Ser/Thr (Tn antigen). The T antigen is a precursor for extended and branched *O*-glycans of largely unknown function. We found that wild-type mice expressed the sialyl-T antigen (NeuAcα2-3Galβ1-3GalNAcα1-Ser/Thr) primarily in endothelial, hematopoietic, and epithelial cells during development. Gene-targeted mice lacking T-synthase instead expressed the nonsialylated Tn antigen in these cells and developed brain hemorrhage that was uniformly fatal by embryonic day 14. T-synthase–deficient brains formed a chaotic microvascular network with distorted capillary lumens and defective association of endothelial cells with pericytes and extracellular matrix.

METHODS IN ENZYMOLOGY, VOL. 416
0076-6879/06 $35.00
DOI: 10.1016/S0076-6879(06)16021-8

These data reveal an unexpected requirement for core 1–derived $O$-glycans during angiogenesis.

## Introduction

Oligosaccharides with GalNAc in $\alpha 1$ linkage to serine or threonine ($O$-glycans) are commonly found on membrane and secreted proteins (Varki *et al.*, 1999). $O$-glycans have a limited number of core structures. The most common is the core 1 disaccharide, which is also the precursor for the branched core 2 trisaccharide (Fig. 1A). Both core 1 and core 2 structures can be further extended and modified into a diverse array of $O$-glycans (core 1–derived $O$-glycans) of mostly unknown function. Some core 1–derived $O$-glycans are components of mucin glycoprotein ligands for the selectins, which initiate leukocyte adhesion to vascular surfaces during infection, tissue injury, and immune surveillance (Fukuda, 2002; McEver, 2002a,b). Other core 1–derived $O$-glycans may limit cell adhesion or modulate T cell function (Fukuda, 2002; Moody *et al.*, 2003).

Formation of the core 1–derived $O$-glycans is controlled by the enzyme core 1 $\beta$1-3-galactosyltransferase (T-synthase), which transfers Gal from UDP-Gal to GalNAc$\alpha$1-Ser/Thr (Tn antigen) to form the core 1 O-glycan (Gal$\beta$1-3GalNAc$\alpha$1-Ser/Thr, T antigen) (Fig. 1A) (Ju *et al.*, 2002a,b). Biochemical evidence and database searches suggest that a single gene, here termed *T-syn*, encodes all T-synthase activity (Ju *et al.*, 2002a,b). Although it was proposed that another gene encodes a second T-synthase (Kudo *et al.*, 2002), this gene actually encodes the chaperone protein Cosmc, which is required for folding and activity of the *T-syn*–encoded T-synthase (Ju and Cummings, 2002). To reveal the functions of core 1–derived $O$-glycans *in vivo*, we disrupted *T-syn* in mice (Xia *et al.*, 2004).

## Generation of $T$-$syn^{-/-}$ Mice

### Construction of Targeting Vector

A bacterial artificial chromosome containing the murine *T-syn* gene was identified by screening a murine embryonic stem cell library (Incyte Genomics) by polymerase chain reaction (PCR). Southern blot analysis showed that murine *T-syn*, like the corresponding human gene (Ju *et al.*, 2002a), has three exons. Exon 1 is 0.3 kb and encodes the ATG translational start site, cytoplasmic domain, transmembrane domain, and stem region of T-synthase. Exon 2 is 0.6 kb and encodes the majority of the catalytic domain. Exon 3 is 0.3 kb and encodes the rest of the catalytic domain and the 3′-untranslated region. Cre/loxP-mediated gene targeting

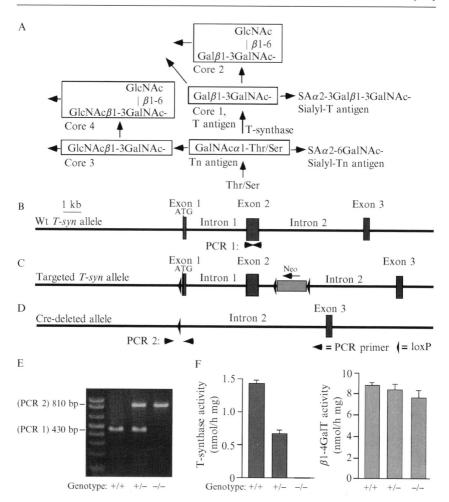

FIG. 1. Generation of T-synthase–deficient mice. (A) Schematic of biosynthesis of the four common *O*-glycan core structures. The cores can be further extended and modified as indicated by additional arrows. The biosynthetic step that T synthase catalyzes is shown. (B) Wild-type (Wt) *T-syn* allele showing the site of PCR 1 used for genotyping. (C) Targeted *T-syn* allele in embryonic stem (ES) cells after homologous recombination with the targeting vector. Exons 1 and 2 and a neomycin cassette (Neo) are flanked by three loxP sites. (D) Deletion of loxP-flanked exons 1 and 2 and Neo in ES cells after *in vitro* Cre-mediated recombination. The site of PCR 2 used for genotyping is indicated. (E) PCR genotyping of DNA from mouse embryos. PCR 1 amplifies a 430-bp fragment from the Wt allele, and PCR 2 amplifies an 810-bp fragment from the Cre-deleted allele. (F) T-synthase and $\beta$1-4-galactosyltransferase ($\beta$1-4Gal-T) activities in E12 tissue extracts. The data represent the mean $\pm$ SD of three independent measurements. (Modified, with permission, from Xia *et al.*, 2004.)

was used to generate the conventional global *T-syn* deletion (Fig. 1B–D). The floxed allele also offers the option to generate conditional and tissue-specific *T-syn* deletions. We engineered a targeting vector with three loxP sites in the *T-syn* allele, which flanked exons 1 and 2 and an inserted *Neo* cassette (Fig. 1B–D).

## Generation of Murine Embryonic Stem Cells with Disrupted T-syn *Allele and Microinjection of Targeted Cells*

The targeting vector was electroporated into CJ7 embryonic stem (ES) cells that were derived from a 129/SvlmJ mouse (gift from Dr. Thomas Sato, Weill Medical College of Cornell University, with permission from Dr. Thomas Gridley, The Jackson Laboratory). ES cell clones with correct homologous recombination were screened by PCR and confirmed by Southern blots. Positive ES cells then were transiently transfected with an expression vector encoding Cre recombinase (gift from Dr. Brian Sauer, Stowers Institute for Medical Research, Kansas City, MO) to delete exons 1 and 2 and the *Neo* cassette to generate conventional *T-syn*–deficient ES cells. Cells confirmed to have a normal karyotype were microinjected into C57BL/6J blastocysts, which were then implanted into pseudopregnant mice. Chimeras among the offspring were bred with C57BL/6J mice to test germline transmission.

## PCR Genotyping

Genotypes of mice were determined by PCR of DNA from adult tails or embryonic tissues. The method for embryo genotyping is given in the following:

Embryonic DNA extraction:
1. A 2- to 4-mm long tail or hind leg of an embryo was collected.
2. The tissues were incubated in 50 $\mu$l lysis buffer (50 m$M$ Tris–HCl, pH 7.5, 20 m$M$ NaCl, 1 m$M$ ethylenediaminetetraacetic acid [EDTA], 0.5% Tween 20, 0.5% NP-40, 200 $\mu$g/ml proteinase K) overnight at 55°.
3. The tissue lysates were boiled for 10 min and then centrifuged at 12,000 rpm for 5 min. Supernatants were used as crude DNA samples for PCR.

PCR:
1. 2 $\mu$l of supernatant was used for a 25-$\mu$l PCR.
2. PCR primers: *T-syn* wild-type allele (primers 5′-TGGGTTAT GACAAGTCCTC3′ and 5′-TCATGTATCCCTGCTTCAC-3′).

*T-syn*$^{-/-}$ allele (primers 5'-GATAAATGTCTTACAGAAGG-3' and 5'-AATACTGTCCTGGGCTATACTACAGTG-3').

3. PCR condition:

Denaturing: Denaturing at 94° for 30 s, annealing at 53° for 30 s, and extension at 72° for 30 s.

Amplification cycles: 30.

Glycosyltransferase Assays

T-synthase activity from murine embryo extracts was measured to verify the *T-syn* deletion and to examine whether *T-syn* is the sole gene encoding core 1β1-3-galactosyltransferase (Ju *et al.*, 2002a,b). The following protocol was used to measure activities of T-synthase and a control enzyme, β1-4-galactosyltransferase.

1. Sample preparation: E12 embryos were homogenized in a 1.5-ml tube containing 200 μl tissue lysis buffer (1 m*M* PMSF, 5 μ*M* benzamidine, and 5 μg/ml leupeptin). The homogenates were centrifuged at 20,000*g* for 30 min at 4°. The supernatants were collected, and the protein concentration in different samples was adjusted to a same level with the BCA assay (Pierce).

2. Assay system:

*Acceptor:* GalNAcα1-O-phenyl (Sigma) for T-synthase; GlcNAc-S-pNp (American Radiolabeled Chemicals, Inc.) for β1-4-galactosyltransferase.

*Donor:* UDP-[$^3$H]Gal (40–60 Ci/mmol. American Radiolabeled Chemicals, Inc.).

*Reaction system:* A 50-μl reaction contains 50 m*M* Tris–HCl, pH 7.0, 2 m*M* GalNAcα1-O-phenyl or GlcNAc-S-pNp, 200 μ*M* UDP-[$^3$H]Gal (60,000–90,000 cpm), 20 m*M* MnCl$_2$, 0.1% Triton X-100, and 100 μg protein from embryo extracts. The reactions were incubated at 37° for 1 h and stopped by addition of 950 μl cold H$_2$O.

*Separation of products from UDP-[$^3$H]Gal:* A 500-mg Sep-Pak C$_{18}$ Cartridge (Water Associates) was activated with 1 ml methanol and equilibrated with 10 ml H$_2$O. The sample was loaded on column, washed once with 10 ml H$_2$O, and then eluted with 1 ml 1-butanol.

*Detection:* The radioactivity of the eluted product was measured by liquid scintillation counting in 10 ml of Scintiverse-BD. A unit of activity is defined as that amount of enzyme transferring 1 nmol of Gal from UDP-Gal to the acceptor GalNAcα1-O-phenyl per hour at 37°.

Our results show that T-synthase activity in tissue extracts was reduced approximately 50% in E12 $T\text{-}syn^{+/-}$ embryos and was eliminated in $T\text{-}syn^{-/-}$ embryos, whereas the activity of the control enzyme $\beta$1-4-galactosyltransferase, which transfers Gal from UDP-Gal to GlcNAc$\beta$1-R, was similar in embryos of all genotypes (Fig. 1F). This result confirms that $T\text{-}syn$ encodes all T-synthase activity, at least through this stage of development. This distinguishes $T\text{-}syn$ from typical multigene families of glycosyltransferases that encode several enzymes with related structures and functions (Lowe and Marth, 2003).

## Characterization of $T\text{-}syn^{-/-}$ Mice

### Disruption of T-syn Causes Fatal Embryonic Hemorrhage

$T\text{-}syn^{+/-}$ mice developed normally. Matings between $T\text{-}syn^{+/-}$ mice generated 364 viable progeny, of which 175 (48%) were male and 189 (52%) were female. Genotyping identified 136 (37%) $T\text{-}syn^{+/+}$ progeny and 228 (63%) $T\text{-}syn^{+/-}$ progeny but did not identify $T\text{-}syn^{-/-}$ progeny. To determine whether deletion of $T\text{-}syn$ on both alleles caused embryonic lethality, we analyzed 293 embryos at embryonic days 9–16 (E9–16) from timed matings of $T\text{-}syn^{+/-}$ mice. Genotyping revealed 78 (27%) $T\text{-}syn^{+/+}$, 142 (48%) $T\text{-}syn^{+/-}$, and 73 (25%) $T\text{-}syn^{-/-}$ progeny (Fig. 1E).

At E9 $T\text{-}syn^{-/-}$ embryos appeared developmentally normal, but thereafter, they developed progressively larger hemorrhages in the brain and spinal cord, with secondary bleeding into the ventricles and spinal canal (Fig. 2). Anemia was frequent, but only after bleeding occurred, and was associated with growth retardation but not with obvious defects in organogenesis. Histological examination revealed no placental abnormalities (unpublished observation). Embryonic hemorrhage invariably preceded death, supporting a casual relationship. All $T\text{-}syn^{-/-}$ embryos died by E14, virtually always at E13 or E14.

### T-syn$^{-/-}$ Embryos Express the Tn Antigen but not the T Antigen

#### Western Blot

To determine whether the levels of T-synthase activity affected core 1 $O$-glycosylation in vivo, we probed blots of embryonic extracts with *Arachis hypogaea* agglutinin (PNA), a lectin that recognizes the nonsialylated T antigen, and with *Helix pomatia* agglutinin (HPA), a lectin that recognizes the nonsialylated Tn antigen (Fig. 3A).

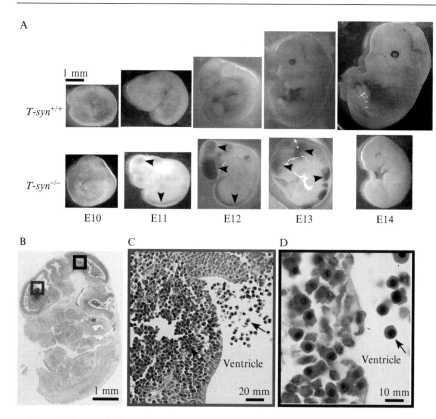

FIG. 2. Targeted disruption of *T-syn* causes fatal embryonic hemorrhage in mice. (A) Comparison of *T-syn*$^{+/+}$ and *T-syn*$^{-/-}$ embryos at different developmental stages. Blood is visible in the hearts of the older *T-syn*$^{+/+}$ embryos. Arrowheads indicate hemorrhage in the brain parenchyma and ventricles, spinal cord, and spinal canal of *T-syn*$^{-/-}$ embryos. Less blood is detected in *T-syn*$^{-/-}$ hearts because of anemia. The E14 *T-syn*$^{-/-}$ embryo is dead and appears pale because blood circulation has ceased. (B) Sagittal section of the E12 *T-syn*$^{-/-}$ embryo. The red and black boxes indicate hemorrhagic lesions, which are shown at higher magnification in (C) and (D). Bleeding is visible in both the brain parenchyma and ventricles. Arrows indicate erythrocytes, which are nucleated at this stage of development. Modified, with permission, from Xia *et al.*, 2004. (See color insert.)

1. Sample preparation: E12 embryos were homogenized in a 1.5-ml tube containing 200 $\mu$l tissue lysis buffer (1 m*M* PMSF, 5 $\mu$*M* benzamidine, and 5 $\mu$g/ml leupeptin). The homogenates were centrifuged at 20,000*g* for 30 min at 4°. The supernatants were collected, and the protein concentration in

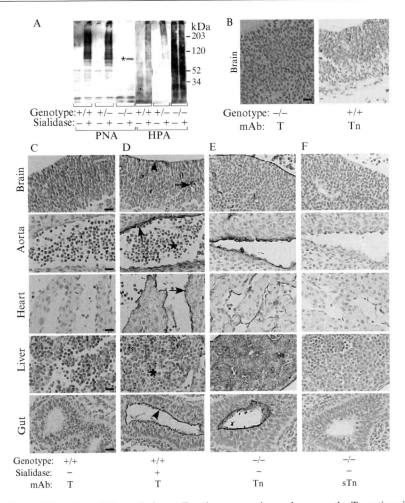

FIG. 3. Disruption of *T-syn* eliminates T-antigen expression and exposes the Tn antigen in murine embryos. (A) Blots of E12 tissue extracts probed with *Arachis hypogaea* agglutinin (PNA), which recognizes nonsialylated core 1 $O$-glycans, and with *Helix pomatia* agglutinin (HPA), which recognizes the nonsialylated Tn antigen. The extracts were incubated with or without sialidase before electrophoresis and blotting. The asterisk indicates the binding of PNA to the added sialidase. The blots are representative of three experiments. (B to F) Immunohistochemical staining of E12 tissue sections with monoclonal antibodies (mAbs) to the T antigen, Tn antigen, or sialyl-Tn (sTn) antigen. The sections in (D) were pretreated with sialidase. Brown reaction product marks sites of antibody binding. Arrows indicate endothelial cells, arrowheads indicate epithelial cells, and asterisks indicate hematopoietic cells. Reproduced, with permission, from Xia *et al.*, 2004. (See color insert.)

different samples was adjusted to the same level with the BCA assay (Pierce).

2. De-sialylation: 200 μg embryonic extract in a 40-μl reaction volume was incubated with 20 mU sialidase from *Arthrobacter ureafaciens* (Roche) in 10 m$M$ Tris–HCl, pH 6.3, overnight at 37°.

3. Western blot: Embryonic extracts (40 μg protein) with or without prior de-sialylation were resolved by sodium dodecylsulfate (SDS)–polyacrylamide gel electrophoresis (PAGE) under reducing conditions and transferred to a nitrocellulose membrane (Bio-Rad). The membrane was blocked with 5% nonfat dry milk and incubated with 2 μg/ml horseradish peroxidase (HRP)–conjugated PNA or 0.25 μg/ml HPA (EY Laboratories) in Tris-buffered saline (TBS) at room temperature for 1 h. Lectin binding was detected with the HighSignal West Pico Chemiluminescent Substrate (Pierce). Binding was detected with HRP-conjugated goat anti-rabbit immunoglobulin G (IgG) (Pierce) using enhanced chemiluminescence (Amersham).

The results indicate that PNA bound to many glycoproteins from $T$-$syn^{+/+}$ and $T$-$syn^{+/-}$ embryos but not from $T$-$syn^{-/-}$ embryos. HPA bound to many glycoproteins from $T$-$syn^{-/-}$ embryos but exhibited little or no binding to glycoproteins from $T$-$syn^{+/+}$ and $T$-$syn^{+/-}$ embryos. Enzymatic de-sialylation of glycoproteins from $T$-$syn^{+/+}$ and $T$-$syn^{+/-}$ embryos was required to expose binding sites for PNA, indicating that sialic acid capped most core 1 $O$-glycans to form the sialyl-T antigen (Fig. 1A). In contrast, de-sialylation of glycoproteins from $T$-$syn^{-/-}$ embryos was not required to expose binding sites for HPA, suggesting that sialic acid did not cap most of the Tn antigen to form the sialyl-Tn antigen (Fig. 1A).

*Immunohistochemical Staining*

To determine which cells expressed core 1 $O$-glycans *in vivo*, we performed immunohistochemical staining of embryonic tissue sections with monoclonal antibodies (mAbs) to the T-antigen, Tn antigen, or sialyl-Tn antigen.

1. Tissue collection: E12 embryos were fixed in 10% formalin. The tissues were then processed and embedded in paraffin; 5-μm sections were used for staining.

2. De-sialylation: Deparaffinized sections were incubated with 0.5 U/ml sialidase from *Arthrobacter ureafaciens* in 10 m$M$ Tris–HCl, pH 6.3, at 37° for 3 h.

3. Immunohistochemical staining: Sections were incubated with mAbs against the T, Tn, or sialyl-Tn antigens (Mandel *et al.*, 1991) (mouse IgG or IgM, gifts from Drs. Ulla Mandel and Henrik Clausen, School of Dentistry, University of Copenhagen, Denmark) or with isotype-matched control

mouse IgG or IgM. Bound antibodies were detected with HRP-conjugated goat-antimouse IgG/IgM (Dako). Alternatively, sections were incubated with peroxidase-conjugated *Maackia amurensis* or *Sambucus nigra* hemagglutinin (EY Laboratories). Immunohistochemical staining was developed with a DAB peroxidase substrate kit (Vector).

The results show that anti-T stained endothelial, hematopoietic, and epithelial cells in tissues of $T$-$syn^{+/+}$ embryos, but only after enzymatic de-sialylation of the tissue sections (Fig. 3C and D, and unpublished observations of multiple sections throughout the entire embryo). By contrast, anti-Tn stained endothelial, hematopoietic, and epithelial cells in tissues of $T$-$syn^{-/-}$ embryos (Fig. 3E). Enzymatic de-sialylation of tissue sections did not enhance binding of anti-Tn to $T$-$syn^{-/-}$ embryos (unpublished observation). Furthermore, anti-sialyl-Tn did not stain $T$-$syn^{-/-}$ embryos (Fig. 3F), although it did stain fixed Jurkat cells, which are known to express the sialyl-Tn antigen (unpublished observation). Anti-T did not stain $T$-$syn^{-/-}$ embryos, and anti-Tn did not stain $T$-$syn^{+/+}$ embryos (Fig. 3B). These results corroborate the lectin blotting data, and they demonstrate that sialic acid modifies the T antigen on $T$-$syn^{+/+}$ embryos but does not appreciably modify the Tn antigen on $T$-$syn^{-/-}$ embryos. Epithelial cells in some adult tissues express core 3 $O$-glycans, which are also formed based on the Tn antigen (Brockhausen and Kuhns, 1997; Iwai *et al.*, 2002; Varki *et al.*, 1999.) (Fig. 1A). However, Fig. 3E demonstrates that $T$-$syn^{-/-}$ embryos do not synthesize sufficient core 3 $O$-glycans to modify all the exposed Tn antigens.

## $T$-$syn^{-/-}$ Embryos have Normal Plasma-Based Blood Coagulation

The $T$-$syn^{-/-}$ embryos exhibited a severe bleeding phenotype. Some murine embryos lacking any one of several coagulation proteins suffer fatal hemorrhage. Coagulation factors V and VIII are modified with $O$-glycans of unknown function (Kaufman, 1998). To investigate whether defective coagulation contributes to the bleeding phenotype, we used the kinetic prothrombin time (kPT) and the kinetic activated partial thromboplastin time (kAPTT) to examine both intrinsic and extrinsic coagulation pathways of E12 embryos.

1. Embryo plasma collection: After severing the umbilical cord, each embryo was placed into a drop of 40 $\mu$l TBS containing 0.19% sodium citrate on a sheet of paraffin film for 5 min. Approximately 3 $\mu$l of embryonic blood was harvested from each embryo. Plasma was obtained after centrifugation at 4000 rpm for 5 min. Each collection was adjusted for similar protein concentration based on the BCA assay (Pierce).

2. Kinetic coagulation assays: Kinetic coagulation assays in 96-well microtiter plates were conducted using modified protocols described for zebrafish (Sheehan *et al.*, 2001); 10 $\mu$l of plasma containing supernatant from each embryo was added to a well containing 3 mg/ml purified human fibrinogen (Calbiochem) in a total volume of 50 $\mu$l. For the kAPTT, 15 $\mu$l partial thromboplastin reagent (Dade Actin, Dade Behring, Inc.) and 8 m$M$ $CaCl_2$ were added. For the kPT, 30 $\mu$l Thromboplastina C Plus (Dade Behring) was added. Clot formation at room temperature was monitored with a kinetic microplate reader (Molecular Devices) set at 405 nm over a 2-h period. Embryo plasma was replaced with TBS as negative control.

3. Coagulation correction assay: 50 $\mu$l pooled murine embryo plasma containing supernatant was mixed with 100 $\mu$l human plasma deficient in factor V or factor VIII (Fisher Scientific). Coagulation triggered by addition of 15 $\mu$l partial thromboplastin reagent and 8 m$M$ $CaCl_2$ was measured on a Start 4 coagulation analyzer (Diagnostica Stago). Pooled normal human plasma diluted to the same protein concentration as the embryo plasma was used as positive control.

Plasma from *T-syn$^{+/+}$* and *T-syn$^{-/-}$* embryos clotted at similar rates as assessed by modified prothrombin times and aPTTs (Fig. 4A and B). Furthermore, plasma from embryos of both genotypes corrected defective clotting of human plasma deficient in factor V or factor VIII (Fig. 4C). These results argue against a general defect in plasma-based blood coagulation.

## T-syn$^{-/-}$ Embryos Develop a Chaotic Microvascular Network

The earliest blood vessels arise by vasculogenesis, the differentiation of precursor cells into endothelial cells that form a primitive capillary plexus. New capillaries then develop by angiogenesis, in which endothelial cells migrate into previously avascular areas. These vessels mature as endothelial cells recruit pericytes, deposit extracellular matrix, and develop an organized branched network (Carmeliet, 2005; Jain, 2005). Mice lacking proteins that affect the recruitment or critical functions of endothelial cells or pericytes exhibit defective vasculogenesis or angiogenesis, and they frequently die from embryonic or perinatal hemorrhage (Carmeliet, 2005; Jain, 2005; McCarty *et al.*, 2005). Using confocal microscopy, we examined microvascular architectures of *T-syn$^{+/+}$* and *T-syn$^{-/-}$* E12 embryos.

1. Sample collection: Embryos were harvested into a 24-well plate containing freshly prepared 4% paraformaldehyde in 0.1 $M$ PBS at room temperature for 90–120 min. The samples were then washed with PBS, cryoprotected with 15% sucrose in PBS at 4° overnight, mounted in TISSUE-TEK O.C.T. compound, and snap-frozen in liquid nitrogen–cooled

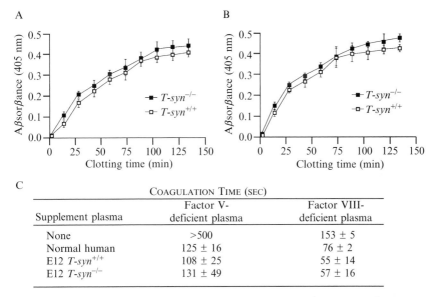

FIG. 4. Plasma-based blood coagulation is equivalent in $T$-$syn^{+/+}$ and $T$-$syn^{-/-}$ embryos. Clotting of individual plasma samples from eight E12 $T$-$syn^{+/+}$ embryos and six E12 $T$-$syn^{-/-}$ embryos (mean $\pm$ standard deviation [SD], $p > .05$) was measured by the kinetic activated partial thromboplastin time (A) and the kinetic prothrombin time (B). (C) The activated partial thromboplastin time of human plasma deficient in factor V or factor VIII was measured with or without supplementation with normal human plasma or with pooled plasma from $T$-$syn^{+/+}$ or $T$-$syn^{-/-}$ embryos. The data represent the mean $\pm$ SD of three independent experiments. (Reproduced, with permission, from Xia *et al.*, 2004.)

isopentane. For whole-mount immunofluorescence, yolk sacs were fixed in 4% paraformaldehyde in 0.1 $M$ PBS at room temperature for 60 min.

2. Cryosection: Thick (100-$\mu$m) cryosections were prepared. The sections were quenched with 0.1 $M$ glycine in PBS for 15 min, washed twice, and blocked with 3% BSA in PBS containing 0.01% saponin for 30 min.

3. Staining: Sections were incubated with rat anti-murine CD31 mAb (1:10 dilution, BD Pharmingen) and rabbit anti-NG2 antibody (1:200 dilution, Chemicon) in 0.3% BSA in PBS with 0.01% saponin or with rabbit anti-laminin antibody (1:500 dilution, Dako) for 1 h at room temperature. Thereafter, the sections were washed in PBS containing 0.01% saponin and incubated with Cy3-conjugated goat anti-rabbit antibody (1:100 dilution, Vector) and biotinylated goat anti-rat IgG absorbed to remove anti-murine IgG (1:100 dilution, Vector), followed by streptavidin conjugated to

fluorescein isothiocyanate (1:100 dilution, Vector). The sections were mounted with VectaShield (Vector). For whole-mount immunofluorescence, yolk sacs were incubated sequentially with rat anti-mouse CD31 mAb, biotinylated goat anti-rat IgG, and streptavidin conjugated to fluorescein isothiocyanate. The yolk sacs were then mounted on slides.

4. Confocal analysis: Sections were analyzed by three-dimensional confocal laser scanning microscopy using a Nikon C1 scanning head mounted on a Nikon ECLIPSE 2000U inverted microscope (Plan Apochromats dry objective lens, 20X, NA 0.75, Nikon Instruments, Dallas, TX). Z-stack images were collected at 1-$\mu$m steps with sequential laser excitation to eliminate bleed-through and with confocal parameters selected to minimize the thickness of the calculated optical section. Volume images from the confocal data sets were processed with IMARIS software (Bitplane AG) for three-dimensional views of the detailed vascular morphology. Images are presented as maximum intensity projections of the z-stacks.

In brains of $T\text{-}syn^{+/+}$ embryos, endothelial cells marked by anti-CD31 antibodies were observed in the perineural plexus, which arises by vasculogenesis (Fig. 5A). Endothelial cells in parenchymal vessels, which develop exclusively by angiogenesis, formed capillaries of relatively uniform diameters with orderly branching patterns. Pericytes marked by antibodies to the proteoglycan NG2 (Ozerdem $et\ al.$, 2001) were associated with endothelial cells in both the perineural plexus and the brain parenchyma (Fig. 5C and E). In brains of $T\text{-}syn^{-/-}$ embryos, endothelial cells and pericytes developed an apparently normal perineural plexus (Fig. 5B, D, and F). However, endothelial cells in the brain parenchyma formed a chaotic microvasculature with irregular capillary diameters and disordered branching, indicating dysregulated angiogenesis (Fig. 5B). Antibodies to NG2 stained the $T\text{-}syn^{-/-}$ brain vessel network, arguing that the vascular defect did not result from failure to recruit pericytes (Fig. 5D and F). Laminin, a basement membrane component, was detected in the microvasculature of both $T\text{-}syn^{+/+}$ and $T\text{-}syn^{-/-}$ brains (Fig. 5G and H). Confocal microscopy did not reveal obvious defects in the microvasculature of other $T\text{-}syn^{-/-}$ embryonic organs.

### T-syn$^{-/-}$ Embryos Have Defective Association of Endothelial Cells with Pericytes and Extracellular Matrix

We used electron microscopy to characterize better the structures of $T\text{-}syn^{+/+}$ and $T\text{-}syn^{-/-}$ brain capillaries. Embryos were fixed with 3% paraformaldehyde and 2% glutaraldehyde in 0.1 $M$ cacodylate, pH 7.2,

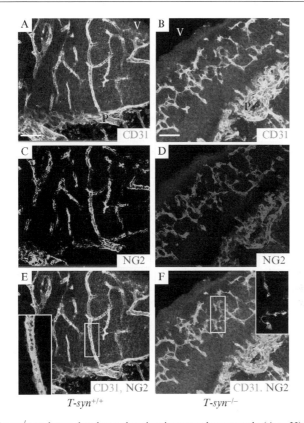

FIG. 5. *T-syn*$^{-/-}$ embryos develop a chaotic microvascular network. (A to H) Visualization of microvessels in E12 hindbrains using maximal intensity projections of z-stacked confocal images. Endothelial cells were stained with antibodies to CD31 (green), pericytes were stained with antibodies to the proteoglycan NG2 (red), and basement membrane was stained with antibodies to laminin (red). *T-syn*$^{+/+}$ embryos form a network of capillaries with uniform diameters and a regular branching pattern (A, C, E, and G), whereas *T-syn*$^{-/-}$ embryos form capillaries with heterogeneous diameters and excessive, irregular branches (B, D, F, and H). Insets in (E) and (F) are thin (5 $\mu$m) optical slices of representative vessels enlarged 2.5-fold, which illustrate the abluminal relationship of the NG2-positive pericytes to the CD31-positive endothelial cells. P, perineural vascular plexus; V, ventricle. Scale bars, 50 $\mu$m unless noted otherwise. Adapted, with permission, from Xia *et al.*, 2004. (See color insert.)

for 2 h, post fixed in 2% osmium tetroxide in 0.1 *M* cacodylate, dehydrated in acetone series, and embedded in EMbed 812 epoxy resin (Electron Microscopy Sciences, Inc.). Thin sections stained with uranyl acetate and lead citrate were examined with a Jeol JEM-1200EX electron microscope.

The electron microscopy images reveal that the endothelial cells of *T-syn⁺/⁺* capillaries were relatively uniform in size and were intimately associated with pericytes (Fig. 6A and E). In sharp contrast, most endothelial cells of *T-syn⁻/⁻* capillaries exhibited distorted shapes with irregular or dilated capillary lumens and with focal cytoplasmic thinning (Fig. 6B–E). Mature interendothelial cell junctions were present, but pericytes typically lacked their normal close apposition to the endothelial cells. Many endothelial cells and pericytes were also surrounded by large spaces, indicating unstable interactions with extracellular matrix or neighboring cells. The morphological abnormalities preceded hemorrhage and tended to be more severe near bleeding lesions. This implies that the structural defects are directly responsible for hemorrhage.

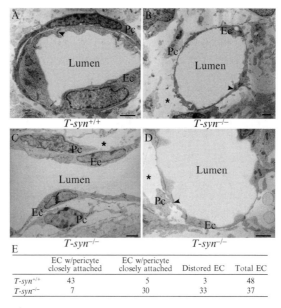

| | EC w/pericyte closely attached | EC w/pericyte closely attached | Distorted EC | Total EC |
|---|---|---|---|---|
| *T-syn⁺/⁺* | 43 | 5 | 3 | 48 |
| *T-syn⁻/⁻* | 7 | 30 | 33 | 37 |

FIG. 6. *T-syn⁻/⁻* embryos have defective capillary structures. (A to D) Electron micrographs of sections from E12 brains. (A) A *T-syn⁺/⁺* capillary has a uniform lumen and a pericyte that is intimately associated with the endothelial cell. (B to D) *T-syn⁻/⁻* capillaries have distended or distorted lumens with attenuated endothelial cell processes. The upper pericyte in (C) is in close contact with an endothelial cell, but other pericytes are partially or completely separated from the endothelial cells. Large empty spaces (asterisks) surround the capillaries. Both *T-syn⁺/⁺* and *T-syn⁻/⁻* capillaries have normal interendothelial cell junctions (arrowheads). Ec, endothelial cell; Pc, pericyte. Scale bars, 2 μm. (E) Sections of capillaries were scored for pericytes closely attached to endothelial cells and for distortions in endothelial cell shape. (Reproduced, with permission, from Xia *et al.*, 2004.)

Conclusion and Future Direction

Blood vessels in the developing brain parenchyma arise exclusively by angiogenesis, (Carmeliet, 2005; Jain, 2003). Thus, the severe vascular defects in $T$-$syn^{-/-}$ embryonic brains indicate a critical and unexpected contribution of core 1–derived $O$-glycans to angiogenesis. The separation of endothelial cells from supporting pericytes and extracellular matrix is a likely mechanism for the capillary fragility in $T$-$syn^{-/-}$ embryos. Targeted deletion of $T$-$syn$ specifically in endothelial cells, pericytes, and surrounding tissues such as neural cells will help address which cell types require $O$-glycans for angiogenesis.

The molecular mechanisms for the defective angiogenesis in $T$-$syn^{-/-}$ mice remain to be determined. The chaotic vascular branching in $T$-$syn^{-/-}$ embryonic brains indicates that the vessels fail to remodel into a hierarchical network. Many molecules involved in the remodeling process are glycoproteins carrying potential sites for $O$-glycosylation, based on preliminary analysis with $O$-glycosylation prediction software (NetOGlyc 3.1) (Julenius et al., 2005). Mice with genetically targeted deficiencies in many of these potentially $O$-glycosylated molecules share bleeding and/or defective angiogenesis phenotypes that resemble those of our $T$-$syn^{-/-}$ mice. For example, mice lacking platelet-derived growth factor B or its receptor fail to recruit pericytes to developing brain vessels and exhibit bleeding at the late embryonic or perinatal period (Lindahl et al., 1997). The phenotype of endoglin (CD105)-deficient mice, a murine model of hereditary hemorrhagic telangiectasia type 1 (HHT1), is similar to that seen in $T$-$syn^{-/-}$ mice (Sorensen et al., 2003). Endoglin is important for vascular maturation. The similarities suggest that $O$-glycans might directly or indirectly affect the functions of one or more of these molecules during vessel development.

Little is known about the synthesis of $O$-glycans during development. A large family of $N$-acetylgalactosaminyltransferases transfers GalNAc from UDP-GalNAc to serine or threonine residues. During murine development, messenger RNAs encoding many of these enzymes are expressed in discrete patterns in multiple tissues, but their functions have not been addressed (Kingsley et al., 2000). Our immunohistochemical data indicate that core 1 $O$-glycans are expressed primarily in endothelial, hematopoietic, and epithelial cells during development. This suggests that these cells express the majority of the Ser/Thr-rich proteins that are $O$-glycosylated to become mucins. Other cells might express fewer or less clustered core 1 $O$-glycans that were not detected by immunohistochemistry. The anti-T antigen staining is sialylation-dependent in $T$-$syn^{-/-}$ embryos, indicating most of the simple core 1 $O$-glycans were $\alpha$2-3-sialylated to form the sialyl-T antigen. In contrast, the negative anti-sialyl-Tn staining in $T$-$syn^{-/-}$

embryos demonstrate that Tn antigen exposed on endothelial, hematopoietic, and epithelial cells of $T\text{-}syn^{-/-}$ embryos was not sialylated to form the sialyl-Tn antigen, as may occur on some malignant cells (Brockhausen and Kuhns, 1997).

Our data provide the first demonstration that normal embryonic development requires an $O$-glycosylation pathway that begins by attachment of GalNAc to serines or threonines. The relatively restricted expression of detectable core 1 $O$-glycans in endothelial, hematopoietic, and epithelial cells may explain why $T\text{-}syn^{-/-}$ embryos do not exhibit multiple developmental defects that cause earlier death. Core 1 $O$-glycans or their derivatives might contribute to physiological or pathological angiogenesis in adults. The expression of core 1 $O$-glycans in epithelial cells deserves further exploration, and other cells might express core 1 or core 2 $O$-glycans adult life. Inducible or cell type–specific deletion of $T\text{-}syn$ may reveal other important functions for core 1 $O$-glycosylation of proteins.

## Acknowledgments

We thank Drs. Richard D. Cummings, Tongzhong Ju, and Florea Lupu for their collaborations. This work was supported by National Institutes of Health grants HL 54502 (R. P. M.), and RR 018758 (R. P. M., and L. X.), and by a Scientist Development Grant from the American Heart Association (L. X.).

## References

Brockhausen, I., and Kuhns, W. (1997). "Glycoproteins and Human Disease." R. G. Landes Company, Austin, TX.

Carmeliet, P. (2005). Angiogenesis in life, disease and medicine. *Nature* **438,** 932–936.

Fukuda, M. (2002). Roles of mucin-type $O$-glycans in cell adhesion. *Biochim. Biophys. Acta* **1573,** 394–405.

Iwai, T., Inaba, N., Naundorf, A., Zhang, Y., Gotoh, M., Iwasaki, H., Kudo, T., Togayachi, A., Ishizuka, Y., Nakanishi, H., and Narimatsu, H. (2002). Molecular cloning and characterization of a novel UDP-GlcNAc:GalNAc-peptide beta1,3-N-acetylglucosaminyltransferase (beta 3Gn-T6), an enzyme synthesizing the core 3 structure of $O$-glycans. *J. Biol. Chem.* **277,** 12802–12809.

Jain, R. K. (2003). Molecular regulation of vessel maturation. *Nat. Med.* **9,** 685–693.

Jain, R. K. (2005). Normalization of tumor vasculature: An emerging concept in antiangiogenic therapy. *Science* **307,** 58–62.

Ju, T., Brewer, K., D'Souza, A., Cummings, R. D., and Canfield, W. M. (2002a). Cloning and expression of human core 1 beta1,3-galactosyltransferase. *J. Biol. Chem.* **277,** 178–186.

Ju, T., and Cummings, R. D. (2002). A unique molecular chaperone Cosmc required for activity of the mammalian core 1 beta 3-galactosyltransferase. *Proc. Natl. Acad. Sci. USA* **99,** 16613–16618.

Ju, T., Cummings, R. D., and Canfield, W. M. (2002b). Purification, characterization, and subunit structure of rat core 1 Beta1, 3-galactosyltransferase. *J. Biol. Chem.* **277,** 169–177.

Julenius, K., Molgaard, A., Gupta, R., and Brunak, S. (2005). Prediction, conservation analysis, and structural characterization of mammalian mucin-type *O*-glycosylation sites. *Glycobiology* **15,** 153–164.

Kaufman, R. J. (1998). Post-translational modifications required for coagulation factor secretion and function. *Thromb. Haemost.* **79,** 1068–1079.

Kingsley, P. D., Hagen, K. G., Maltby, K. M., Zara, J., and Tabak, L. A. (2000). Diverse spatial expression patterns of UDP-GalNAc:Polypeptide *N*-acetylgalactosaminyl-transferase family member mRNAs during mouse development. *Glycobiology* **10,** 1317–1323.

Kudo, T., Iwai, T., Kubota, T., Iwasaki, H., Takayma, Y., Hiruma, T., Inaba, N., Zhang, Y., Gotoh, M., Togayachi, A., and Narimatsu, H. (2002). Molecular cloning and characterization of a novel UDP-Gal:GalNAc(alpha) peptide beta 1,3-galactosyltransferase (C1Gal-T2), an enzyme synthesizing a core 1 structure of O-glycan. *J. Biol. Chem.* **277,** 47724–47731.

Lindahl, P., Johansson, B. R., Leveen, P., and Betsholtz, C. (1997). Pericyte loss and microaneurysm formation in PDGF-B-deficient mice. *Science* **277,** 242–245.

Lowe, J. B., and Marth, J. D. (2003). A genetic approach to Mammalian glycan function. *Annu. Rev. Biochem.* **72,** 643–691.

Mandel, U., Petersen, O. W., Sorensen, H., Vedtofte, P., Hakomori, S., Clausen, H., and Dabelsteen, E. (1991). Simple mucin-type carbohydrates in oral stratified squamous and salivary gland epithelia. *J. Invest. Dermatol.* **97,** 713–721.

McCarty, J. H., Lacy-Hulbert, A., Charest, A., Bronson, R. T., Crowley, D., Housman, D., Savill, J., Roes, J., and Hynes, R. O. (2005). Selective ablation of alpha integrins in the central nervous system leads to cerebral hemorrhage, seizures, axonal degeneration and premature death. *Development* **132,** 165–176.

McEver, R. P. (2002a). P-selectin and PSGL-1: Exploiting connections between inflammation and venous thrombosis. *Thromb. Haemost.* **87,** 364–365.

McEver, R. P. (2002b). Selectins: Lectins that initiate cell adhesion under flow. *Curr. Opin. Cell Biol.* **14,** 581–586.

Moody, A. M., North, S. J., Reinhold, B., Van Dyken, S. J., Rogers, M. E., Panico, M., Dell, A., Morris, H. R., Marth, J. D., and Reinherz, E. L. (2003). Sialic acid capping of CD8beta core 1-*O*-glycans controls thymocyte-major histocompatibility complex class I interaction. *J. Biol. Chem.* **278,** 7240–7246.

Ozerdem, U., Grako, K. A., Dahlin-Huppe, K., Monosov, E., and Stallcup, W. B. (2001). NG2 proteoglycan is expressed exclusively by mural cells during vascular morphogenesis. *Dev. Dyn.* **222,** 218–227.

Sheehan, J., Templer, M., Gregory, M., Hanumanthaiah, R., Troyer, D., Phan, T., Thankavel, B., and Jagadeeswaran, P. (2001). Demonstration of the extrinsic coagulation pathway in Teleostei: Identification of zebrafish coagulation factor VII. *Proc. Natl. Acad. Sci. USA* **98,** 8768–8773.

Sorensen, L. K., Brooke, B. S., Li, D. Y., and Urness, L. D. (2003). Loss of distinct arterial and venous boundaries in mice lacking endoglin, a vascular-specific TGFbeta coreceptor. *Dev. Biol.* **261,** 235–250.

Varki, A., Cummings, R., Esko, J., Freeze, H., Hart, G., and Marth, J. (1999). "Essentials of Glycobiology." Cold Spring Harbor Press, Cold Spring Harbor, New York.

Xia, L., Ju, T., Westmuckett, A., An, G., Ivanciu, L., McDaniel, J. M., Lupu, F., Cummings, R. D., and McEver, R. P. (2004). Defective angiogenesis and fatal embryonic hemorrhage in mice lacking core 1-derived *O*-glycans. *J. Cell. Biol.* **164,** 451–459.

## [22]   Roles of Mucin-Type O-Glycans Synthesized by Core2β1,6-N-Acetylglucosaminyltransferase

*By* MINORU FUKUDA

### Abstract

Core 2 branched O-linked oligosaccharides (O-glycans) represent the first example of onco-developmental antigens in mucin-type O-glycans. Core 2 branched O-glycans are expressed in immature T lymphocytes (cortical thymocytes), disappear on mature T lymphocytes (medullary thymocytes) and T lymphocytes in the peripheral blood, and appear again in activated T lymphocytes, leukemic cells, and other cancer cells. Core 2 branched O-linked oligosaccharides are synthesized by Core2β1,6-N-acetylglucosaminyltransferase (Core2GlcNAcT). The first cloned Core2GlcNAcT-1 has been inactivated in mice through homologous recombination, and mutants show significantly reduced leukocyte rolling on E-, P-, and L-selectin–coated plates. Moreover, mutant mice exhibit an impaired peritoneal inflammatory response associated with reduced neutrophil infusion. By contrast, lymphocyte homing to secondary lymphoid organs is only marginally compromised. These results combined indicate that Core2GlcNAcT-1 plays a major role in leukocyte trafficking and distinguish leukocyte trafficking to inflamed sites from lymphocyte homing to secondary lymphoid organs.

### Overview

Oligosaccharides attached to mucin-type glycoproteins are usually terminated with N-acetylgalactosamines (GalNAc) at the reducing terminal. This reducing end is linked to the hydroxyl group of serine or threonine residues in a polypeptide (Schachter and Brockhausen, 1992). Initially, it was thought that these so-called mucin-type O-linked oligosaccharides or mucin-type O-glycans were only attached to mucins, mucus proteins secreted from the linings of the respiratory duct and digestive tract. However, a major plasma membrane glycoprotein in human red cells, glycoprotein A, was found to have a long stretch of polypeptide enriched with serine and threonine residues, which are glycosylated (Tomita *et al.*, 1978). Subsequently, when a major glycoprotein of leukocytes, leukosialin (CD43), was characterized, its extracellular domain was found not only to be enriched with serine and threonine residues but also to contain almost 70 residues of N-acetylgalactosamine per molecule (Carlsson and Fukuda,

METHODS IN ENZYMOLOGY, VOL. 416
0076-6879/06 $35.00
DOI: 10.1016/S0076-6879(06)16022-X

1986; Killeen *et al.*, 1987; Pallant *et al.*, 1989). It is estimated that approximately one in three amino acid residues in the extracellular domain of leukosialin is glycosylated by mucin-type $O$-glycans.

At the same time, it was found that the structure of mucin-type $O$-glycans attached to leukosialin differs among that of different cell types expressing that protein (Carlsson *et al.*, 1986). Early on, such cell-type–specific expression was found to depend on whether a particular cell expressed core 2 branched $O$-glycans, Gal$\beta$1→4GlcNAc$\beta$1→6(Gal$\beta$1→3)GalNAc$\alpha$-Ser/Thr, as opposed to core 1 branched $O$-glycans, Gal$\beta$1→3GalNAc$\alpha$-Ser/Thr (Carlsson *et al.*, 1986). These initial observations are highly relevant to several research fields. First, in various cancer cells including those found in leukemia and chorionic carcinoma, the amount of core 2 branched $O$-linked oligosaccharides is increased compared to non-transformed cells (Amano *et al.*, 1988; Brockhausen *et al.*, 1991; Saitoh *et al.*, 1991; Yousefi *et al.*, 1991). In all of these cancer cells examined, this increase in core 2 branched $O$-glycans resulted from increased levels of core 2 $\beta$1,6-$N$-acetylglucosaminyltransferase, which forms the core 2 branch (Fig. 1). Increases in core 2 branched $O$-glycans are highly correlated with tumor progression, such as venous and lymphatic invasion of colon and lung adenocarcinoma (Machida *et al.*, 2001; Shimodaira *et al.*, 1997). Increases in core 2 branched $O$-glycans in tumor cells are apparently due to increased transcription of the gene encoding Core2Glc NAcT-1.

An increase in core 2 branched $O$-glycans is also a prominent feature of certain autoimmune diseases, such as rheumatoid arthritis and immune deficiency (Higgins *et al.*, 1991; Piller *et al.*, 1991). One antibody raised against lymphocytes in the synovial fluid of rheumatoid arthritis patients, T305 (Fox *et al.*, 1983), was found to react specifically with core 2 branched $O$-glycans (Sportsman *et al.*, 1985). It was also found that T305 binds preferentially to core 2 branched $O$-linked oligosaccharides attached to leukosialin (CD43) (Bierhuizen *et al.*, 1994). Using T305 to detect transfected cells expressing core 2 branched $O$-glycans, we used an expression-cloning strategy to identify cDNA-encoding Core2GlcNAcT (Bierhuizen and Fukuda, 1992). T305 was also used to demonstrate that Wiskott–Aldrich syndrome is associated with increases in core 2 branched $O$-glycans. In Wiskott–Aldrich syndrome, patients display immature platelets and immature lymphocytes, causing recurrent infections. Wiskott–Aldrich syndrome is due to a mutation in the gene encoding Wiskott–Aldrich syndrome protein (WASP) (Derry *et al.*, 1994). Studies show that WASP complexes with calcium and integrin binding protein (CIB) (Tsuboi, 2002). This complex then binds to $\alpha$II$\beta$3 integrin to activate platelets. Mutated WASP seen in WASP patients is expressed at lower levels or exhibits a lower affinity for CIB than wild-type WASP. Inefficient binding reduces complex formation

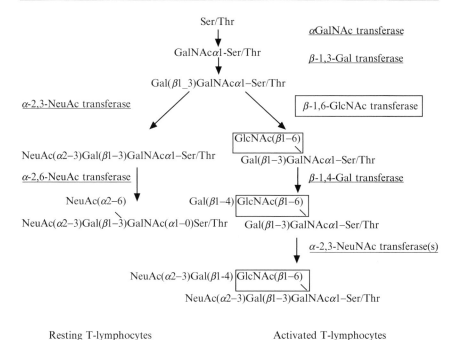

Resting T-lymphocytes          Activated T-lymphocytes

FIG. 1. Biosynthesis of *O*-glycans in T lymphocytes before and after activation. Resting T lymphocytes lack Core2GlcNAcT, thus synthesizing the tetrasaccharide (bottom left). In activated T lymphocytes, Core2GlcNAcT-1 is expressed and the branched hexasaccharide (bottom right) is formed. (Used, with permission, from Piller *et al.*, 1988.)

required for $\alpha II\beta 3$ activation, resulting in impaired $\alpha II\beta 3$ activation (Tsuboi *et al.*, 2006). Overexpression of Core2GlcNAcT-1 in T-lymphocytes in transgenic mice resulted in reduced immune responses by both T and B lymphocytes, which was associated with reduced adhesion of these lymphocytes, indicating that immunodeficiency is due in part to impaired immune cell interaction (Tsuboi and Fukuda, 1997, 1998).

In parallel to these studies, it was found that human resting T lymphocytes express unbranched core 1 oligosaccharides, while activated T lymphocytes express core 2 branched oligosaccharides (Piller *et al.*, 1988) (Fig. 1). This discovery was the first report to correlate a change in mucin-type *O*-glycans with a functional change in cells expressing those oligosaccharides. This study was followed by another critical finding that immature T lymphocytes, thymocytes in the cortical layer of the thymus, express core 2 branched *O*-linked oligosaccharides, while mature thymocytes in the thymus medulla express almost undetectable amounts of core 2 branched

$O$-glycans. Consistently, transcripts encoding Core2GlcNAcT-1 were detected in the cortical layer of the thymus but were undetectable in the medulla (Baum *et al.*, 1995). This low expression level of core 2 branched $O$-glycans continues in circulating T lymphocytes until T lymphocytes are stimulated. Expression of core 2 branched $O$-glycans activated in T lymphocytes apparently triggers apoptosis of T lymphocytes to avoid an excessive response of activated T lymphocytes (Priatel *et al.*, 2000). Considering that core 2 branched $O$-glycans are ectopically expressed in leukemic cells, these findings collectively identified core 2 branched $O$-glycans as an onco-developmental antigen (Fukuda, 1996), the first instance for which mucin-type $O$-glycans have been identified as such.

Immediately after the discovery of core 2 branched $O$-glycans as cell-type–specific carbohydrates, it was shown that core 2 branched $O$-glycans carry Lewis X and sialyl Lewis X oligosaccharides in neutrophils (Fukuda *et al.*, 1986). These findings illustrate two critical attributes of core 2 branched $O$-glycans. First, core 2 branched $O$-glycans are cell-type specific. For example, red cells lack the oligosaccharides (Thomas and Winzler, 1969), while granulocytes express them, as noted earlier. Secondly, core 2 branched oligosaccharides provide critical underlying backbone structures onto which functional oligosaccharides are attached. In mucin-type $O$-glycans of granulocytes, sialyl Lewis and Lewis X are exclusively attached to core 2 branched oligosaccharides (Fig. 2). Sialyl Lewis X was discovered as a selectin ligand (Lowe *et al.*, 1990; Phillips *et al.*, 1990; Walz *et al.*, 1990), and such identification heavily relied on structural data showing that human granulocytes express sialyl Lewis X (Fukuda *et al.*, 1986). These findings indicate that core 2 branching in $O$-linked oligosaccharides is critical to provide $N$-acetyllactosamine, on which functional groups can be added.

The presence of sialyl Lewis and sialyl Lewis A has been shown to be positively correlated with tumor progression (Fukuda, 1996; Fukushima *et al.*, 1984; Inaba *et al.*, 2003; Itzkowitz *et al.*, 1988; Ohyama *et al.*, 1999). The presence of sialyl Lewis X on core 2 branched $O$-glycans then raises the question, between core 2 branched $O$-glycans and sialyl Lewis X, which is more highly correlated with tumor progression? In detailed studies of lung and colonic carcinomas, the presence of core 2 branched $O$-glycans and Core2GlcNAcT-1 transcripts was independently measured in relation to tumor progression. Because sialyl Lewis X and sialyl Lewis A are selectin ligands, it was assumed that these ligands play the critical roles in cancer cell attachment to endothelial cells, distant from primary tumor origin. Surprisingly, it was found that the presence of core 2 branched $O$-glycans or Core2GlcNAcT-1 transcripts has a higher correlation with tumor progression than the presence of sialyl Lewis X or sialyl Lewis A (Machida *et al.*, 2001; Shimodaira *et al.*, 1997). This observation was supported by a finding that lung

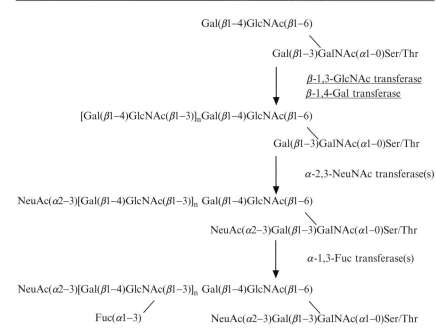

FIG. 2. Biosynthesis of sialyl Lewis X on core 2 branched O-glycans. N-Acetyllactosaminyl chain extended from the GlcNAcβ1→6 linkage in core 2 can be further modified by α1→3 fucosyltransferase, forming sialyl Lewis X termini. Core 2 branched O-glycans form poly-N-acetyllactosamine extension at lower levels than N-glycans. (From Maemura and Fukuda, 1992.)

colonization of sialyl Lewis X-expressing B16 cells in E- and P-selectin-deficient mice is as efficient as that seen in wild-type mice, suggesting that molecules other than E- and P-selectin play a role in recognizing carbohydrates expressed on cancer cells (Zhang et al., 2002). Interestingly, single nucleotide polymorph is correlated with the progression of prostate carcinoma and valine at the codon 152 of Core2GlcNAcT-1 and is highly correlated with the increased risk of prostate cancer than leucine residue (Wang et al., 2005). These findings should now stimulate new efforts to understand and determine how expression of core 2 branched O-glycans leads to tumor progression.

### Generation of Core2GlcNAcT-1 Knockout Mice

Core2GlcNAcT-1 genomic DNA was isolated from a mouse 124/SVJ genomic DNA library using the mouse Core2GlcNAcT-1 cDNA as a probe. cDNA sequences encoding the neomycin resistance gene and the thymidine kinase (tk) gene were inserted into the 3' noncoding region of Core2GlcNAcT-1 genomic DNA. This chimeric genomic DNA was cloned into the pflox vector so that 5'-loxP site is in the 5'-untranslated region of the Core2GlcNAcT-1 gene, while 3'-loxP site flanks the neo-tk cassette gene (Ellies *et al.*, 1998). After linearizing the targeting vector with Not1, it was introduced into R1ES cells via electroporation. Transfected ES cells were selected in the presence of 150 $\mu$g/ml of active G418. ES cell clones obtained were then transiently transfected with supercoiled Cre expression plasmid by electrophoresis. Transfectants were selected for resistance to ganciclovir. Following selection, isolated ES cells were analyzed for the presence of loxP sites using Southern blot analysis. After recombination, only one loxP site remains in Core2GlcNAcT-1$^\Delta$, in which the GlcNAcT-1 gene is deleted, while two loxP sites are present in Core2GlcNAcT-1F, which can be used to delete the gene in a tissue-specific manner using transgenic mouse expressing Cre-recombinase driven by a tissue-specific promoter.

## Peritoneal Inflammation

Among various assays performed, recruitment of neutrophils to peritoneal exudates was reduced in Core2GlcNAcT-1 knockout mice injected intraperitoneally with 1 ml of 3% thioglycollate. Various times after the injection, mice were sacrificed and the peritoneal cavities lavaged with 10 ml of ice-cold phosphate-buffered saline (PBS) containing 1% bovine serum albumin (BSA) and 0.5 m$M$ ethylenediaminetetraacetic acid (EDTA). Red cells were removed by hypotonic lysis, and leukocytes were counted manually using a hemocytometer. After cytocentrifugation specimens were stained with Leukostat (Sigma) and neutrophils were counted. Peripheral exudates were also stained with granulocyte-specific antibody Gr-1 and analyzed by flow cytometry (Broide *et al.*, 2002; Ellies *et al.*, 1998).

## Selectin•IgM Chimera Binding to HEV

Interaction of selectin and selectin carbohydrate ligands is not strong as antibody–antigen interaction. In particular, an interaction between L-selectin and L-selectin ligands is weak and requires shear force (Finger *et al.*, 1996). To overcome this problem, cDNA encoding L-selectin•IgM chimeric protein was developed using E-selectin•IgM cDNA as a template (Maly *et al.*, 1996). In this system, L-selectin works as a pentamer, which exponentially increases its affinity to ligands. In many of the L-selectin studies, L-selectin•IgM chimera protein is recommended to detect L-selectin ligands on high endothelial venules (HEVs). Similarly cDNA encoding P-selectin•IgM can be constructed. The generation of cDNA encoding L-selectin•IgM chimera, and production of L-selectin•IgM chimera and its use are fully described in Chapter 18.

Leukocyte Rolling

To evaluate the role of carbohydrates on leukocytes during inflammation, leukocytes are rolled on plates coated with E-, P- or L-selectin•IgG chimeric proteins. For these experiments, selectin IgG chimeric protein is recommended. Soluble E-, P-, and E-selectin chimeric protein is produced by transient transfection of pcDNA I-selectin•IgG in Cos-1 cells.

A cDNA-encoding soluble human L-selectin•IgG was amplified by polymerase chain reaction (PCR) using L-selectin cDNA (Tedder et al., 1989) as a template. The 5′-primer contains an EcoRI site flanking sequences corresponding to a region 46 bp upstream from the initiation methionine codon, and the 3′-primer encodes antisense L-selectin sequence from codon 336 to 341 with a flanking BamHI site. The amplified cDNA fragment is ligated upstream of a BamHI-Xba fragment of the human IgG hinge plus constant region, derived from CDM8-CD4•IgG chimeric vector (Aruffo et al., 1990) then cloned into pcDNAI, resulting in pcDNAI-L-selectin•IgG (Sueyoshi et al., 1994).

It is critical to use a BamHI site to join the two cDNAs because that sequence encodes a proline as the joint amino acid. Since proline usually perturbs $\alpha$-helical, $\beta$-sheet, or other ordered structures, two proteins joined as a chimeric protein behave independently, each maintaining its separate function. Moreover, a proline-rich sequence in the hinge region of the IgG sequence also allows formation of a flexible joint. In an almost identical manner, pcDNAI-E-selectin•IgG and pcDNA1-P-selectin•IgG vectors can be constructed.

Soluble mouse E-, P-, and L-selectin•IgG chimeric proteins are coated onto polystyrene dishes and assembled in a parallel chamber. To achieve reasonable leukocyte rolling, E-, P-, and L-selectin•IgG are coated at different densities. The following concentrations have been used in representative experiments: E-selectin, P-selectin, and L-selectin at 63, 1469, and 2840 molecules/$\mu$m$^2$, respectively (Ellies et al., 1998). Neutrophils from wild-type and Core2GlcNAcT-1$^{-/-}$ or FucT-VII$^{-/-}$ mice are prepared at a concentration of $1 \times 10^8$/ml in 0.2% BSA in HSS with calcium and magnesium free cell-dissociation solution (Life Technologies, Inc., Gaithersburg, MD) titrated with 10 m$M$ HEPES (pH 7.2). The calcium concentration is adjusted to 2 m$M$ immediately before infusion of the neutrophils into the flow chamber for 30 s at 5 dnm/cm$^2$ using a syringe pump (KD Scientific, Inc., Boston, MA). After infusion is stopped, wall shear forces are increased to 0.19 dyn/cm$^2$ and doubled every 2 min without interrupting the flow. Fields of neutrophils are observed microscopically using a 10× objective and the results are recorded on VCR tape.

Image analysis is performed using the Scion version of the NIH Image program (Scion Corporation, Frederick, MD). The number of cells remaining adherent under static conditions after 2 min at each specific shear force is determined by counting the adherent cells and dividing this number by the number of adherent cells observed preceding the initiation of flow. One hundred percent represents cells in the observed field after static adhesion, before initiation of the lowest shear flow rate, 0.19 dyn/cm$^2$. Alternatively, we can plot the number of cells bound, as the initial number of neutrophils infused is constant.

## Lymphocyte Rolling

To determine the roles of oligosaccharides as L-selectin ligands in HEVs, lymphocyte rolling on GlyCAM-1–coated polystyrene plates can be employed. GlyCAM-1 exclusively contains mucin-type O-linked oligosaccharides and represents HEV-derived L-selectin ligands (Lasky et al., 1992). GlyCAM is a secretory glycoprotein, which is apparently continuously secreted into the bloodstream. GlyCAM-1, thus, can be isolated from serum from mice of various genetic backgrounds (Singer and Rosen, 1996). Specifically, mouse serum is extracted with 4 volumes of chloroform-methanol (2:1) in a glass-stoppered bottle. After centrifugation at 2500g for 15 min, the upper (aqueous) phase is collected and boiled to reduce the volume of the aqueous phase. Once the volume of serum is reduced to one-half the original quantity, the solution is cooled. A clear supernatant is obtained after centrifugation at 12,000g for 15 min and then dialyzed against PBS to remove the residual methanol. After centrifugation at 12,000g for 15 min, the supernatant is reconstituted to the starting volume by addition of PBS. The sample is then applied to plates coated with anti-GlyCAM-1 antibodies. Usually, a GlyCAM-1 sample derived from 10 μl of serum is applied twice (6 h and overnight), or three times (6 h, overnight, and 6 h) on polystyrene plates (Kawashima et al., 2005). Antibodies are purified using a synthetic peptide synthesized to correspond to the Gly-CAM-1 sequence, and the peptide is conjugated to UltraLink Biosupport Medium (Pierce). Bound antibodies are eluted by ImmunoPure IgG Elution Buffer (Pierce) and then titrated to pH 7.0 by adding 1 M NaOH. Resultant antibodies are coated at a concentration of 20 or 50 μg/ml.

Lymphocytes are added to the flow chamber and assayed as described earlier in this chapter. Based on velocity, rolling cells can be classified into two groups: rolling lymphocytes that roll at least two cell diameters for more than 0.5 s below critical velocity and transiently tethered lymphocytes with a high rolling velocity near that of free-floating cells and those that exhibit multiple pauses in less than 0.5 s (Kawashima *et al.*, 2005).

Lymphocyte rolling assays showed that Core2GlcNAcT-1 contributes to all E-, P-, and L-selectin ligands on neutrophils. Among the three selectin ligand levels, P- and L-selectin ligands are most significantly reduced (Ellies *et al.*, 1998). In these assays, it is also possible that different densities of selectins, in particular P- and L-selectins, might yield different results. This is because certain threshold concentrations of P- and L-selectin chimeric proteins are necessary to achieve sufficient rolling. On the other hand, a subtle decrease in selectin ligands may be detected more easily when suboptimal concentrations of these selectins are coated on plates.

Future Perspectives

Core2GlcNAcT-1 was the first Core2$\beta$1,6-$N$-acetylglucosaminyltransferase cloned, and since then many studies have investigated its function. However, studies indicate that two additional enzymes, Core2GlcNAcT-2 and Core2GlcNAcT-3, are encoded in the mouse and human genome (Schwientek *et al.*, 2000; Yeh *et al.*, 1999).

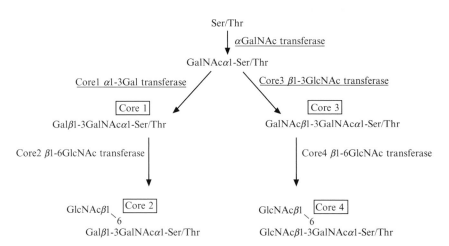

FIG. 3. Biosynthesis of *O*-glycan core structures. The initial step of *O*-glycosylation is the addition of GalNAc to serine (or threonine) of a glycoprotein. Addition of $\beta$1,3-linked galactose and $\beta$1,3-linked *N*-acetylglucosamine results in formation of core 1 and core 3. Core 2 and core 4 are formed by addition of *N*-acetylglucosamine in reactions catalyzed by enzymes unique to each reaction (Schachter and Brockhausen, 1992). Core 2 and core 4 structures are further modified by galactosylation, sialylation, fucosylation, sulfation or elongation reactions.

While Core2GlcNAcT-1 is fairly ubiquitously expressed, Core2Glc-NAc6ST-2 is predominantly expressed in mucin-synthesizing cells, such as gastric cells in the small and large intestines (Yeh *et al.*, 1999). Indeed, Huang *et al.* (2006) showed that Core2GlcNAcT-2 is down-regulated in colorectal cancer and its forced expression causes growth inhibition of colon cancer cells such as HCT 116 and SC 480. These findings indicate that Core2GlcNAcT-2 functions as tumor suppressor, and it is now important to determine whether this is the case for large numbers of patients with colorectal cancer and gastric cancer. Core2GlcNAcT-2 can form core 4 branched *O*-glycans when core 3 forming enzyme is present (Iwai *et al.*, 2005; Yeh *et al.*, 1999). It is, thus, possible that the expression of core 4 oligosaccharides may be critical for Core2GlcNAcT-2 to function as a tumor suppressor (Fig. 3).

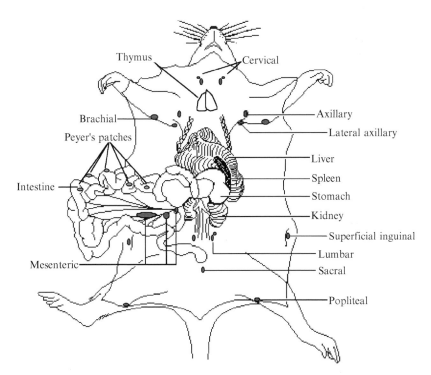

FIG. 4. Lymph nodes and Peyer's patches of mouse. Axillary, lateral axillary, brachial cervical, lumbar, sacral, superficial inguinal, and popliteal are the peripheral lymph nodes. Mesenteric lymph nodes and Peyer's patches are also shown. They appear yellow-white small tissues with a round shape (see also Chapter 18). (Modified, with permission, from Reeves and Reeves, 1991.)

Core2GlcNAcT-3, on the other hand, is expressed in thymus and T lymphocytes (Schwientek *et al.*, 2000). Merzaban *et al.* (2005) showed that Core2GlcNAcT-3 plays a significant role in P-selectin ligand formation in activated CD8 T lymphocytes. In CD8 T lymphocytes, Core2GlcNAcT-3 compensates for Core2GlcNAcT-1 and the role of Core2GlcNAcT-3 is more significant in Core2GlcNAcT-1 knockout mice. Mice deficient in Core2GlcNAcT-2 or Core2GlcNAcT-3 have been generated in Dr. Jamey Marth's laboratory at the University of California, San Diego. The novel aspect of *in vivo* functions of both Core2GlcNAcT-2 and Core2GlcNAcT-3 should be revealed by phenotypic analysis of these mice.

Core2GlcNAcT-1 knockout mice exhibit only marginal decreases in L-selectin ligands of HEVs in the peripheral and mesenteric lymph nodes (Fig. 4 describes the location of lymph nodes). Analysis of Core2GlcNAcT-1 knockout mice identified the remaining L-selectin ligands as those based on extended core 1–based L-selectin ligands (Yeh *et al.*, 2001). In particular, the MECA-79 antibody was found to bind 6-sulfated $N$-acetyllactosamine attached to core 1 structure, $\mathrm{Gal}\beta1{\rightarrow}4(\mathrm{sulfo}{\rightarrow}6)$ $\mathrm{GlcNAc}\beta1{\rightarrow}3\mathrm{Gal}\beta1{\rightarrow}3$ $\mathrm{GalNAc}\alpha1{\rightarrow}R$ and its sialylated, fucosylated form, 6-sulfo sialyl Lewis sialic $\mathrm{acid}\alpha2{\rightarrow}3\mathrm{Gal}\beta1{\rightarrow}4(\mathrm{Fuc}\alpha1{\rightarrow}3[\mathrm{sulfo}{\rightarrow}6])$ $\mathrm{GlcNAc}\beta1{\rightarrow}3\mathrm{Gal}\beta1{\rightarrow}3\mathrm{GalNAc}$ $\alpha1{\rightarrow}R$. Not only the knockout studies determined the function of oligosaccharides by abolishing their expression, but they also revealed novel functions based on newly identified oligosaccharides, thus further expanding the direction of this research. We expect that we will see more of this exemplified discovery revealing novel functions of carbohydrates by studying knockout mice.

## Acknowledgments

The author thanks colleagues and collaborators for generating relevant results, Elise Lamar for critical reading of the manuscript, and Aleli Morse for organizing and editing the manuscript. The work in my laboratory was supported by National Institutes of Health (NIH) grants CA33000, CA48737, and CA91932.

## References

Amano, J., Nishimura, R., Mochizuki, M., and Kobata, A. (1988). Comparative study of the mucin-type sugar chains of human chorionic gonadotropin present in the urine of patients with trophoblastic diseases and healthy pregnant women. *J. Biol. Chem.* **263**, 1157–1165.

Aruffo, A., Stamenkovic, I., Melnick, M., Underhill, C. B., and Seed, B. (1990). CD44 is the principal cell surface receptor for hyaluronate. *Cell* **61**, 1303–1313.

Baum, L. G., Pang, M., Perillo, N. L., Wu, T., Delegeane, A., Uittenbogaart, C. H., Fukuda, M., and Seilhamer, J. J. (1995). Human thymic epithelial cells express an endogenous lectin, galectin-1, which binds to core 2 *O*-glycans on thymocytes and T lymphoblastoid cells. *J. Exp. Med.* **181,** 877–887.

Bierhuizen, M. F., and Fukuda, M. (1992). Expression cloning of a cDNA encoding UDP-GlcNAc:Gal beta 1–3-GalNAc-R (GlcNAc to GalNAc) beta 1–6GlcNAc transferase by gene transfer into CHO cells expressing polyoma large tumor antigen. *Proc. Natl. Acad. Sci. USA* **89,** 9326–9330.

Bierhuizen, M. F., Maemura, K., and Fukuda, M. (1994). Expression of a differentiation antigen and poly-N-acetyllactosaminyl *O*-glycans directed by a cloned core 2 beta-1,6-N-acetylglucosaminyltransferase. *J. Biol. Chem.* **269,** 4473–4479.

Brockhausen, I., Kuhns, W., Schachter, H., Matta, K. L., Sutherland, D. R., and Baker, M. A. (1991). Biosynthesis of *O*-glycans in leukocytes from normal donors and from patients with leukemia: Increase in O-glycan core 2 UDP-GlcNAc:Gal beta 3 GalNAc alpha-R (GlcNAc to GalNAc) beta(1–6)-N-acetylglucosaminyltransferase in leukemic cells. *Cancer Res.* **51,** 1257–1263.

Broide, D. H., Miller, M., Castaneda, D., Nayar, J., Cho, J. Y., Roman, M., Ellies, L. G., and Sriramarao, P. (2002). Core 2 oligosaccharides mediate eosinophil and neutrophil peritoneal but not lung recruitment. *Am. J. Physiol. Lung Cell. Mol. Physiol.* **282,** L259–L266.

Carlsson, S. R., and Fukuda, M. (1986). Isolation and characterization of leukosialin, a major sialoglycoprotein on human leukocytes. *J. Biol. Chem.* **261,** 12779–12786.

Carlsson, S. R., Sasaki, H., and Fukuda, M. (1986). Structural variations of O-linked oligosaccharides present in leukosialin isolated from erythroid, myeloid, and T-lymphoid cell lines. *J. Biol. Chem.* **261,** 12787–12795.

Derry, J. M., Ochs, H. D., and Francke, U. (1994). Isolation of a novel gene mutated in Wiskott-Aldrich syndrome. *Cell* **78,** 635–644.

Ellies, L. G., Tsuboi, S., Petryniak, B., Lowe, J. B., Fukuda, M., and Marth, J. D. (1998). Core 2 oligosaccharide biosynthesis distinguishes between selectin ligands essential for leukocyte homing and inflammation. *Immunity* **9,** 881–890.

Finger, E. B., Puri, K. D., Alon, R., Lawrence, M. B., von Andrian, U. N., and Springer, T. A. (1996). Adhesion through L-selectin requires a threshold hydrodynamic shear. *Nature* **379,** 266–269.

Fox, R. I., Hueniken, M., Fong, S., Behar, S., Royston, I., Singhal, S. K., and Thompson, L. (1983). A novel cell surface antigen (T305) found in increased frequency on acute leukemia cells and in autoimmune disease states. *J. Immunol.* **131,** 762–767.

Fukuda, M. (1996). Possible roles of tumor-associated carbohydrate antigens. *Cancer Res.* **56,** 2237–2244.

Fukuda, M., Carlsson, S. R., Klock, J. C., and Dell, A. (1986). Structures of *O*-linked oligosaccharides isolated from normal granulocytes, chronic myelogenous leukemia cells, and acute myelogenous leukemia cells. *J. Biol. Chem.* **261,** 12796–12806.

Fukushima, K., Hirota, M., Terasaki, P. I., Wakisaka, A., Togashi, H., Chia, D., Suyama, N., Fukushi, Y., Nudelman, E., and Hakomori, S. (1984). Characterization of sialosylated Lewisx as a new tumor-associated antigen. *Cancer Res.* **44,** 5279–5285.

Higgins, E. A., Siminovitch, K. A., Zhuang, D. L., Brockhausen, I., and Dennis, J. W. (1991). Aberrant O-linked oligosaccharide biosynthesis in lymphocytes and platelets from patients with the Wiskott-Aldrich syndrome. *J. Biol. Chem.* **266,** 6280–6290.

Huang, M. C., Chen, H. Y., Huang, H. C., Huang, J., Liang, J. T., Shen, T. L., Lin, N. Y., Ho, C. C., Cho, I. M., and Hsu, S. M. (2006). C2GnT-M is downregulated in colorectal cancer and its re-expression causes growth inhibition of colon cancer cells. *Oncogene* **25,** 3267–3276.

Inaba, Y., Ohyama, C., Kato, T., Satoh, M., Saito, H., Hagisawa, S., Takahashi, T., Endoh, M., Fukuda, M. N., Arai, Y., and Fukuda, M. (2003). Gene transfer of alpha1,

3-fucosyltransferase increases tumor growth of the PC-3 human prostate cancer cell line through enhanced adhesion to prostatic stromal cells. *Int. J. Cancer* **107,** 949–957.

Itzkowitz, S. H., Yuan, M., Fukushi, Y., Lee, H., Shi, Z. R., Zurawski, V., Jr., Hakomori, S., and Kim, Y. S. (1988). Immunohistochemical comparison of Lea, monosialosyl Lea (CA 19–9), and disialosyl Lea antigens in human colorectal and pancreatic tissues. *Cancer Res.* **48,** 3834–3842.

Iwai, T., Kudo, T., Kawamoto, R., Kubota, T., Togayachi, A., Hiruma, T., Okada, T., Kawamoto, T., Morozumi, K., and Narimatsu, H. (2005). Core 3 synthase is down-regulated in colon carcinoma and profoundly suppresses the metastatic potential of carcinoma cells. *Proc. Natl. Acad. Sci. USA* **102,** 4572–4577.

Kawashima, H., Petryniak, B., Hiraoka, N., Mitoma, J., Huckaby, V., Nakayama, J., Uchimura, K., Kadomatsu, K., Muramatsu, T., Lowe, J. B., and Fukuda, M. (2005). N-acetylglucosamine-6-*O*-sulfotransferases 1 and 2 cooperatively control lymphocyte homing through L-selectin ligand biosynthesis in high endothelial venules. *Nat. Immunol.* **6,** 1096–1104.

Killeen, N., Barclay, A. N., Willis, A. C., and Williams, A. F. (1987). The sequence of rat leukosialin (W3/13 antigen) reveals a molecule with O-linked glycosylation of one third of its extracellular amino acids. *EMBO J.* **6,** 4029–4034.

Lasky, L. A., Singer, M. S., Dowbenko, D., Imai, Y., Henzel, W. J., Grimley, C., Fennie, C., Gillett, N., Watson, S. R., and Rosen, S. D. (1992). An endothelial ligand for L-selectin is a novel mucin-like molecule. *Cell* **69,** 927–938.

Lowe, J. B., Stoolman, L. M., Nair, R. P., Larsen, R. D., Berhend, T. L., and Marks, R. M. (1990). ELAM-1–dependent cell adhesion to vascular endothelium determined by a transfected human fucosyltransferase cDNA. *Cell* **63,** 475–484.

Machida, E., Nakayama, J., Amano, J., and Fukuda, M. (2001). Clinicopathological significance of core 2 beta1,6-*N*-acetylglucosaminyltransferase messenger RNA expressed in the pulmonary adenocarcinoma determined by *in situ* hybridization. *Cancer Res.* **61,** 2226–2231.

Maemura, K., and Fukuda, M. (1992). Poly-*N*-acetyllactosaminyl *O*-glycans attached to leukosialin. The presence of sialyl Le(x) structures in *O*-glycans. *J. Biol. Chem.* **267,** 24379–24386.

Maly, P., Thall, A., Petryniak, B., Rogers, C. E., Smith, P. L., Marks, R. M., Kelly, R. J., Gersten, K. M., Cheng, G., Saunders, T. L., Camper, S. A., Camphausen, R. T., Sullivan, F. X., Isogai, Y., Hindsgaul, O., von Andrian, U. H., and Lowe, J. B. (1996). The $\alpha(1,3)$ fucosyltransferase Fuc-TVII controls leukocyte trafficking through an essential role in L-, E-, and P-selectin ligand biosynthesis. *Cell* **86,** 643–653.

Merzaban, J. S., Zuccolo, J., Corbel, S. Y., Williams, M. J., and Ziltener, H. J. (2005). An alternate core 2 beta1,6-*N*-acetylglucosaminyltransferase selectively contributes to P-selectin ligand formation in activated CD8 T cells. *J. Immunol.* **174,** 4051–4059.

Ohyama, C., Tsuboi, S., and Fukuda, M. (1999). Dual roles of sialyl Lewis X oligosaccharides in tumor metastasis and rejection by natural killer cells. *EMBO J.* **18,** 1516–1525.

Pallant, A., Eskenazi, A., Mattei, M. G., Fournier, R. E., Carlsson, S. R., Fukuda, M., and Frelinger, J. G. (1989). Characterization of cDNAs encoding human leukosialin and localization of the leukosialin gene to chromosome 16. *Proc. Natl. Acad. Sci. USA* **86,** 1328–1332.

Phillips, M. L., Nudelman, E., Gaeta, F. C., Perez, M., Singhal, A. K., Hakomori, S., and Paulson, J. C. (1990). ELAM-1 mediates cell adhesion by recognition of a carbohydrate ligand, sialyl-Lex. *Science* **250,** 1130–1132.

Piller, F., Le Deist, F., Weinberg, K. I., Parkman, A., and Fukuda, M. (1991). Altered *O*-glycans synthesis in lymphocytes from patients with Wiskott-Aldrich syndrome. *J. Exp. Med.* **173,** 1501–1510.

Piller, F., Piller, V., Fox, R. I., and Fukuda, M. (1988). Human T-lymphocyte activation is associated with changes in O-glycan biosynthesis. J. Biol. Chem. **263,** 15146–15150.

Priatel, J. J., Chui, D., Hiraoka, N., Simmons, C. J., Richardson, K. B., Page, D. M., Fukuda, M., Varki, N. M., and Marth, J. D. (2000). The ST3Gal-I sialyltransferase controls CD8+ T lymphocyte homeostasis by modulating O-glycan biosynthesis. Immunity **12,** 273–283.

Reeves, J. P., and Reeves, P. A. (1991). "Removal of Lymphoid Organs, Current Protocols in Immunology," Section III, Unit, 1.9, John Wiley & Sons, New York.

Saitoh, O., Piller, F., Fox, R. I., and Fukuda, M. (1991). T-lymphocytic leukemia expresses complex, branched O-linked oligosaccharides on a major sialoglycoprotein, leukosialin. Blood **77,** 1491–1499.

Schachter, H., and Brockhausen, I. (1992). The biosynthesis of serine (threonine)-N-acetylgalactosamine-linked carbohydrate moieties. In "Glycoconjugates: Composition, Structure and Function" (H. J. Allen and E. C. Kisailus, eds.), pp. 263–332. Marcel Dekker, New York.

Schwientek, T., Yeh, J. C., Levery, S. B., Keck, B., Merkx, G., van Kessel, A. G., Fukuda, M., and Clausen, H. (2000). Control of O-glycan branch formation. Molecular cloning and characterization of a novel thymus-associated core 2 beta1,6-n-acetylglucosaminyltransferase. J. Biol. Chem. **275,** 11106–11113.

Shimodaira, K., Nakayama, J., Nakamura, N., Hasebe, O., Katsuyama, T., and Fukuda, M. (1997). Carcinoma-associated expression of core 2 beta-1,6-N-acetylglucosaminyltransfer-ase gene in human colorectal cancer: Role of O-glycans in tumor progression. Cancer Res. **57,** 5201–5206.

Singer, M. S., and Rosen, S. D. (1996). Purification and quantification of L-selectin-reactive GlyCAM-1 from mouse serum. J. Immunol. Methods **196,** 153–161.

Sportsman, J. R., Park, M. M., Cheresh, D. A., Fukuda, M., Elder, J. H., and Fox, R. I. (1985). Characterization of a membrane surface glycoprotein associated with T-cell activation. J. Immunol. **135,** 158–164.

Sueyoshi, S., Tsuboi, S., Sawada-Hirai, R., Dang, U. N., Lowe, J. B., and Fukuda, M. (1994). Expression of distinct fucosylated oligosaccharides and carbohydrate-mediated adhesion efficiency directed by two different alpha-1,3-fucosyltransferases. Comparison of E- and L-selectin-mediated adhesion. J. Biol. Chem. **269,** 32342–32350.

Tedder, T. F., Isaacs, C. M., Ernst, T. J., Demetri, G. D., Adler, D. A., and Disteche, C. M. (1989). Isolation and chromosomal localization of cDNAs encoding a novel human lymphocyte cell surface molecule, LAM-1. Homology with the mouse lymphocyte homing receptor and other human adhesion proteins. J. Exp. Med. **170,** 123–133.

Thomas, D. B., and Winzler, R. J. (1969). Structural studies on human erythrocyte glycoproteins. Alkali-labile oligosaccharides. J. Biol. Chem. **244,** 5943–5946.

Tomita, M., Furthmayr, H., and Marchesi, V. T. (1978). Primary structure of human erythrocyte glycophorin A. Isolation and characterization of peptides and complete amino acid sequence. Biochemistry **17,** 4756–4770.

Tsuboi, S. (2002). Calcium integrin-binding protein activates platelet integrin alpha IIbbeta 3. J. Biol. Chem. **277,** 1919–1923.

Tsuboi, S., and Fukuda, M. (1997). Branched O-linked oligosaccharides ectopically expressed in transgenic mice reduce primary T-cell immune responses. EMBO J. **16,** 6364–6373.

Tsuboi, S., and Fukuda, M. (1998). Overexpression of branched O-linked oligosaccharides on T cell surface glycoproteins impairs humoral immune responses in transgenic mice. J. Biol. Chem. **273,** 30680–30687.

Tsuboi, S., Nonoyama, S., and Ochs, H. D. (2006). Wiskott-Aldrich syndrome protein is involved in $\alpha$IIb$\beta$3-mediated cell adhesion. EMBO Rep. **7,** 506–511.

Walz, G., Aruffo, A., Kolanus, W., Bevilacqua, M., and Seed, B. (1990). Recognition by ELAM-1 of the sialyl-Lex determinant on myeloid and tumor cells. *Science* **250**, 1132–1135.

Wang, L., Mitoma, J., Tsuchiya, N., Narita, S., Horikawa, Y., Habuchi, T., Imai, A., Ishimura, H., Ohyama, C., and Fukuda, M. (2005). An A/G polymorphism of core 2 branching enzyme gene is associated with prostate cancer. *Biochem. Biophys. Res. Commun.* **331**, 958–963.

Yeh, J. C., Hiraoka, N., Petryniak, B., Nakayama, J., Ellies, L. G., Rabuka, D., Hindsgaul, O., Marth, J. D., Lowe, J. B., and Fukuda, M. (2001). Novel sulfated lymphocyte homing receptors and their control by a Core1 extension beta 1,3-*N*-acetylglucosaminyltransferase. *Cell* **105**, 957–969.

Yeh, J. C., Ong, E., and Fukuda, M. (1999). Molecular cloning and expression of a novel beta-1, 6-*N*-acetylglucosaminyltransferase that forms core 2, core 4, and I branches. *J. Biol. Chem.* **274**, 3215–3221.

Yousefi, S., Higgins, E., Daoling, Z., Pollex-Kruger, A., Hindsgaul, O., and Dennis, J. W. (1991). Increased UDP-GlcNAc:Gal beta 1–3GalNAc-R (GlcNAc to GalNAc) beta-1, 6-*N*-acetylglucosaminyltransferase activity in metastatic murine tumor cell lines. Control of polylactosamine synthesis. *J. Biol. Chem.* **266**, 1772–1782.

Zhang, J., Nakayama, J., Ohyama, C., Suzuki, M., Suzuki, A., Fukuda, M., and Fukuda, M. N. (2002). Sialyl Lewis X-dependent lung colonization of B16 melanoma cells through a selectin-like endothelial receptor distinct from E- or P-selectin. *Cancer Res.* **62**, 4194–4198.

# [23]    Analysis of Leukocyte Rolling *In Vivo* and *In Vitro*

By Markus Sperandio, John Pickard,
Sunil Unnikrishnan, Scott T. Acton, and Klaus Ley

## Abstract

Leukocyte rolling is an important step for the successful recruitment of leukocytes from blood to tissues mediated by a specialized group of glycoproteins termed *selectins*. Because of the dynamic process of leukocyte rolling, binding of selectins to their respective counter-receptors (selectin ligands) needs to fulfill three major requirements: (1) rapid bond formation, (2) high tensile strength, and (3) fast dissociation rates. These criteria are perfectly met by selectins, which interact with specific carbohydrate determinants on selectin ligands. This chapter describes the theoretical background, technical requirements, and analytical tools needed to quantitatively assess leukocyte rolling *in vivo* and *in vitro*. For the *in vivo* setting, intravital microscopy allows the observation and recording of leukocyte rolling under different physiological and pathological conditions in almost every organ. Real-time and off-line analysis tools help to assess geometric, hemodynamic, and rolling parameters. Under *in vitro*

METHODS IN ENZYMOLOGY, VOL. 416                                    0076-6879/06 $35.00
Copyright 2006, Elsevier Inc. All rights reserved.                     DOI: 10.1016/S0076-6879(06)16023-1

conditions, flow chamber assays such as parallel plate flow chamber systems have been the mainstay to study interactions between leukocytes and adhesion molecules under flow. In this setting, adhesion molecules are immobilized on plastic, in a lipid monolayer, or presented on cultured endothelial cells on the chamber surface. Microflow chambers are available for studying leukocyte adhesion in the context of whole blood and without blood cell isolation. The microscopic observation of leukocyte rolling in different *in vivo* and *in vitro* settings has significantly contributed to our understanding of the molecular mechanisms responsible for the stepwise extravasation of leukocytes into inflamed tissues.

## The Multistep Model of Leukocyte Recruitment

Leukocyte recruitment as seen by intravital microscopy starts with the capture of free-flowing leukocytes to the vessel wall, followed by leukocyte rolling along the endothelial surface layer (Butcher, 1991; Ley, 1989; Springer, 1995). Both capture and rolling are mediated by selectins. During rolling, leukocytes establish intimate contact with the endothelium, which gives endothelial-bound chemokines the opportunity to interact with specific chemokine receptors expressed on leukocytes. This in turn leads to the activation of leukocyte integrins, which results in firm adhesion of the leukocyte on the endothelium (Butcher, 1991; Ley, 2002; Springer, 1995). Additional groups of adhesion and signaling molecules then mediate leukocyte transmigration into tissue (Muller, 2003).

## Selectins, Selectin Ligands, and Glycosyltransferases

Selectins are professional mediators of leukocyte rolling in inflamed microvessels of nonlymphoid tissues and in high endothelial venules (HEVs) of lymphatic tissues (Bargatze *et al.*, 1995). Three selectins have been identified in humans, mice, and other mammals: P-, E-, and L-selectin (Vestweber and Blanks, 1999). P-selectin is stored in secretory granules called *Weibel–Palade bodies* of endothelial cells and in $\alpha$-granules of platelets. P-selectin can be rapidly mobilized to the cell surface upon activation where it serves as rolling receptor during inflammation (Sperandio and Ley, 2005). E-selectin expression on activated endothelium is induced by several pro-inflammatory cytokines such as tumor necrosis factor-$\alpha$ (TNF-$\alpha$) or interleukin-1 (IL-1). Its expression is controlled by nuclear factor-$\kappa$B (NF-$\kappa$B) and by translational mechanisms (Kraiss *et al.*, 2003; Read *et al.*, 1995). L-selectin is constitutively expressed on most leukocytes and shed from their surface following leukocyte activation. L-selectin is the predominant rolling receptor during naive lymphocyte homing to lymphatic tissue (Ley and Kansas, 2004).

All selectins belong to the C-type lectin family of glycoproteins that recognize and bind specific carbohydrate determinants on selectin ligands in a calcium-dependent fashion. These interactions are only loosely dependent on the underlying protein backbone but require the proper post-translational glycosylation of selectin ligands. This may be the reason why many selectin ligand proteins have been identified, but few have been validated *in vivo* (Vestweber and Blanks, 1999). P-selectin Glycoprotein Ligand-1 (PSGL-1) has been demonstrated to function as a selectin ligand under *in vivo* conditions (Briskin *et al.*, 1993; Sako *et al.*, 1993). PSGL-1 serves as major P- and L-selectin ligand and important E-selectin ligand during inflammation (Sperandio *et al.*, 2003; Xia *et al.*, 2002). Mucosal Addressin Cell Adhesion Molecule-1 (MAdCAM-1) has been proposed to mediate L-selectin–dependent rolling in HEVs of gut-associated lymphatic tissue (Berlin *et al.*, 1993).

Several glycosyltransferases have been identified to contribute to the synthesis of functional selectin ligands (Lowe, 2002). Among them are core 2 $N$-acetylglucosaminyltransferase (Ellies *et al.*, 1998), $\alpha1$–3 fucosyltransferases FucT-VII and -IV (Maly *et al.*, 1996; Weninger *et al.*, 2000), $\alpha2$–3 sialyltransferase ST3Gal-IV (Ellies *et al.*, 2002), and $\beta1$–4 galactosyltransferase-I (Asano *et al.*, 2003). Investigations in mice lacking these glycosyltransferases revealed a broad range of rolling defects reaching from subtle changes in leukocyte rolling velocity in ST3Gal-IV$^{-/-}$ mice (Ellies *et al.*, 2002) to the loss of almost all leukocyte rolling in FucT-VII$^{-/-}$ mice (Homeister *et al.*, 2001; Weninger *et al.*, 2000). An additional modification consisting of the sulfation of a crucial tyrosine at the N-terminus of PSGL-1 mediated by distinct tyrosylsulfotransferases enhances P-selectin binding to PSGL-1 (Ouyang *et al.*, 1998; Ouyang and Moore, 1998; Pouyani and Seed, 1995). Carbohydrate sulfation catalyzed by two specific carbohydrate sulfotransferases, GlcNAc6ST-1 and GlcNAc6ST-2, has been demonstrated to confer maximum affinity for L-selectin ligands in HEVs of peripheral lymph nodes (Hiraoka *et al.*, 1999; Kawashima *et al.*, 2005; Uchimura *et al.*, 2005; Van Zante *et al.*, 2003).

Biomechanical Aspects of Leukocyte Rolling

Leukocyte rolling is a remarkable process made possible by the unique properties of selectins and selectin ligands. From the biomechanical point of view, rolling can be considered a state of dynamic equilibrium whereby rapid bond formation at the leading edge is balanced by rapid bond breakage at the trailing edge of the leukocyte–endothelial contact zone

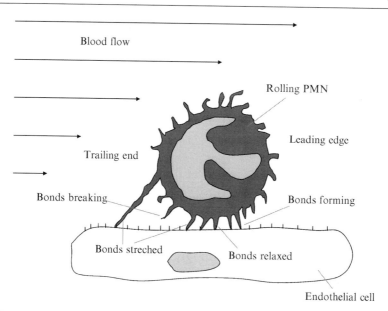

Fig. 1. Leukocyte rolling is a dynamic process whereby bonds between selectins and selectin ligands are constantly formed at the leading edge. This is followed by a short period in which bonds are in a relaxed state until they eventually become stretched and rupture at the trailing end.

(Fig. 1). As leukocytes are exposed to shear stress during rolling, force is loaded on the bonds formed between selectins and their ligands (Fig. 1). To support binding, both selectins and their ligands must be associated with the cytoskeleton (Pavalko *et al.*, 1995). Selectin bonds are unique in that their binding strength increases with the pulling force so that the bond locks even more tightly when blood flows over the leukocyte (catch bonds) (Marshall *et al.*, 2003; Yago *et al.*, 2004). However, at higher pulling forces, the selectin bonds eventually break (Smith *et al.*, 2002). These properties contribute significantly to the creation of an effective system that recruits free-flowing leukocytes to the endothelial wall and prepares them (during rolling) for subsequent adhesion and transmigration.

Intravital Microscopy to Study Leukocyte Rolling *In Vivo*

The direct microscopic observation of flowing blood *in vivo* was reported for the first time by Malpighi in 1661 who found "almost an infinite number of particles" moving in capillaries of the frog lung. Since then many modifications have enabled scientists to investigate microcirculatory phenomena

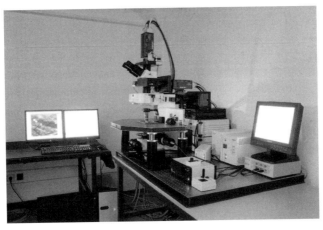

FIG. 2. Side view of a modern intravital microscope. The upright microscope is mounted onto a motorized *xy* stage that is placed on a stable desk. Several light sources and detectors are connected to the system enabling video and/or digital recording of experiments. Filters and dichroic mirrors can be inserted at different positions in the illumination and detection pathways providing a versatile system that allows a whole variety of different microscopy techniques and experiments (Courtesy LaVision BioTec, Bielefeld, Germany.)

including leukocyte rolling and adhesion in a variety of different models and setups (Ley, 2001; Ley *et al.*, 1987; Pries, 1988). Multiphoton laser scanning microscopy has been added to the armamentarium of intravital microscopy (Mempel *et al.*, 2004) and may one day be used to study leukocyte rolling.

Figure 2 shows the side view of a modern intravital microscope. The different components of the setup are listed in Table I and are discussed later in this chapter. The upright microscope is fixed to a motorized *xy*-stage that is mounted on a stable optical table (Fig. 2). Hence, to examine the sample, the whole microscope with its different light sources and detectors (Table I) can be moved while the sample remains fixed in its position.

The illumination devices include conventional light sources (e.g., halogen lamp) for transillumination or reflection microscopy, as well as stroboscopic flash lamp systems (e.g., Rapp Optoelectronic, Hamburg, Germany) and continuous light Hg or Xe sources for epifluorescence microscopy applications (Slaaf *et al.*, 1986). The choice of the light source depends on the particular fluorochromes and their excitation wavelengths. The mercury lamp offers some advantages for fluorochromes that are best excited at the peaks of the Hg spectrum. In contrast, Xe lamps have an almost continuous spectrum reaching from ultraviolet over the whole visible range, which favors its use for applications in which different fluorochromes are used. For epifluorescence applications, pulsed illumination systems offer

TABLE I

COMPONENTS OF A MODERN INTRAVITAL MICROSCOPE

| Component | Details | Characteristics |
|---|---|---|
| Motorized xy-stage | | Moves the whole microscope while the sample is resting |
| Upright microscope | | Microscope type that allows the adaptation of various light sources and detectors |
| Optical table | | Stable, absorption of vibrations and shocks |
| Light source | 1. Halogen lamp | Transmission/reflection experiments |
| | 2. Continuous light from Xe- or Hg-lamp | Fluorescence excitation of various fluorochromes |
| | 3. Stroboscopic flash lamp system | Fluorescence excitation offering several advantages over continuous light sources |
| Detection device | CCD-camera with analog or digital output signal | Fast CCD camera for digital recording (frame rates up to 100 fps) connected to high-end computer |
| Objective lens | Standard (salt) water immersion: ×10, ×20, ×40, ×60 with high numerical aperture | Long working distance Large field of view with high resolution |
| Condensor | Important for transillumination microscopy applications | Köhler illumination, integration of contrast enhancing techniques (phase contrast, DIC, darkfield) |
| Fast Filter Wheel | | Multicolor fluorescence |
| On-line blood flow velocity measurement device | Digital temporal cross-correlation of intensity profiles collected by two adjacent photodiodes (Pries, 1988) | Easy-to-use application, powerful bright field lamp necessary |
| Video-timer | For videomicroscopy only | Important component for off-line analysis |
| Superfusion system and thermo-controlled heating pad | Different superfusion solutions are described (f.e. Klitzman and Duling, 1979) | Thermocontrolled environment |
| Animal stage | Custom-made from Plexiglass | Support of tissue, reservoirs for superfusion |
| Micromanipulator for microinjections | Micromanipulators with x-y-z axis and inclination | |

significant advantages over continuous light sources. Most importantly, because of the long pause after each flash (the duration of one pulse is typically only a few microseconds), the phototoxic effects exerted on the living tissue under observation are significantly decreased. In addition, a proper synchronization of the flash duration and frequency to the readout cycle of the detection devices (digital camera or videocamera) enables (automated) off-line analysis of several hemodynamic parameters including blood flow velocity, leukocyte rolling velocity, and even entire velocity profiles (Long *et al.*, 2004; Pries, 1988). Of note, placing two consecutive flashes one shortly before frame transfer and the second shortly after frame transfer produces a pair of consecutive images recorded only a few milliseconds apart (Pries, 1988). This method can be used to assess high blood flow velocities (f.e. in arterioles) that are too fast to be picked up by conventional analysis. For detection devices, many intravital microscopists use CCD cameras with analog output signal. Powerful CCD cameras are capable of recording and saving short movies directly to the computer in real time with frame transfer rates and image resolution exceeding video quality (up to 100 fps, pixel size 5–10 $\mu$m). Although these completely digitized systems will slowly replace video recordings, it should be kept in mind that digital recording of dynamic processes such as leukocyte rolling requires the use of fast cameras that should reach frame rates of at least 20 fps. In fact, high-speed computer processors with large RAM ($>2$ GB) and hard drive capacity (100+ GB) are necessary to handle and store recorded sequences. The transfer of stored movie clips to mobile hard drives or DVD is still time consuming when compared to the routine handling of videotapes. Currently, no digital camera can produce color images at the sensitivity and speed needed for real-time fluorescence imaging.

To enable the continuous observation of microcirculatory phenomena by intravital microscopy, surgically exteriorized living tissue has to be kept warm and moist. This is achieved by superfusing the tissue with thermo-controlled (37°) bicarbonate buffered salt solution as described (Klitzman and Duling, 1979). To make the superfusion, 100 ml of solution I (NaCl 292.9 g, KCl 13.3 g, $CaCl_2$ 11.2 g, and $MgCl_2$ 7.7 g dissolved in 3.8 liters of deionized water) is added into a 1000-ml cylinder filled with 800 ml deionized water. Thereafter, 100 ml of solution II ($NaHCO_3$ 57.5 g dissolved in 3.8 liters of deionized water) is added. After mixing the final solution, the superfusion is equilibrated with $N_2$ 95%/$CO_2$ 5% using a foam disperser. Besides its usefulness for keeping the living tissue under standardized conditions during the experiment, the superfusion fluid when used in conjunction with a (salt) water objective also offers several advantages concerning optical properties. Having water between the tissue and the objective helps to stabilize the optical path and increases numerical aperture. As a whole variety of different water immersion objectives are on the market, the choice of the right objective should be based not only on the magnification and the numerical aperture but also on a small objective nose providing easy access to the tissue. For most intravital microscopy applications, we use a ×40 objective with a numerical aperture of 0.8 and a narrow objective nose (Carl Zeiss, Jena, Germany).

## Transillumination Microscopy

Transillumination intravital microscopy has been used for more than 150 years to investigate leukocyte rolling *in vivo* (Wagner, 1839). Several animal models have been described. In view of the fact that the generation of knockout animals is largely restricted to mice, studies on leukocyte recruitment and rolling have mainly focused on mouse models. For the observation of leukocyte rolling during inflammation, the murine cremaster muscle preparation has become a frequently used model (Baez, 1973). The muscle is relatively thin and sufficiently vascularized. Inflammation can be induced by surgical trauma or by intrascrotal injection of pro-inflammatory cytokines (Jung *et al.*, 1998; Ley *et al.*, 1995). In contrast to the rat mesentery, which has been used for studying leukocyte rolling, the mesentery of the mouse has very few microvessels, which prevents its widespread application for studying leukocyte rolling *in vivo*.

Intravital microscopic experiments can be recorded by a videocamera connected to a video recorder. Leukocyte rolling and adhesion, as well as geometric and hemodynamic parameters such as vessel diameter, vessel segment length, blood flow velocity, rolling velocity, pause times, acceleration, and deceleration, are analyzed after the experiment. Centerline blood flow velocities during transillumination microscopy experiments can be measured using the dual slit photometric method originally designed by Wayland and Johnson (1967), consisting of a dual photodiode that is connected to a digital on-line temporal intensity cross-correlation program (Circusoft Instrumentation, Hockessin, USA). However, information content is limited, because only one vessel is analyzed at a time. A more advanced technology is spatial correlation with dual flash, which allows the measurement of leukocyte velocity in all vessels in the field of view (Pries, 1988).

The most recent development is microparticle image velocimetry ($\mu$PIV) that yields entire flow profiles, local viscosity, shear rate, and details of near-wall effects caused by endothelial surface molecules (Long *et al.*, 2004; Smith *et al.*, 2003). An automated image processing software extracting all these parameters from a short recording is expected in 2007.

Using fluorescence microscopy, centerline blood flow velocity can be determined after intravenous injection of 1 $\mu$m diameter fluorescent YG microspheres (Polysciences, Warrington, PA) by measuring frame-to-frame displacement of single beads located in the center of the vessel. Wall shear rates ($\gamma_w$) are estimated as 4.9 (8 $v_b$/d), where $v_b$ is the mean blood flow velocity and $d$ the diameter of the vessel. The constant 4.9 is a mean empirical correction factor obtained from recently described velocity profiles measured in microvessels *in vivo* using PIV (Long *et al.*, 2004; Smith *et al.*, 2003).

Fluorescence Microscopy

The observation of leukocyte rolling by fluorescence microscopy has become an essential technical tool for the dissection of lymphocyte homing *in vivo*. In addition, intravital fluorescence microscopy studies have also been instrumental in elucidating the role of leukocytes in ischemia/reperfusion injury of internal organs. Since its first description by von Andrian (1996), the peripheral lymph node model has become a very popular model to study leukocyte rolling in lymphatic tissue. To investigate lymphocyte trafficking into lymphatic tissue of the gut, intravital fluorescence microscopy of Peyer's patches has been used (Miura *et al.*, 1987). Intravital microscopy of the skin and in the dorsal skin fold chamber are attractive models to study leukocyte rolling during inflammation and allergic reactions (Lehr *et al.*, 1993; Radeke *et al.*, 2005). Leukocyte recruitment during ischemia reperfusion injury has been studied by intravital microscopy in several organs such as the liver, intestine, and brain (Liu and Kubes, 2003). Interestingly, these studies revealed that adhesion molecules involved in leukocyte rolling may differ considerably between organs.

To visualize the cells of interest during intravital fluorescence microscopy, the cells must be labeled with a fluorescent dye. This can be achieved by isolating leukocyte subsets under *in vitro* conditions and injecting them during the *in vivo* experiment. Magnetic bead negative selection is a widespread tool to isolate a sufficient number of lymphocyte subsets. For neutrophil isolation, Percoll gradient separation has been successfully applied not only in humans but also in mice (Stark *et al.*, 2005). For blood neutrophils, bead-based isolation techniques are used (Cotter *et al.*, 2001). However, *ex vivo* isolated neutrophils do not roll normally when injected back into mice, showing that isolation procedures cause subtle activation. *In vivo* staining of leukocytes can easily be achieved by injection of rhodamine 6G or acridine orange. This method is convenient but rarely allows investigating rolling of leukocyte subsets (Janssens *et al.*, 1989). Several green fluorescent protein (GFP)–transgenic mice have become available where leukocyte subsets including neutrophils and lymphocytes express GFP under control of a subset specific promoter (Faust *et al.*, 2000; Jung *et al.*, 2000; Manjunath *et al.*, 1999; Singbartl *et al.*, 2001; Unutmaz *et al.*, 2000).

Microparticle Image Velocimetry

Complex flows often encountered in the microvasculature, like flow around adherent and rolling leukocytes, are poorly understood because of the lack of accurate quantitative data, often leading to the assumption of a parabolic profile in analysis (Sugii *et al.*, 2002; Tangelder *et al.*, 1986). Methods like laser Doppler velocimetry and tracking of red blood cells (RBCs) are limited by poor spatial resolution (Sugii *et al.*, 2002; Tangelder *et al.*, 1986). Also, near-wall velocity profiles are needed to obtain accurate wall shear stress data. Evidence of an endothelial surface layer that impedes plasma flow requires accurate near-wall velocity measurements to understand the forces involved in leukocyte adhesion to the vessel wall (Smith *et al.*, 2003). These requirements demand high spatial resolution methods.

Particle image velocimetry (PIV) is an optical technique for velocity field measurement and visualization, widely used in traditional fields of fluid mechanics like aerodynamics and combustion. In this method, the flow is first seeded with tracer particles. It is illuminated by a plane of light at two different time points, separated by a known time interval, and the event is recorded photographically on a single image or on a succession of images. Spatial correlation techniques are then used to obtain the two-dimensional velocity vectors. PIV was first extended for microfluidics in a Hele-Shaw flow around an elliptic cylinder, using 100–300 nm fluorescent particles as tracers under epi-illumination (Santiago *et al.*, 1998). These authors were able to obtain a spatial resolution of $6.9 \times 6.9 \times 1.5 \ \mu$m. Koutsiaris *et al.* in 1999 demonstrated the feasibility of using $\mu$PIV in glass capillaries and the microcirculation. The development of advanced postprocessing techniques such as a combined PIV-particle tracking velocimetry (PTV) algorithm (Devasenathipathy *et al.*, 2003; Keane *et al.*, 1995), coupled with the advent of high-speed digital cameras, has led to a considerable improvement in spatial and temporal resolution (Shinohara *et al.*, 2004; Sugii *et al.*, 2005). Combined with intravital fluorescence microscopy, this technique lends itself well to *in vivo* animal studies because of the noninvasive nature of interrogation.

In $\mu$PIV, unlike conventional PIV, the flow field is volume illuminated and the plane of measurement is determined by the depth of field of the microscope optics. Hence, out-of-plane particles appear progressively blurred as we move away from the focus and add to background noise. To define the measurement plane, it is useful to specify an intensity threshold (as a fraction of the peak in-focus intensity) below which we assume the particle does not contribute to the measurement. This measurement plane is dependent on numerical aperture, refractive index of the medium in which the objective is immersed, collected light wavelength, and particle diameter (Meinhart et al., 2000; Nguyen and Wereley, 2002). An important consideration in $\mu$PIV is the size of the particle injected in the flow. In microlevel flows, the particles themselves influence the flow, especially near the wall, so appropriate theoretical analysis should be used to correct the obtained experimental values (Smith et al., 2003). Brownian motion and surface effects also introduce errors in measurement and need to be accounted for in postprocessing algorithms (Wereley and Meinhart, 2005).

### Off-Line Analysis Tools: What Is Needed to Assess Leukocyte Rolling *In Vivo*?

In the microcirculation, leukocytes marginate to outer blood flow layers, a phenomenon that is predominantly observed in venules. But even under conditions of maximal margination, leukocytes still move at about 50% of the centerline velocity in small venules (Gaehtgens et al., 1985). In contrast, rolling leukocytes travel with velocities that are at least one order of magnitude lower than average blood flow velocity, demonstrating the presence of adhesive interactions between leukocytes and the endothelium. Theoretical considerations based on hydrodynamic velocity near the wall (Goldman et al., 1967) led to the introduction of the term *critical velocity*, which defines that the slowest velocity free-flowing leukocytes can reach close to the vessel wall (see also Table II for definitions). Below critical velocity, adhesive interactions between leukocytes and endothelial cells have to be assumed (Ley and Gaehtgens, 1991). Critical velocity ($v_{crit}$) *in vivo* is dependent on blood flow velocity, leukocyte diameter, and vessel diameter. $v_{crit}$ can be estimated from mean blood flow velocity) ($v_{mean}$) in cylindrical vessels as $v_{crit} = v_{mean} \times \varepsilon \times (2 - \varepsilon)$, where $\varepsilon$ is the cell to vessel diameter ratio (Ley and Gaehtgens, 1991). This gives an easily determined estimate and is accurate enough for many studies. However, better estimates became possible with $\mu$PIV (Long et al., 2004). Leukocyte rolling velocities in inflamed cremaster muscle venules usually range from 1 to 100 $\mu$m/s, while blood flow velocities in postcapillary venules (diameter 20–50 $\mu$m) range from 0.5 to 10 mm/s.

Using videotape recordings or digital movies, leukocyte rolling can be quantitatively assessed as leukocyte rolling flux (RF), which is the number of leukocytes crossing an imaginary line across the observed vessel (Table II). Leukocyte rolling flux is easily measured; however, it does not take into

TABLE II

PARAMETERS IMPORTANT FOR THE ASSESSMENT OF LEUKOCYTE ROLLING BY INTRAVITAL MICROSCOPY

| Parameter | Definition | Unit |
| --- | --- | --- |
| Leukocyte rolling flux | Number of rolling leukocytes passing a perpendicular line placed across the observed vessel in one minute | n/min |
| Leukocyte rolling flux fraction | Rolling flux divided by the complete leukocyte flux | % |
| Leukocyte rolling velocity | Movement of leukocytes in intermittent and/or continuous contact with the endothelial surface | $\mu$m/s |
| Critical velocity | Minimal velocity of free-flowing leukocytes close to the vessel wall without contact to the wall | $\mu$m/s |
| Vessel diameter | Maximal orthogonal linear distance between endothelial cells on opposite sides of vessel | $\mu$m |
| Centerline blood flow velocity | Converted into mean blood flow velocity by multiplying with an empirical correction factor (Lipowsky and Zweifach, 1978) | $\mu$m/s |
| Wall shear rate | $\gamma_w = 4.9 \, (8v_{bl}/d_{vessel})$ ($\gamma_w$: wall shear rate) (Long et al., 2004; Smith et al., 2003) | $(\text{s}^{-1})$ |
| Wall shear stress | $\tau_w = \eta \, \gamma_w$ ($\tau_w$: wall shear stress, $\eta$: viscosity) | $(\text{dyn/cm}^2)$ |

account differences in systemic leukocyte count. This may become important in experiments in which antibody treatments are conducted. For example, injection of the L-selectin blocking monoclonal antibody (mAb) MEL-14 causes a significant drop in systemic neutrophil count (Sperandio *et al.*, 2001). In addition, some adhesion molecule–deficient mice such as CD18$^{-/-}$ mice or P-/E-selectin$^{-/-}$ mice have markedly elevated systemic leukocyte counts (Bullard, 2001). To overcome this problem, many groups assess leukocyte rolling as rolling flux fraction (RFF), which is the number of rolling cells divided by all leukocytes passing a perpendicular line in a certain time. To quantify the number of free-flowing leukocytes in postcapillary venules, fluorescent *in vivo* labeling of all leukocytes as described earlier can be used to count the number of free-flowing leukocytes. As a rough approximation, the number of free-flowing leukocytes can be calculated by multiplying systemic leukocyte count (cells/$\mu$l) and microvascular blood flow ($\mu$l/min). Microvascular blood flow can be obtained from measuring vascular diameter ($\mu$m) and blood flow velocity ($\mu$m/s). Vascular diameter can easily be measured off-line using digital image processing systems (Norman, 2001; Pries, 1988). Blood flow can be deduced from centerline RBC velocity measurements using a dual photodiode and a digital cross-correlation program as described earlier. These centerline velocities can be converted to mean blood flow velocities by multiplying with an empirical factor of approximately 0.625 (Lipowsky and Zweifach, 1978).

Leukocyte rolling velocities are traditionally obtained from video or digital recordings by manually determining the position of hundreds of rolling leukocytes using a frame-by-frame analysis method, a tedious task (Jung *et al.*, 1996). Advancements in imaging and computational techniques, in conjunction with the need to improve and accelerate the assessment of microvascular parameters, have spawned promising new automated approaches that have the additional advantage of eliminating observer bias and allow the possibility of real-time velocity assessment.

The accuracy of these algorithms is largely dependent on image quality, which can be highly variable between animals and even between vessels within the same animal. Rolling leukocytes often move in and out of the focal plane, on top of one another, and under occluding structures causing changes in intensity, cell shape, and edge sharpness. Image clutter, a result of the layered tissue, further complicates the analysis environment providing numerous interfering objects and imaging artifacts. Furthermore, animal respiration and muscle contractions result in gross tissue movement, which must be identified and corrected for. Flow chamber and microcapillary tube experiments, on the other hand, have few of these problems, and customary image processing techniques have demonstrated good results. Spatial filtering detection and cross-correlation, sum-of-absolute-differences, and centroid-based tracking algorithms have been used for automated particle and cell tracking analysis (Debeir *et al.*, 2005; DiVietro *et al.*, 2001; Ghosh and Webb, 1994; Kusumi *et al.*, 1993). Image quality determines the best algorithm for a particular application (Cheezum *et al.*, 2001), and, unfortunately, the lower image quality of *in vivo* data make these methods unsuitable for intravital experiments (Acton *et al.*, 2002; Ray *et al.*, 2002).

*In vivo* data analysis typically begins with an image enhancement step, which attempts to remove the respiratory motion artifacts, image noise, and clutter. To this end, frame differencing (optical flow) and cross-correlation of image edges have been effectively used (Acton *et al.*, 2002; Eden *et al.*, 2005; Goobic *et al.*, 2005). Vessel identification can be integrated into this step to improve processing speed and accuracy (Jinshan and Acton, 2004). The first automatically acquired *in vivo* rolling velocities used spatiotemporal image analysis (Sato *et al.*, 1997); however, this analysis was confined to the imaged vessel wall and lacked exact position determinations (Eden *et al.*, 2005). More advanced techniques exploit large intensity differences, which delineate cell borders, for template-based detection methods, which attempt to find leukocyte shaped objects (Dong *et al.*, 2005; Gang *et al.*, 2005; Mukherjee *et al.*, 2004). Tracking detects cells using deformable models, specifically parametric active contours, which have shown better results than correlation or centroid trackers (Jinshan, 2002). Initially applied to *in vitro* cells (Leymarie and Levine, 1993), the inclusion of *a priori* size, shape, and velocity information have allowed active contour extension to *in vivo* data (Ray and Acton, 2004; Ray *et al.*, 2002). Fully integrated systems using the above methodology have been demonstrated after respiratory motion correction (Goobic *et al.*, 2005), moving field of view registration (Dunne *et al.*, 2004), and by real-time detection and tracking on a multiprocessor system (Jellish, 2005). Additionally, a neural-network decision framework, termed *CellTrack*, was used on color data for cell detection and tracking (Eden *et al.*, 2005). While none of these systems achieves 100% accuracy, a reliable computer-based tracking and data assessment system offers a rapid, highly accurate, and consistent alternative to the cumbersome and observer-biased manual work necessary to assess leukocyte rolling *in vivo*.

Flow Chamber Assays

Leukocyte rolling *in vivo* takes place in a complex environment, which is exposed to constant changes in hemodynamic and geometric parameters such as wall shear stress, volume flow rate, tube hematocrit, and vessel diameter. In addition, the vessel wall is covered by endothelial cells and a thick surface layer (Vink and Duling, 1996) hosting numerous adhesion molecules at variable site densities even along the same vessel segment, leading to local changes in rolling velocity or adhesion dynamics (Damiano *et al.*, 1996; Jung and Ley, 1997). To create a more reductionist system in which interactions between different adhesion molecules can be studied on a molecular level and under clearly defined flow conditions, several flow chamber assays have been developed (Lawrence, 2001). These dynamic *in vitro* assays helped to identify the shear threshold for selectin-mediated rolling (Finger *et al.*, 1996; Lawrence *et al.*, 1997), the ability of $\alpha_4$-integrins to initiate lymphocyte tethering and rolling in the absence of selectins (Berlin *et al.*, 1995), and the role of chemokines in leukocyte arrest (Campbell *et al.*, 1998; Shamri *et al.*, 2005).

Parallel plate flow chambers are among the most popular flow chambers used to study leukocyte rolling *in vitro*. They were described as early as 1912 (Burrows, 1912) but were used for the investigation of leukocyte rolling only in the late 1980s (Lawrence *et al.*, 1987). Commercially available parallel plate flow chamber systems (www.glycotech.com) consist of two flat plates separated by a gasket, which forms a geometrically defined narrow channel between the two plates. An inflow and outflow port are used to enable the perfusion of the cell suspension through the chamber. An additional port generates a negative pressure, which keeps both plates together. To control for the flow of the cell suspension through the flow chamber, a high-precision perfusion pump is used. By knowing the geometry of the inner chamber, the flow rate of the perfusion pump $Q$, and the viscosity $\eta$, wall shear stress can be calculated as

$$\tau_w = (3\eta Q)/(2ba^2), \tag{1}$$

where $b$ is chamber width, and $a$ is channel half height. To observe leukocyte rolling, the flow chamber is placed under an inverted microscope, which is connected to a videocamera or digital camera for recording the experiment. Two popular assays to investigate leukocyte rolling in the flow chamber are the leukocyte attachment and leukocyte detachment assay. In the attachment assay, leukocytes are perfused through the chamber at constant flow (wall shear stress: 0.5–1.5 dyn/cm$^2$), leading to a gradual increase in the number of rolling leukocytes over time until a maximum is reached. In the detachment assay, leukocytes are infused into the chamber and flow is ceased for a little while. This gives leukocytes the chance to bind to immobilized proteins or endothelial cells. Flow is then reinitiated at very low shear stress levels (f.e. 0.05 dyn/cm$^2$) and increased until all rolling leukocytes detach from the surface. The off-line analysis of *in vitro* leukocyte rolling assays can be accomplished by counting the number of rolling cells per field of view and per time period. Leukocyte rolling velocity can be assessed by measuring the distance leukocytes have traveled in a certain time. Several automatic tracking systems have been developed to automatically quantify the number and the velocity of rolling cells *in vitro* (Mangan *et al.*, 2005; Rao *et al.*, 2002).

## Autoperfused Flow Chamber

Although flow chambers have been an important tool to study leukocyte rolling *in vivo*, there are several limitations such as the large dead volume of fluid within the flow system ($\sim$300 $\mu$l in parallel plate flow chambers), necessitating the isolation of large quantities of cells. This prevented widespread use of flow chambers not only in the mouse system but also in experiments in which rare human leukocyte populations were investigated. The availability of genetically engineered mice that offer ideal conditions to study adhesive interactions between distinct pairs/groups of adhesion molecules stimulated the development of new flow systems, which can be conducted with small sample volumes. Smith *et al.* (2004) described an *ex vivo* autoperfused flow chamber system in the mouse that consists of a rectangular glass capillary ($200 \times 2000$ $\mu$m) where adhesion molecules of choice can be immobilized. After immobilization the flow chamber is placed under an upright microscope and connected via PE-tubing to the jugular vein and the carotid artery of the mouse. Because of the pressure difference between the arterial and venous side, blood is continuously driven through the flow chamber. Additional tubing connected to the venous and the arterial side is used to continuously measure the pressure drop ($\Delta$P) along the glass capillary. From this measurement in conjunction with the known geometry of the chamber, wall shear stress ($\tau_w$) can be determined exactly and independent of any assumptions about blood viscosity as

$$\tau_w = (\Delta P D_h)/(4L), \qquad (2)$$

where $L$ is the length of the capillary and $D_h$ the hydraulic diameter. Hydraulic diameter can be calculated using

$$D_h = (4A)/(p), \qquad (3)$$

where $A$ is the channel cross sectional area and $p$ the wetted perimeter of the cross section (Smith *et al.*, 2004). Adhesive interactions in the flow chamber between blood constituents (i.e., leukocytes) and immobilized proteins on the glass surface can then be observed and recorded (Smith

TABLE III
TOOLS TO MANIPULATE LEUKOCYTE ROLLING

| Mode of manipulation | Impaired function | | References |
|---|---|---|---|
| Genetically engineered mice targeting: | | | |
| 1. Adhesion molecules | | | |
| Selectins: | P-selectin | Rolling | Mayadas et al., 1993 |
| | E-selectin | Slow rolling | Labow et al., 1994 |
| | L-selectin | Tethering | Arbones et al., 1994 |
| Selectin ligands | PSGL-1 | Rolling | Xia et al., 2002; Yang et al., 1999 |
| 2. Post-translational modification of selectin ligands | | | |
| Glycosyltransferases: | FucT-VII | All selectins | Maly et al., 1996 |
| | FucT-IV | E-selectin rolling | Weninger et al., 2000 |
| | Core2 GlcNAcT | All selectins | Ellies et al., 1998 |
| | ST3Gal-IV | E- and L-selectin | Ellies et al., 2002 |
| | $\beta$1-4GalT | P-selectin | Asano et al., 2003 |
| Carbohydrate | GlcNAc6ST-1 | L-selectin | Uchimura et al., 2004 |
| sulfotransferases | GlcNAc6ST-2 | L-selectin | Hemmerich et al., 2001 |
| Tyrosylsulfotransferases | TST | P-selectin | Ouyang and Moore, 1998 |
| Blocking antibodies against selectins and selectin ligands: | | | |
| Anti P-selectin | RB40.34 ($\alpha$ mouse) | | Bosse and Vestweber, |
| | G1 ($\alpha$ human) | | 1994; Hamburger and McEver, 1990 |
| Anti E-selectin | 9A9, RME-1 ($\alpha$ mouse) | | Bevilacqua et al., 1987; |
| | H18/7 ($\alpha$ human) | | Norton et al., 1993; Walter et al., 1997 |
| Anti L-selectin | MEL-14 ($\alpha$ mouse) | | Gallatin et al., 1983; |
| | DREG-56 ($\alpha$ human) | | Kishimoto et al., 1990 |
| Anti PSGL-1 | 4RA10 ($\alpha$ mouse) | | Frenette et al., 2000; |
| | PL-1, KPL-1 ($\alpha$ human) | | Li et al., 1996; Snapp et al., 1998 |

et al., 2004). This flow chamber was further miniaturized to a cross-section of $20 \times 200$ $\mu$m (Chesnutt et al., 2006).

Because of the small volume in the system, the rectangular glass capillaries can also be used for in vitro rolling studies by simply connecting the glass capillary to a perfusion pump. This provides superior optical conditions over round glass capillaries, which have also been used for leukocyte rolling studies (Nandi et al., 2000). If appropriate condensers, objectives, and additional accessories are chosen, microflow chamber assays with cell suspensions can also be performed on the upright microscope using contrast enhancing techniques such as phase contrast, darkfield, or differential interference contrast (DIC) microscopy.

## Manipulation of Selectin-Dependent Leukocyte Rolling

Most leukocyte rolling is mediated by selectins binding to carbohydrate structures on selectin ligands. Manipulation of these interactions can be achieved on several levels (Table III) (Sperandio *et al.*, 2004). Adhesion molecules can be eliminated directly (gene targeted disruption of selectins or selectin ligands) or indirectly (f.e. glycosyltransferase knockout mice). At the protein level, a variety of functional blocking mAbs against selectins and selectin ligands exist and are commercially available (Table III). These antibodies can be applied during *in vivo* and *in vitro* experiments and offer the advantage that rolling can be studied before and after antibody injection and adaptive processes common in knockout mice are absent. On the other hand, incomplete blocking of leukocyte rolling either by an insufficient dose or by specificity of the used antibody has to be considered. Structural mimetics of selectin ligands have been described to block selectin ligand function under *in vitro* conditions (Kogan *et al.*, 1998). This may open a new and interesting avenue to manipulate selectin-dependent rolling *in vivo* (Aydt and Wolff, 2002).

## Conclusion

The observation of leukocyte rolling *in vivo* and *in vitro* has become an indispensable tool for the understanding of how leukocytes find their way out of the vasculature into tissue. Using intravital microscopy, combined with the advancements in genetic engineering and digital imaging, has made the mouse an ideal organism to unravel the molecular mechanisms necessary for leukocyte extravasation. The development of innovative microscopy setups such as multiphoton laser scanning intravital microscopy may further speed up the gain of knowledge on leukocyte recruitment *in vivo*, which will also include the observation of leukocyte subsets traveling through the different compartments of the organism.

## Acknowledgment

We thank Volker Andresen, Ph.D. for helpful discussion.

## References

Acton, S. T., Wethmar, K., and Ley, K. (2002). Automatic tracking of rolling leukocytes *in vivo*. *Microvasc. Res.* **63,** 139–148.

Arbones, M. L., Ord, D. C., Ley, K., Ratech, H., Maynard-Curry, C., Otten, G., Capon, D. J., and Tedder, T. F. (1994). Lymphocyte homing and leukocyte rolling and migration are impaired in L-selectin-deficient mice. *Immunity* **1,** 247–260.

Asano, M., Nakae, S., Kotani, N., Shirafuji, N., Nambu, A., Hashimoto, N., Kawashima, H., Hirose, M., Miyasaka, M., Takasaki, S., and Iwakura, Y. (2003). Impaired selectin-ligand biosynthesis and reduced inflammatory responses in beta-1,4-galactosyltransferase-I-deficient mice. *Blood* **102**, 1678–1685.

Aydt, E., and Wolff, G. (2002). Development of synthetic pan-selectin antagonists: A new treatment strategy for chronic inflammation in asthma. *Pathobiology* **70**, 297–301.

Baez, S. (1973). An open cremaster muscle preparation for the study of blood vessels by *in vivo* microscopy. *Microvasc. Res.* **5**, 384–394.

Bargatze, R. F., Jutila, M. A., and Butcher, E. C. (1995). Distinct roles of L-selectin and integrins $\alpha_4\beta_7$ and LFA-1 in lymphocyte homing to Peyer's patch-HEV *in situ*: The multistep model confirmed and refined. *Immunity* **3**, 99–108.

Berlin, C., Bargatze, R. F., Campbell, J. J., von Andrian, U. H., Szabo, M. C., Hasslen, S. R., Nelson, R. D., Berg, E. L., Erlandsen, S. L., and Butcher, E. C. (1995). $\alpha_4$ integrins mediate lymphocyte attachment and rolling under physiologic flow. *Cell* **80**, 413–422.

Berlin, C., Berg, E. L., Briskin, M. J., Andrew, D. P., Kilshaw, P. J., Holzmann, B., Weissman, I. L., Hamann, A., and Butcher, E. C. (1993). $\alpha_4\beta_7$ integrin mediates lymphocyte binding to the mucosal vascular addressin MAdCAM-1. *Cell* **74**, 185–195.

Bevilacqua, M. P., Pober, J. S., Mendrick, D. L., Cotran, R. S., and Gimbrone, M. A., Jr. (1987). Identification of an inducible endothelial-leukocyte adhesion molecule. *Proc. Natl. Acad. Sci. USA* **84**, 9238–9242.

Bosse, R., and Vestweber, D. (1994). Only simultaneous blocking of the L- and P-selectin completely inhibits neutrophil migration into mouse peritoneum. *Eur. J. Immunol.* **24**, 3019–3024.

Briskin, M. J., McEvoy, L. M., and Butcher, E. C. (1993). MAdCAM-1 has homology to immunoglobulin and mucin-like adhesion receptors and to IgA1. *Nature* **363**, 461–464.

Bullard, D. C. (2001). Knockout mice in inflammation research. *In* "Physiology of Inflammation" (K. Ley, ed.), pp. 381–401. Oxford University Press, New York.

Burrows, M. T. (1912). A method of furnishing a continuous supply of new medium to a tissue culture *in vitro*. *Anat. Rec.* **6**, 141–144.

Butcher, E. C. (1991). Leukocyte-endothelial cell recognition – Three (or more) steps to specificity and diversity. *Cell* **67**, 1033–1036.

Campbell, J. J., Hedrick, J., Zlotnik, A., Siani, M. A., Thompson, D. A., and Butcher, E. C. (1998). Chemokines and the arrest of lymphocytes rolling under flow conditions. *Science* **279**, 381–384.

Cheezum, M. K., Walker, W. F., and Guilford, W. H. (2001). Quantitative comparison of algorithms for tracking single fluorescent particles. *Biophys. J.* **81**, 2378–2388.

Chesnutt, B. C., Smith, D. F., Raffler, N. A., Smith, M. L., White, E. J., and Ley, K. (2006). Induction of LFA-1-dependent neutrophil rolling on ICAM-1 by engagement of E-selectin. *Microcirculation* **13**, 99–109.

Cotter, M. J., Norman, K. E., Hellewell, P. G., and Ridger, V. C. (2001). A novel method for isolation of neutrophils from murine blood using negative immunomagnetic separation. *Am. J Pathol.* **159**, 473–481.

Damiano, E. R., Westheider, J., Tözeren, A., and Ley, K. (1996). Variation in the velocity, deformation, and adhesion energy density of leukocytes rolling within venules. *Circ. Res.* **79**, 1122–1130.

Debeir, O., Van Ham, P., Kiss, R., and Decaestecker, C. (2005). Tracking of migrating cells under phase-contrast video microscopy with combined mean-shift processes. *IEEE Trans. Med. Imaging* **24**, 697–711.

Devasenathipathy, S., Santiago, B., Wereley, S. T., Meinhart, C. D., and Takehara, K. (2003). Particle imaging techniques for microfabricated fluidic systems. *Exp. Fluids* **34**, 504–514.

DiVietro, J. A., Smith, M. J., Smith, B. R., Petruzzelli, L., Larson, R. S., and Lawrence, M. B. (2001). Immobilized IL-8 triggers progressive activation of neutrophils rolling *in vitro* on P-selectin and intercellular adhesion molecule-1. *J. Immunol.* **167**, 4017–4025.

Dong, G., Ray, N., and Acton, S. T. (2005). Intravital leukocyte detection using the gradient inverse coefficient of variation. *IEEE Trans. Med. Imaging* **24**, 910–924.

Dunne, J. L., Goobic, A. P., Acton, S. T., and Ley, K. (2004). A novel method to analyze leukocyte rolling behavior *in vivo*. *Biol. Proced. Online.* **6**, 173–179.

Eden, E., Waisman, D., Rudzsky, M., Bitterman, H., Brod, V., and Rivlin, E. (2005). An automated method for analysis of flow characteristics of circulating particles from *in vivo* video microscopy. *IEEE Trans. Med. Imaging* **24**, 1011–1024.

Ellies, L. G., Sperandio, M., Underhill, G. H., Yousef, J., Smith, M., Priatel, J. J., Kansas, G. S., Ley, K., and Marth, J. (2002). Sialyltransferase specificity in selectin ligand formation. *Blood* **100**, 3618–3625.

Ellies, L. G., Tsuboi, S., Petryniak, B., Lowe, J. B., Fukuda, M., and Marth, J. D. (1998). Core 2 oligosaccharide biosynthesis distinguishes between selectin ligands essential for leukocyte homing and inflammation. *Immunity* **9**, 881–890.

Faust, N., Varas, F., Kelly, L. M., Heck, S., and Graf, T. (2000). Insertion of enhanced green fluorescent protein into the lysozyme gene creates mice with green fluorescent granulocytes and macrophages. *Blood* **96**, 719–726.

Finger, E. B., Puri, K. D., Alon, R., Lawrence, M. B., von Andrian, U. H., and Springer, T. A. (1996). Adhesion through L-selectin requires a threshold hydrodynamic shear. *Nature* **379**, 266–269.

Frenette, P. S., Denis, C. V., Weiss, L., Jurk, K., Subbarao, S., Kehrel, B., Hartwig, J. H., Vestweber, D., and Wagner, D. D. (2000). P-Selectin glycoprotein ligand 1 (PSGL-1) is expressed on platelets and can mediate platelet-endothelial interactions *in vivo*. *J. Exp. Med.* **191**, 1413–1422.

Gaehtgens, P., Ley, K., Pries, A. R., and Müller, R. (1985). Mutual interaction between leukocytes and microvascular blood flow. *In* "White Cell Rheology and Inflammation" (K. Mebmer and F. Hammersen, eds.), pp. 15–28. Basel.

Gallatin, W. M., Weissman, I. L., and Butcher, E. C. (1983). A cell-surface molecule involved in organ-specific homing of lymphocytes. *Nature* **304**, 30–34.

Gang, D., Ray, N., and Acton, S. T. (2005). Intravital leukocyte detection using the gradient inverse coefficient of variation. *IEEE Trans. Med. Imaging* **24**, 910–924.

Ghosh, R. N., and Webb, W. W. (1994). Automated detection and tracking of individual and clustered cell surface low density lipoprotein receptor molecules. *Biophys. J.* **66**, 1301–1318.

Goldman, A. J., Cox, R. G., and Brenner, B. M. (1967). Slow viscous motion of a sphere parallel to a plane wall - I Motion through a quiescent fluid. *Chem. Eng. Sci.* **22**, 637–651.

Goobic, A. P., Tang, J., and Acton, S. T. (2005). Image stabilization and registration for tracking cells in the microvasculature. *IEEE Trans. Biomed. Eng.* **52**, 287–299.

Hamburger, S. A., and McEver, R. P. (1990). GMP-140 mediates adhesion of stimulated platelets to neutrophils. *Blood* **75**, 550–554.

Hemmerich, S., Bistrup, A., Singer, M. S., Van Zante, A., Lee, J. K., Tsay, D., Peters, M., Carminati, J. L., Brennan, T. J., Carver-Moore, K., Leviten, M., Fuentes, M. E., Ruddle, N. H., and Rosen, S. D. (2001). Sulfation of L-selectin ligands by an HEV-restricted sulfotransferase regulates lymphocyte homing to lymph nodes. *Immunity* **15**, 237–247.

Hiraoka, N., Petryniak, B., Nakayama, J., Tsuboi, S., Suzuki, M., Yeh, J. C., Izawa, D., Tanaka, T., Miyasaka, M., Lowe, J. B., and Fukuda, M. (1999). A novel, high endothelial venule-specific sulfotransferase expresses 6-sulfo sialyl lewis(x), an L-selectin ligand displayed by CD34. *Immunity* **11,** 79–89.

Homeister, J. W., Thall, A. D., Petryniak, B., Maly, P., Rogers, C. E., Smith, P. L., Kelly, R. J., Gersten, K. M., Askari, S. W., Cheng, G., Smithson, G., Marks, R. M., Misra, A. K., Hindsgaul, O., von Andrian, U. H., and Lowe, J. B. (2001). The alpha(1,3)fucosyl-transferases FucT-IV and FucT-VII exert collaborative control over selectin-dependent leukocyte recruitment and lymphocyte homing. *Immunity* **15,** 115–126.

Janssens, C. J. J. G., Reneman, R. S., Slaaf, D. W., and Tangelder, G. J. (1989). Differentiation *in vivo* of leukocytes rolling in mesenteric venules of anaesthetized rabbits. *J. Physiol. (London)* **420,** 135P.

Jellish, S. (2005). Real-time video detection and tracking of rolling leukocytes [MS Thesis]. University of Virginia, Charlottesville.

Jinshan, T., and Acton, S. T. (2004). Vessel boundary tracking for intravital microscopy via multiscale gradient vector flow snakes. *IEEE Trans. Biomed. Eng.* **51,** 316–324.

Jinshan, T., Gang, D., Ray, N., and Acton, S. T. (2002). Evaluation of intravital tracking algorithms. The 2002 45th Midwest Symposium on Circuits and Systems, 1.I-3.

Jung, S., Aliberti, J., Graemmel, P., Sunshine, M. J., Kreutzberg, G. W., Sher, A., and Littman, D. R. (2000). Analysis of fractalkine receptor CX(3)CR1 function by targeted deletion and green fluorescent protein reporter gene insertion. *Mol. Cell Biol.* **20,** 4106–4114.

Jung, U., Bullard, D. C., Tedder, T. F., and Ley, K. (1996). Velocity difference between L-selectin and P-selectin dependent neutrophil rolling in venules of the mouse cremaster muscle *in vivo*. *Am. J. Physiol.* **271,** H2740–H2747.

Jung, U., and Ley, K. (1997). Regulation of E-selectin, P-selectin and ICAM-1 expression in mouse cremaster muscle vasculature. *Microcirculation* **4,** 311–319.

Jung, U., Ramos, C. L., Bullard, D. C., and Ley, K. (1998). Gene-targeted mice reveal importance of L-selectin-dependent rolling for neutrophil adhesion. *Am. J. Physiol.* **274,** H1785–H1791.

Kawashima, H., Petryniak, B., Hiraoka, N., Mitoma, J., Huckaby, V., Nakayama, J., Uchimura, K., Kadomatsu, K., Muramatsu, T., Lowe, J. B., and Fukuda, M. (2005). N-acetylglucosamine-6-O-sulfotransferases 1 and 2 cooperatively control lymphocyte homing through L-selectin ligand biosynthesis in high endothelial venules. *Nat. Immunol.* **6,** 1096–1104.

Keane, R. D., Adrian, R. J., and Zhang, Y. (1995). Super-resolution particle imaging velocimetry. *Meas. Sci. Technol.* **6,** 754–768.

Kishimoto, T. K., Jutila, M. A., and Butcher, E. C. (1990). Identification of a human peripheral lymph node homing receptor: A rapidly down-regulated adhesion molecule. *Proc. Natl. Acad. Sci. USA* **87,** 2244–2248.

Klitzman, B., and Duling, B. R. (1979). Microvascular hematocrit and red cell flow in resting and contracting striated muscle. *Am. J. Physiol.* **237,** H481–H490.

Kogan, T. P., Dupre, B., Bui, H., McAbee, K. L., Kassir, J. M., Scott, I. L., Hu, X., Vanderslice, P., Beck, P. J., and Dixon, R. A. (1998). Novel synthetic inhibitors of selectin-mediated cell adhesion: Synthesis of 1,6-bis[3-(3-carboxymethylphenyl)-4-(2-alpha-D-mannopyranosyloxy)phenyl]hexane (TBC1269). *J. Med. Chem.* **41,** 1099–1111.

Koutsiaris, A. G., Mathioulakis, D. K., and Tsangaris, S. (1999). Microscope PIV for velocity-field measurement of particle suspensions flowing inside glass capillaries. *Meas. Sci. Technol.* **10,** 1037–1046.

Kraiss, L. W., Alto, N. M., Dixon, D. A., McIntyre, T. M., Weyrich, A. S., and Zimmerman, G. A. (2003). Fluid flow regulates E-selectin protein levels in human endothelial cells by inhibiting translation. *J. Vasc. Surg.* **37,** 161–168.

Kusumi, A., Sako, Y., and Yamamoto, M. (1993). Confined lateral diffusion of membrane receptors as studied by single particle tracking (nanovid microscopy). Effects of calcium-induced differentiation in cultured epithelial cells. *Biophys. J.* **65,** 2021–2040.

Labow, M. A., Norton, C. R., Rumberger, J. M., Lombard-Gillooly, K. M., Shuster, D. J., Hubbard, J., Bertko, R., Knaack, P. A., Terry, R. W., Harbison, M. L., Kontgen, F., Stewart, C. L., McIntyre, K. W., Will, P. C., Burns, D. K., and Wolitzky, B. A. (1994). Characterization of E-selectin-deficient mice: Demonstration of overlapping function of the endothelial selectins. *Immunity* **1,** 709–720.

Lawrence, M. B. (2001). *In vitro* flow models of leukocyte adhesion. *In* "Physiology of Inflammation" (K. Ley, ed.), pp. 204–221. Oxford University Press, New York.

Lawrence, M. B., Kansas, G. S., Ghosh, S., Kunkel, E. J., and Ley, K. (1997). Threshold levels of fluid shear promote leukocyte adhesion through selectins (CD62L,P,E). *J. Cell Biol.* **136,** 717–727.

Lawrence, M. B., McIntire, L. V., and Eskin, S. G. (1987). Effect of flow on polymorphonu-clear leukocyte/endothelial cell adhesion. *Blood* **70,** 1284–1290.

Lehr, H. A., Leunig, M., Menger, M. D., Nolte, D., and Messmer, K. (1993). Dorsal skinfold chamber technique for intravital microscopy in nude mice. *Am. J. Pathol.* **143,** 1055–1062.

Ley, K. (1989). Granulocyte adhesion to microvascular and cultured endothelium. *Studia Biophys.* **34,** 179–184.

Ley, K. (2001). Leukocyte recruitment as seen by intravital microscopy. *In* "Physiology of Inflammation" (K. Ley, ed.), pp. 303–337. Oxford University Press, New York.

Ley, K. (2002). Integration of inflammatory signals by rolling neutrophils. *Immunol Rev.* **186,** 8–18.

Ley, K., Bullard, D. C., Arbones, M. L., Bosse, R., Vestweber, D., Tedder, T. F., and Beaudet, A. L. (1995). Sequential contribution of L- and P-selectin to leukocyte rolling *in vivo*. *J. Exp. Med.* **181,** 669–675.

Ley, K., and Gaehtgens, P. (1991). Endothelial, not hemodynamic differences are responsible for preferential leukocyte rolling in venules. *Circ. Res.* **69,** 1034–1041.

Ley, K., and Kansas, G. S. (2004). Selectins in T-cell recruitment to non-lymphoid tissues and sites of inflammation. *Nat. Rev. Immunol.* **4,** 325–335.

Ley, K., Pries, A. R., and Gaehtgens, P. (1987). A versatile intravital microscope design. *Int. J. Microcirc. Clin. Exp.* **6,** 161–167.

Leymarie, F., and Levine, M. D. (1993). Tracking deformable objects in the plane using an active contour model. Anonymous. *IEEE Trans Pattern, Analysis, and Machine Intelligence* **15,** 617–634.

Li, F., Erickson, H. P., James, J. A., Moore, K. L., Cummings, R. D., and McEver, R. P. (1996). Visualization of P-selectin glycoprotein ligand-1 as a highly extended molecule and mapping of protein epitopes for monoclonal antibodies. *J. Biol. Chem.* **271,** 6342–6348.

Lipowsky, H. H., and Zweifach, B. W. (1978). Application of the "two-slit" photometric technique to the measurement of microvascular volumetric flow rates. *Microvasc. Res.* **15,** 93–101.

Liu, L., and Kubes, P. (2003). Molecular mechanisms of leukocyte recruitment: Organ-specific mechanisms of action. *Thromb. Haemost.* **89,** 213–220.

Long, D. S., Smith, M. L., Pries, A. R., Ley, K., and Damiano, E. R. (2004). Microviscometry reveals reduced blood viscosity and altered shear rate and shear stress profiles in microvessels after hemodilution. *Proc. Natl. Acad. Sci. USA* **101,** 10060–10065.

Lowe, J. B. (2002). Glycosylation in the control of selectin counter-receptor structure and function. *Immunol Rev.* **186,** 19–36.

Maly, P., Thall, A. D., Petryniak, B., Rogers, C. E., Smith, P. L., Marks, R. M., Kelly, R. J., Gersten, K. M., Cheng, G., Saunders, T. L., Camper, S. A., Camphausen, R. T., Sullivan,

F. X., Isogai, Y., Hindsgaul, O., von Andrian, U. H., and Lowe, J. B. (1996). The $\alpha(1,3)$ fucosyltransferase Fuc-TVII controls leukocyte trafficking through an essential role in L-, E-, and P-selectin ligand biosynthesis. *Cell* **86,** 643–653.

Mangan, P. R., O'quinn, D., Harrington, L., Bonder, C. S., Kubes, P., Kucik, D. F., Bullard, D. C., and Weaver, C. T. (2005). Both Th1 and th2 cells require p-selectin glycoprotein ligand-1 for optimal rolling on inflamed endothelium. *Am. J. Pathol.* **167,** 1661–1675.

Manjunath, N., Shankar, P., Stockton, B., Dubey, P. D., Lieberman, J., and von Andrian, U. H. (1999). A transgenic mouse model to analyze CD8(+) effector T cell differentiation *in vivo. Proc. Natl. Acad. Sci. USA* **96,** 13932–13937.

Marshall, B. T., Long, M., Piper, J. W., Yago, T., McEver, R. P., and Zhu, C. (2003). Direct observation of catch bonds involving cell-adhesion molecules. *Nature* **423,** 190–193.

Mayadas, T. N., Johnson, R. C., Rayburn, H., Hynes, R. O., and Wagner, D. D. (1993). Leukocyte rolling and extravasation are severely compromised in P selectin–deficient mice. *Cell* **74,** 541–554.

Meinhart, C. D., Wereley, S. T., and Gray, M. H. B. (2000). Volume illumination for two-dimensional particle image velocimetry. *Meas. Sci. Technol.* **11,** 809–814.

Mempel, T. R., Scimone, M. L., Mora, J. R., and vand on Andrian, U. H. (2004). *In vivo* imaging of leukocyte trafficking in blood vessels and tissues. *Curr. Opin. Immunol.* **16,** 406–417.

Miura, S., Asakura, H., and Tsuchiya, M. (1987). Dynamic analysis of lymphocyte migration into Peyer's patches of rat small intestine. *Lymphology* **20,** 252–256.

Mukherjee, D. P., Ray, N., and Acton, S. T. (2004). Level set analysis for leukocyte detection and tracking. *IEEE Trans. Image Process* **13,** 562–572.

Muller, W. A. (2003). Leukocyte-endothelial-cell interactions in leukocyte transmigration and the inflammatory response. *Trends Immunol.* **24,** 327–334.

Nandi, A., Estess, P., and Siegelman, M. H. (2000). Hyaluronan Anchoring and Regulation on the Surface of Vascular Endothelial Cells Is Mediated through the Functionally Active Form of CD44. *J. Biol. Chem.* **275,** 14939–14948.

Nguyen, N.-T., and Wereley, S. T. (2002). "Fundamentals and Applications of Microfluidics." Artech House, Norwood, MA.

Norman, K. E. (2001). An effective and economical solution for digitizing and analyzing video recordings of the microcirculation. *Microcirculation* **8,** 243–249.

Norton, C. R., Rumberger, J. M., Burns, D. K., and Wolitzky, B. A. (1993). Characterization of murine E-selectin expression *in vitro* using novel anti-mouse E-selectin monoclonal antibodies. *Biochem. Biophys. Res. Commun.* **195,** 250–258.

Ouyang, Y. B., Lane, W. S., and Moore, K. L. (1998). Tyrosylprotein sulfotransferase-purification and molecular cloning of an enzyme that catalyzes tyrosine O-sulfation, a common posttranslational modification of eukaryotic proteins. *Proc. Natl. Acad. Sci. USA* **95,** 2896–2901.

Ouyang, Y. B., and Moore, K. L. (1998). Molecular cloning and expression of human and mouse tyrosylprotein sulfotransferase-2 and a tyrosylprotein sulfotransferase homologue in *Caenorhabditis elegans. J. Biol. Chem.* **273,** 24770–24774.

Pavalko, F. M., Walker, D. M., Graham, L., Goheen, M., Doerschuk, C. M., and Kansas, G. S. (1995). The cytoplasmic domain of L-selectin interacts with cytoskeletal proteins via $\alpha$-actinin: Receptor positioning in microvilli does not require interaction with a-actinin. *J. Cell Biol.* **129,** 1155–1164.

Pouyani, T., and Seed, B. (1995). PSGL-1 recognition of P-selectin is controlled by a tyrosine sulfation consensus at the PSGL-1 amino terminus. *Cell* **83,** 333–343.

Pries, A. R. (1988). A versatile video image analysis system for microcirculatory research. *Int. J. Microcirc. Clin. Exp.* **7,** 327–345.

Radeke, H. H., Ludwig, R. J., and Boehncke, W. H. (2005). Experimental approaches to lymphocyte migration in dermatology *in vitro* and *in vivo*. *Exp. Dermatol.* **14**, 641–666.

Rao, R. M., Haskard, D. O., and Landis, R. C. (2002). Enhanced recruitment of Th2 and CLA-negative lymphocytes by the S128R polymorphism of E-selectin. *J. Immunol.* **169**, 5860–5865.

Ray, N., and Acton, S. T. (2004). Motion gradient vector flow: An external force for tracking rolling leukocytes with shape and size constrained active contours. *IEEE Trans. Med. Imaging* **23**, 1466–1478.

Ray, N., Acton, S. T., and Ley, K. (2002). Tracking leukocytes *in vivo* with shape and size constrained active contours. *IEEE Trans. Med. Imaging* **21**, 1222–1235.

Read, M. A., Neish, A. S., Luscinskas, F. W., Palombella, V. J., Maniatis, T., and Collins, T. (1995). The proteasome pathway is required for cytokine-induced endothelial-leukocyte adhesion molecule expression. *Immunity* **2**, 493–506.

Sako, D., Chang, X.-J., Barone, K. M., Vachino, G., White, H. M., Shaw, G., Veldman, G. M., Bean, K. M., Ahern, T. J., Furie, B., Cumming, D. A., and Larsen, G. R. (1993). Expression cloning of a functional glycoprotein ligand for P-selectin. *Cell* **75**, 1179–1186.

Santiago, J. G., Wereley, S. T., Meinhart, C. D., Beebe, D. J., and Adrian, R. J. (1998). A particle image velocimetry system for microfluidics. *Exp. Fluids* **25**, 316–319.

Sato, Y., Chen, J., Zoroofi, R. A., Harada, N., Tamura, S., and Shiga, T. (1997). Automatic extraction and measurement of leukocyte motion in microvessels using spatiotemporal image analysis. *IEEE Trans. Biomed. Eng.* **44**, 225–236.

Shamri, R., Grabovsky, V., Gauguet, J. M., Feigelson, S., Manevich, E., Kolanus, W., Robinson, M. K., Staunton, D. E., von Andrian, U. H., and Alon, R. (2005). Lymphocyte arrest requires instantaneous induction of an extended LFA-1 conformation mediated by endothelium-bound chemokines. *Nat. Immunol.* **6**, 497–506.

Shinohara, K., Sugii, S., Aota, A., Hibara, A., Tokeshi, M., Kitamori, T., and Okamoto, K. (2004). High-speed micro-PIV measurements of transient flow in microfluidic devices. *Meas. Sci. Technol.* **15**, 1965–1970.

Singbartl, K., Thatte, J., Smith, M. L., Wethmar, K., Day, K., and Ley, K. (2001). A CD2-green fluorescence protein-transgenic mouse reveals very late antigen-4-dependent CD8+ lymphocyte rolling in inflamed venules. *J. Immunol.* **166**, 7520–7526.

Slaaf, D. W., Jongsma, F. H. M., Tangelder, G. J., and Reneman, R. S. (1986). Characteristics of optical systems for intravital microscopy. *In* "Microcirculatory Technology" (C. H. Baker and W. L. Nastuk, eds.), pp. 211–228. Academic Press, New York.

Smith, M. L., Long, D. S., Damiano, E. R., and Ley, K. (2003). Near-wall micro-PIV reveals a hydrodynamically relevant endothelial surface layer in venules *in vivo*. *Biophys. J.* **85**, 637–645.

Smith, M. L., Smith, M. J., Lawrence, M. B., and Ley, K. (2002). Viscosity-independent velocity of neutrophils rolling on p-selectin *in vitro* or *in vivo*. *Microcirculation* **9**, 523–536.

Smith, M. L., Sperandio, M., Galkina, E. V., and Ley, K. (2004). Autoperfused mouse flow chamber reveals synergistic neutrophil accumulation through P-selectin and E-selectin. *J. Leukoc. Biol.* **76**, 985–993.

Snapp, K. R., Ding, H., Atkins, K., Warnke, R., Luscinskas, F. W., and Kansas, G. S. (1998). A novel P-selectin glycoprotein ligand-1 (PSGL-1) monoclonal antibody recognizes an epitope within the tyrosine sulfate motif of human PSGL-1 and blocks recognition of both P- and L-selectin. *Blood* **91**, 154–164.

Sperandio, M., Forlow, S. B., Thatte, J., Ellies, L. G., Marth, J. D., and Ley, K. (2001). Differential requirements for core2 glucosaminyltransferase for endothelial L-selectin ligand function *in vivo*. *J. Immunol.* **167**, 2268–2274.

Sperandio, M., and Ley, K. (2005). The physiology and pathophysiology of P-selectin. *Mod. Asp. Immunobiol.* **15**, 24–26.

Sperandio, M., Linderkamp, O., and Leo, A. (2004). Blocking leukocyte rolling: Does it have a role in disease prevention? *Vasc. Dis. Prev.* **1,** 185–195.

Sperandio, M., Smith, M. L., Forlow, S. B., Olson, T. S., Xia, L., McEver, R. P., and Ley, K. (2003). P-selectin Glycoprotein Ligand-1 mediates L-selectin-dependent leukocyte rolling in venules. *J. Exp. Med.* **197,** 1355–1363.

Springer, T. A. (1995). Traffic signals on endothelium for lymphocyte recirculation and leukocyte emigration. *Annu. Rev. Physiol.* **57,** 827–872.

Stark, M. A., Huo, Y., Burcin, T. L., Morris, M. A., Olson, T. S., and Ley, K. (2005). Phagocytosis of apoptotic neutrophils regulates granulopoiesis via IL-23 and IL-17. *Immunity* **22,** 285–294.

Sugii, Y., Nishio, S., and Okamoto, K. (2002). In vivo PIV measurement of red blood cell velocity field in microvessels considering mesentery motion. *Physiol. Meas.* **23,** 403–416.

Sugii, Y., Okuda, R., Okamoto, K., and Madarame, H. (2005). Velocity measurement of both red blood cells and plasma of in vitro blood flow using high-speed micro PIV technique. *Meas. Sci. Technol.* **16,** 1126–1130.

Tangelder, G. J., Slaaf, D. W., Muijtjens, A. M. M., Arts, T., oude Egbrink, M. G. A., and Reneman, R. S. (1986). Velocity profiles of blood platelets and red blood cells flowing in arterioles of the rabbit mesentery. *Circ. Res.* **59,** 505–514.

Uchimura, K., Gauguet, J. M., Singer, M. S., Tsay, D., Kannagi, R., Muramatsu, T., von Andrian, U. H., and Rosen, S. D. (2005). A major class of L-selectin ligands is eliminated in mice deficient in two sulfotransferases expressed in high endothelial venules. *Nat. Immunol.* **6,** 1105–1113.

Uchimura, K., Kadomatsu, K., El Fasakhany, F. M., Singer, M. S., Izawa, M., Kannagi, R., Takeda, N., Rosen, S. D., and Muramatsu, T. (2004). N-acetylglucosamine 6-O-sulfotransferase-1 regulates expression of L-selectin ligands and lymphocyte homing. *J. Biol. Chem.* **279,** 35001–35008.

Unutmaz, D., Xiang, W., Sunshine, M. J., Campbell, J., Butcher, E., and Littman, D. R. (2000). The primate lentiviral receptor Bonzo/STRL33 is coordinately regulated with CCR5 and its expression pattern is conserved between human and mouse. *J. Immunol.* **165,** 3284–3292.

Van Zante, A., Gauguet, J. M., Bistrup, A., Tsay, D., von Andrian, U. H., and Rosen, S. D. (2003). Lymphocyte-HEV interactions in lymph nodes of a sulfotransferase-deficient mouse. *J. Exp. Med.* **198,** 1289–1300.

Vestweber, D., and Blanks, J. E. (1999). Mechanisms that regulate the function of the selectins and their ligands. *Physiol. Rev.* **79,** 181–213.

Vink, H., and Duling, B. R. (1996). Identification of distinct luminal domains for macromolecules, erythrocytes, and leukocytes within mammalian capillaries. *Circ. Res.* **79,** 581–589.

von Andrian, U. H. (1996). Intravital microscopy of the peripheral lymph node microcirculation in mice. *Microcirculation* **3,** 287–300.

Wagner, R. (1839). "Erläuterungstafeln zur Physiologie und Entwicklungsgeschichte." Leopold Voss, Leipzig.

Walter, U. M., Ayer, L. M., Manning, A. M., Frenette, P. S., Wagner, D. D., Hynes, R. O., Wolitzky, B. A., and Issekutz, A. C. (1997). Generation and characterization of a novel adhesion function blocking monoclonal antibody recognizing both rat and mouse e-selectin. *Hybridoma* **16,** 355–361.

Wayland, H., and Johnson, P. C. (1967). Erythrocyte velocity measurement in microvessels by two-slit photometric method. *J. Appl. Physiol.* **22,** 333–337.

Weninger, W., Ulfman, L. H., Cheng, G., Souchkova, N., Quackenbush, E. J., Lowe, J. B., and von Andrian, U. H. (2000). Specialized contributions by $\alpha(1,3)$-fucosyltransferase-IV and FucT-VII during leukocyte rolling in dermal microvessels. *Immunity* **12,** 665–676.

Wereley, S. T., and Meinhart, C. D. (2004). Micron resolution particle image velocimetry. *In* "Microscale Diagnostic Techniques" (K. S. Breuer, ed.), pp. 60–65. Springer Publishing, Heidelberg, Germany.

Xia, L., Sperandio, M., Yago, T., McDaniel, J. M., Cummings, R. D., Pearson-White, S., Ley, K., and McEver, R. P. (2002). P-selectin glycoprotein ligand-1-deficient mice have impaired leukocyte tethering to E-selectin under flow. *J. Clin. Invest.* **109,** 939–950.

Yago, T., Wu, J., Wey, C. D., Klopocki, A. G., Zhu, C., and McEver, R. P. (2004). Catch bonds govern adhesion through L-selectin at threshold shear. *J. Cell Biol.* **166,** 913–923.

Yang, J., Hirata, T., Croce, K., Merrill-Skoloff, G., Tchernychev, B., Williams, E., Flaumenhaft, R., Furie, B., and Furie, B. C. (1999). Targeted gene disruption demonstrates that PSGL-1 is required for P-Selectin mediated but not E-Selectin mediated neutrophil rolling and migration. *J. Exp. Med.* **190,** 1769–1782.

# [24]  Cell Type–Specific Roles of Carbohydrates in Tumor Metastasis

*By* Shihao Chen and Minoru Fukuda

## Abstract

Protein- or lipid-bound glycans have been shown to play important roles in many biological processes; their functional diversity is due primarily to the multiple linkages, branching patterns, and terminal modifications seen in these glycans. Furthermore, one particular glycan may play different roles depending on the biological system. Sialyl Lewis X oligosaccharides, prototypic ligands for E-, L-, and P-selectins, are essential for naive lymphocyte homing to secondary lymphoid organs. They have also been implicated clinically in tumor metastasis and poor prognosis of cancer patients. In this chapter, we describe the protocol for the formation of lung tumor after intravenous injection of B16 melanoma cells. In our study, B16 melanoma cells formed more tumors when the cells were transfected with fucosyltransferase-III to express sialyl Lewis X. In the second experimental protocol, we describe metastatic tumor formation at the draining lymph node after primary tumor was formed at the footpad. In this experimental system, natural killer (NK) cell recruitment to the draining lymph node was found to be critical to suppress tumor metastasis and this tumor suppression is dependent on L-selectin–mediated trafficking of NK cells to the lymph nodes.

METHODS IN ENZYMOLOGY, VOL. 416
0076-6879/06 $35.00
DOI: 10.1016/S0076-6879(06)16024-3

## Overview

Cancer is a group of many related diseases in which cells become abnormal and grow out of control. The site where cancer cells first form is called the *primary cancer* or *primary tumor*. However, cancer cells can break away from a primary tumor and migrate to a distant site where they form a new or metastatic tumor. Metastasis occurs when (1) tumor cells are shed from a primary tumor and invade a local tissue across the basement membrane; (2) cells enter into the bloodstream or lymphatic circulation and travel to a distant site; (3) cells escape the circulation at that distant site (extravasation), invade the parenchyma of a new organ, and form a secondary tumor (Fidler, 2003). This process is considered "inefficient," because most cells disseminated from a primary tumor cannot successfully complete all the steps to give rise to metastatic tumors. Most circulating tumor cells are killed by host immune cells or trapped in a capillary bed in another organ. The rate-limiting step for metastasis has been considered the survival of tumor cells in circulation and their subsequent lodging on endothelial cells at the distant site. Anchorage of tumor cells to endothelial cells is partly mediated by carbohydrate-dependent adhesion and can be assayed using at least three experimental paradigms.

In the first, cancer cells in single cell suspensions are intravenously (i.v.) injected through a mouse tail vein, and tumor formation is evaluated in the lung (Elkin and Vlodavsky, 2001). The lung is the primary organ for i.v. injected tumor cells to colonize, simply because the lung capillary bed is the first to be encountered by injected tumor cells, and its narrow fine structures are conducive to form tumors. Such assays clearly do not involve migration of cells from a primary tumor and, therefore, are not bona fide assays for tumor metastasis. Thus, they are best considered representative of lung tumor focus formation. Employing this model, we have determined the role of sialyl Lewis X oligosaccharides in melanoma lung colonization (Ohyama *et al.*, 1999, 2002). Mouse B16 melanoma cells are negative for sialyl Lewis X expression but become positive after transfection with human fucosyltransferase-III (B16-FTIII cells). We injected parental B16 cells and B16-FTIII cells i.v. into syngeneic C57BL/6 mice. Parental B16 melanoma cells formed a small number of tumor foci; B16-FTIII cells, on the other hand, formed a significantly increased number of tumor foci. This increase was inhibited by preinjection of sialyl Lewis X oligosaccharides, showing that increased tumor formation is due to the forced expression of sialyl Lewis X on tumor cells. This experimental protocol is described in detail later in this chapter and in Fig. 1. Similarly, a report showed that forced expression of core 3 synthesizing enzyme, $\beta$3GlcNAcT-6, on colonic cancer cells resulted in reduced lung tumor formation, implicating that the core 3 structure GlcNAc$\beta$1→3 GalNAc$\alpha$1→Ser/Thr may function as a tumor suppressor (Iwai *et al.*, 2005). Finally, rat glioma C6 cells, which are negative for polysialic acid, barely invaded tissues of the corpus callosum when inoculated in the caudate putamen of brains of adult C57BL/6 mice via a stereotaxic device. On the other hand, similarly injected C6 cells transfected with a polysialyltransferase enabling expression of polysialic acid did invade corpus callosum tissues (Suzuki *et al.*, 2005).

In a second paradigm of tumor cell migration, a primary tumor is first allowed to form and then tumor formation is measured at sites distant from the primary tumor. Cancer cells must complete all necessary migration steps described earlier to form secondary tumor colonies, so this assay more closely mimics tumor metastasis *in vivo*. In one example, when inoculated subcutaneously (s.c.), only polysialic acid–positive small cell lung carcinoma cells (NCI-H69) formed intracutaneous tumors (Scheidegger *et al.*, 1994), while polysialic acid–negative cells did not. In this case, the distance between the primary and metastatic tumor was small; nonetheless, the authors, with

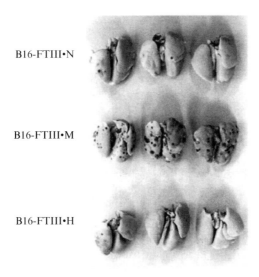

B16-FTIII•N

B16-FTIII•M

B16-FTIII•H

FIG. 1. Representative lungs of BALB/c mice after intravenous injection of B16 melanoma cells transfected with human fucosyltransferase-III (B16-FTIII). B16-FTIII•N, B16 cells that did not express sialyl Lewis X oligosaccharides; B16-FTIII•M, B16 cells expressing moderate amounts of sialyl Lewis X; B16-FTIII•H, B16 cells expressing high levels of sialyl Lewis X. Lungs were photographed 3 wk after injection. (Used, with permission, from Ohyama *et al.*, 1999.)

some reservations, regarded this migration as tumor metastasis. In another example, colonic carcinoma cells, which efficiently form metastasis at the liver, were established by Morikawa *et al.* (1988). This mouse model is one system suited to determine the roles of cell surface carbohydrates in tumor metastasis, because the liver is a major target for metastasizing colonic carcinoma cells in humans. We have found that highly metastatic colonic carcinoma cell lines expressed more cell surface poly-*N*-acetyllactosamine carried on lamp-1 *N*-glycans than poorly metastatic colonic carcinoma cells (Saitoh *et al.*, 1992).

As a third approach to evaluating tumor cell migration, we employed lymph node metastasis of B16 melanoma cells (Chen *et al.*, 2005). In this assay, tumor cells were inoculated into the right footpad to form metastatic tumors in the draining popliteal lymph node, leaving the left popliteal lymph node to serve as a control. Tumor metastasis is suppressed by NK cell recruitment to tumor-bearing lymph nodes, a process facilitated by interactions between L-selectin on NK cells and L-selectin ligands on endothelial cells. When the aforementioned assay was carried out in mutant mice deficient in L-selectin or L-selectin ligands, more aggressive tumors formed in the draining lymph nodes. Moreover, reconstitution in the mutant mice with wild-type, but not L-selectin–deficient NK cells, greatly suppressed lymph node tumor formation. Because this study represents an important experimental model in studying tumor cell–immune cell interaction, the procedure of this assay is described as the second protocol.

"Experimental" Model of Bloodborne Melanoma Pulmonary
  Tumor Formation

In this model, we investigated the function of sialyl Lewis X oligosaccharides in tumor survival and colonization. By intravenous injection, most tumor cells are initially arrested in the lung tissue, simply because the lung capillary bed is the first encountered. Thus, the lung is the primary organ to form tumor colonies, and such tumor formation is evaluated by a tail vein assay.

*Forced Expression of Sialyl Lewis x in Melanoma Cells*

Mouse or human melanoma cells (B16-F1 or MeWo), which are negative for sialyl Lewis X expression, are maintained in DME high glucose medium (Irvine Scientific) supplemented with 10% fetal calf serum (FCS). Tumor cells are transfected by electroporation with cDNA encoding human fucosyltransferase-III in the pcDNA3 vector (pcDNA3-FTIII) and selected in the presence of 800 $\mu$g/ml of G418 (GIBCO/BRL). Isolated colonies are dissociated into monodispersed cells using an enzyme-free cell dissociation solution (Hanks' based) purchased from Cell and Molecular Technologies (Phillipsburg, NJ) and incubated with 10 $\mu$g/ml of anti-sialyl Lewis X antibody (CSLEX-1; Becton Dickinson) followed by fluorescent isothiocyanate (FITC)–conjugated goat affinity-purified F (ab')2 fragments specific to mouse immunoglobulin M (IgM) (Cappel). Stained cells are separated by cell sorting into three populations: (1) cells negative for sialyl Lewis X expression, (2) cells expressing moderate amounts of sialyl Lewis X, and (3) cells strongly positive for sialyl Lewis X.

*Tumor Inoculation*

*Inoculation of B16-F1 transfectants:* $1 \times 10^5$ cells in 100 $\mu$l of serum-free DME are injected into the tail vein of 6- to 8-wk-olds of C57BL/6, SCID (C.B.-17) and beige (C.B.-17scid-beige) mice. After 2 wk (for beige mice) or 3 wk (for C57BL/6 and SCID mice), mice are euthanized, lungs are fixed with Bouin's solution, and tumor foci are counted under a dissecting microscope.

*Inoculation of MeWo transfectants:* $5 \times 10^6$ cells (>90% viability) are suspended in 100 $\mu$l of serum-free RPMI medium 1640 and injected into the tail vein of BALB/c nude (nu/nu) or beige mice (C.B-17 scid-beige; 6–8 wk, female). After 2 (beige mice) or 3 (nude mice) wk, mice are killed, lungs are fixed with Bouin's solution, and tumor foci are counted under a dissecting microscope.

The results showed that formation of lung tumor foci by melanoma cells is highly increased by moderate expression of sialyl Lewis X oligosaccharides (Fig. 1). Cells with high sialyl Lewis X expression formed tumors only in NK cell–deficient beige mice and not in wild-type mice.

"Spontaneous" Model for Melanoma Lymph Node Metastasis

In this model, tumor cells are subcutaneously inoculated to allow formation of a primary tumor at the inoculation site. Disseminating tumor cells then enter the lymphatic circulation to form metastases in regional lymph nodes. Our results showed that L-selectin on NK cells and L-selectin ligands on endothelial cells facilitate migration of NK cells to lymph nodes from the bloodstream. This process is crucial to suppress metastatic formation in lymph nodes.

## Tumor Inoculation

The mouse melanoma cell line B16-F10 (ATCC CRL-6475) is cultured in DME high glucose medium (Irvine Scientific) supplemented with 10% FCS. Cells are harvested in cell dissociation buffer (Specialty Media), washed three times with phosphate-buffered saline (PBS), and resuspended in serum-free DME high glucose medium. Two million B16-F10 cells in 20 $\mu$l are inoculated s.c. into the right hind footpad of wild-type, $Rag-1^{-/-}$, L-selectin$^{-/-}$, and fucosyltransferase-IV$^{-/-}$/VII$^{-/-}$ mice.

## Relative Tumor Burden

Ten days after tumor inoculation, primary tumors at the right footpad are visible (animals failing to exhibit tumors are sacrificed). Right (tumor-draining) and left (control) popliteal lymph nodes are collected in PBS. Metastases are melanotic and occur preferentially in marginal sinuses and spread superficially to subcortical regions of the draining lymph node. Lymph nodes are photographed, and the tumor area measured using NIH ImageJ software. The relative tumor burden is expressed as the percentage of tumor areas over the widest cross-sectional area of the lymph nodes. Data are statistically analyzed using the Prism Mann Whitney test. A $p$ value less than .05 is considered statistically significant.

## Isolation of NK Cells

Single cell suspensions are prepared by mechanical dissociation of spleens and then filtered through a 70-$\mu$m cell strainer. Splenocytes are treated with 17 m$M$ Tris–HCl buffer (pH 7.4) containing 0.83% NH$_4$Cl to lyse red blood cells, washed once with PBS, and resuspended in DME high glucose medium containing 5% FCS (DME-5) ($10^8$ cells/ml). Cells are applied to a nylon fiber column (Wako) pre-equilibrated with 20 ml of PBS warmed to 37°, followed by 20 ml of warmed DME-5. Once samples have entered the nylon fiber bed, 1 ml of DME-5 is added carefully to keep the column from drying. After incubation at 37° for 1 h, nonadherent cells are eluted from columns by carefully adding 5 ml of warmed DME-5. Eluted cells are further negatively sorted using a mouse NK cell isolation kit (Miltenyi Biotec) according to the manufacturer's protocols. In brief, 10 $\mu$l of a cocktail of biotin-labeled anti-CD3, anti-CD14, anti-CD19, anti-CD36, and anti-IgE antibodies is added to $10^7$ cells resuspended in 40 $\mu$l of PBS containing 0.5% bovine serum albumin (BSA) and 2 m$M$ ethylenediamine-tetraacetic acid (EDTA) (Buffer B). After incubation for 10 min at 6–12°, 20 $\mu$l of anti-biotin–conjugated magnetic beads is added and incubated for another 15 min. Cells are then washed once with the addition of 1–2 ml of Buffer B and subjected to magnetic sorting. NK cells are collected in the non-bound fraction and resuspended in PRMI-10 for intravenous injection, or PBS with 2% FCS at $10^7$ cells/ml for fluorescent dye labeling.

## Suppression of Tumor Formation in L-Selectin–Deficient Mice

NK cells were isolated from 48 wild-type C57BL/6 mice as described earlier, and purified NK cells (6–10 × $10^6$ cells/mouse) were injected i.v. into four L-selectin–deficient mice one day (Day 1) after B16 tumors were inoculated. A similar number of NK cells were adoptively transferred to the same mice on Days 4 and 7, and the popliteal lymph nodes were monitored for tumor formation on Day 10.

## Flow Cytometry for NK Cells

Single cell suspensions are prepared from control and draining popliteal lymph nodes by mechanical dissociation in ice-cold RPMI-1640 medium (Irvine Scientific) supplemented with 5% FCS, $10^{-5}$ $M$ $\beta$-mercaptoethanol ($\beta$-ME, Sigma-Aldrich), 100 $\mu$g/ml streptomycin and 100 U/ml of penicillin (Invitrogen), 1X sodium pyruvate, and 1X nonessential amino acids (RPMI-10). Cells are filtered through a 70-$\mu$m cell strainer (BD Labware) and resuspended in 100 $\mu$l of PBS containing 1% BSA, 0.01% sodium azide, and 4 $\mu$g/ml of anti-FcR mAb 2.4G2 (mouse BD Fc BlockTM, BD Pharmingen). After incubation for 10 min in ice, FITC-conjugated anti-CD3 (1A17) and phycoerythrin (PE)-conjugated anti-NK1.1 (PK136) are added at a concentration of 10 $\mu$g/ml and cells are further incubated on ice for another 30 min. Stained cells are washed two times with ice-cold PBS containing 2% FCS and 0.01% sodium azide (Buffer A). Washed cells are resuspended in Buffer A containing 0.5 $\mu$g/ml PI and analyzed on a FACSort or FACScanto flow cytometer (BD Biosciences). NK cells are identified as $CD3^-NK1.1^+$. Data are analyzed with FlowJo software.

## NK Cell Adoptive Transfer

Purified wild-type or mutant NK cells in PBS with 2% FCS are labeled with either 1 $\mu M$ CFSE or 2.5 $\mu M$ SNARF-1 (Molecular Probes) at 37° for 10 min. After three washes with RPMI-10, cells are resuspended in ice-cold serum-free RPMI-1640 medium. A mixture containing an approximately equal number of wild-type C57BL/6 and mutant NK cells (1–5 × $10^6$) is injected i.v. into the tail vein of wild-type C57BL/6 mice. After 3, 18, or 42 h, peripheral blood is collected from the posterior vena cava, and the

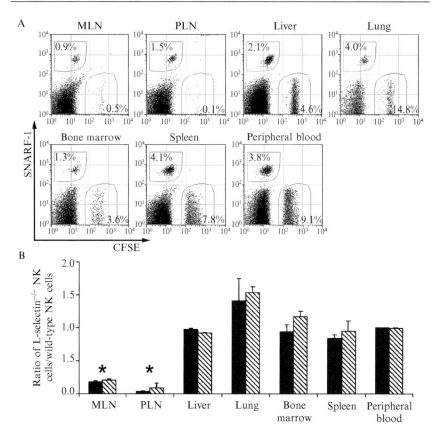

FIG. 2. L-selectin–mediated natural killer (NK) cell trafficking to lymph nodes. (A) Wild-type and L-selectin–deficient NK cells were labeled with SNARF-1 and CFSE, respectively, or vice versa. A mixture was injected intravenously into the tail vein of a wild-type C57BL/6 mouse. After 18 or 42 h, various tissues were harvested and single cell suspensions were stained with anti-NK1.1 monoclonal antibody and subjected to flow cytometric analysis. MLN, mesenteric lymph nodes; PLN, peripheral lymph nodes. Only NK1.1[+] (gated) lymphocytes are shown 42 h after transfer. The numbers in dot plots indicate the percentages of transferred NK cells among all NK1.1[+] cells. (B) Wild-type NK cells were labeled with SNARF-1 and L-selectin–deficient NK cells were labeled with CFSE. Closed and hatched bars represent ratios of L-selectin–deficient cells to wild-type cells 18 and 42 h after transfer, respectively. The data shown represent the average $\pm$ standard deviation of three independent experiments. Asterisk (*) indicates $p < .001$ versus the peripheral blood. (Modified, with permission, from Chen *et al.*, 2005.)

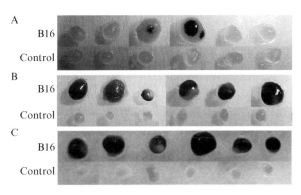

FIG. 3. Tumor metastasis in L-selectin$^{-/-}$ and FucT-IV$^{-/-}$/VII$^{-/-}$ mice. B16-F10 melanoma cells were inoculated subcutaneously into the right hind footpads. Development of tumor foci was examined and photographed 10 days later. Tumor formation was examined in 22 C57BL/6 (A), 26 L-selectin$^{-/-}$ (B), and 14 FucT-IV$^{-/-}$/VII$^{-/-}$ (C) mice. Representative results derived from six mice from each group are shown. (Modified, with permission, from Chen *et al.*, 2005.) (See color insert.)

mouse perfused with PBS containing 10 U/ml heparin (Sigma-Aldrich). Single-cell suspensions of mesenteric lymph nodes, peripheral lymph nodes, spleen, and peripheral blood are prepared as described earlier. Bone marrow cells from one femur and one tibia are flushed with RPMI-10. The liver and lung are minced and digested with 50 U/ml collagenase (Roche Applied Sciences) and 0.01% DNase I (Sigma-Aldrich) at 37° for 1 h. Tissue homogenates are then forced through a 70-$\mu$m filter and centrifuged at 300$g$ for 5 min. Cell pellets are resuspended in 20 ml (for liver) or 8 ml (for lung) of 35% Percoll in PBS (Amersham Biosciences) containing 200 U/ml heparin and overlaid onto 67% Percoll in PBS. Samples are centrifuged at 600$g$ for 20 min at room temperature without braking. Mononuclear cells are collected at the interface and washed three times with PBS before antibody staining. The frequency of adoptively transferred NK cells (NK1.1$^+$CFSE$^+$ or NK1.1$^+$SNARF-1$^+$) in various organs of the recipients is determined by flow cytometry. The relative efficiency of migration to peripheral organs is expressed as the ratio of L-selectin$^{-/-}$ NK cells over wild-type NK cells, which is further normalized by comparison with the same ratio calculated from peripheral blood because it serves as a reservoir for NK cells before their entry into organs.

In our study, the results obtained in the aforementioned system showed that (1) NK cells, but not T or B cells, play a major role in suppressing tumor formation in lymph nodes (Fig. 2); (2) NK cells are recruited to tumor-bearing lymph node or regional lymph node where an immune response is evoked; (3) NK cell recruitment depends on interaction between L-selectin expressed on NK cells and L-selectin ligands expressed on endothelial cells (Fig. 3); and (4) NK cells derived from wild-type mice rescue suppression of tumor formation in L-selectin–deficient mutant mice (Chen *et al.*, 2005).

## Acknowledgments

We thank Dr. Elise Lamar for critically reading the manuscript. The work in our laboratory was supported by National Institutes of Health (NIH) grants CA33000, CA33895, CA48737, and CA71932.

## References

Chen, S., Kawashima, H., Lowe, J. B., Lanier, L. L., and Fukuda, M. (2005). Suppression of tumor formation in lymph nodes by L-selectin–mediated natural killer cell recruitment. *J. Exp. Med.* **202,** 1679–1689.

Elkin, M., and Vlodavsky, I. (2001). Tail vein assay of cancer metastasis. *Current Protocols Cell Biol.* 19.2.1–19.2.7.

Fidler, I. J. (2003). The pathogenesis of cancer metastasis: The 'seed and soil' hypothesis revisited. *Nat. Rev. Cancer* **3,** 453–458.

Iwai, T., Kudo, T., Kawamoto, R., Kubota, T., Togayachi, A., Hiruma, T., Okada, T., Kawamoto, T., Morozumi, K., and Narimatsu, H. (2005). Core 3 synthase is down-regulated in colon carcinoma and profoundly suppresses the metastatic potential of carcinoma cells. *Proc. Natl. Acad. Sci. USA* **102,** 4572–4577.

Morikawa, K., Walker, S. M., Jessup, J. M., and Fidler, I. J. (1988). *In vivo* selection of highly metastatic cells from surgical specimens of different primary human colon carcinomas implanted into nude mice. *Cancer Res.* **48,** 1943–1948.

Ohyama, C., Tsuboi, S., and Fukuda, M. (1999). Dual roles of sialyl Lewis X oligosaccharides in tumor metastasis and rejection by natural killer cells. *EMBO J.* **18,** 1516–1525.

Ohyama, C., Kanto, S., Kato, K., Nakano, O., Arai, Y., Kato, T., Chen, S., Fukuda, M. N., and Fukuda, M. (2002). Natural killer cells attack tumor cells expressing high levels of sialyl Lewis x oligosaccharides. *Proc. Natl. Acad. Sci. USA* **99,** 13789–13794.

Saitoh, O., Wang, W. C., Lotan, R., and Fukuda, M. (1992). Differential glycosylation and cell surface expression of lysosomal membrane glycoproteins in sublines of a human colon cancer exhibiting distinct metastatic potentials. *J. Biol. Chem.* **267,** 5700–5711.

Scheidegger, E. P., Lackie, P. M., Papay, J., and Roth, J. (1994). *In vitro* and *in vivo* growth of clonal sublines of human small cell lung carcinoma is modulated by polysialic acid of the neural cell adhesion molecule. *Lab. Invest.* **70,** 95–106.

Suzuki, M., Nakayama, J., Suzuki, A., Angata, K., Chen, S., Sakai, K., Hagihara, K., Yamaguchi, Y., and Fukuda, M. (2005). Polysialic acid facilitates tumor invasion by glioma cells. *Glycobiology* **15,** 887–894.

# Author Index

# Subject Index

## A

α1,4-*N*-Acetylgalactosamine
  crystal studies of ligand binding, 3, 9
  isothermal titration calorimetry of
      donor sugar binding
    experimental design, 6
    materials, 11
    protein expression and purification, 9–11
    running conditions, 11
    UDP-GalNAc studies, 6–9
    UDP-GlcNAc studies, 6–9
Arylsulfatases
  functions, 244–245
  human enzymes
    assay in conditioned media, 246–248
    endoglucosamine-6-sulfatase assay,
        248–249
    enzyme-linked immunosorbent assay
        of heparin–vascular endothelial
        growth factor interaction
        modulation, 249, 251–252
    preparation, 245–246
  types, 244

## B

BIAcore, *see* Surface plasmon resonance
BLAST, *in silico* cloning of
    glycosyltransferase genes, 91
Blood group B galactosyltransferase,
    donor sugar binding studies with
    nuclear magnetic resonance
  protein shifts, 23, 25
  sugar donor shifts, 19–21

## C

Cancer, *see* Tumors
Carbohydrate mimetics, *see* Phage display
Carbohydrate–protein interactions, *see*
    Isothermal titration calorimetry; Nuclear
    magnetic resonance; X-ray crystallography

Chemokine–glycosaminoglycan interactions,
    *see* Chondroitin sulfate
Chinese hamster ovary cells
  carbohydrate-modifying enzyme
      transfection studies
    complementary DNA identification, 293
    flow cytometry for oligosaccharide
        analysis, 299–301
    multiple gene expression, 298–299
    overview, 293–295
    plasmid preparation and cell culture, 295
    stable expression, 296, 298
    transient expression, 296
  glycosaminoglycan mutants
    biosynthesis, 205–206
    characterization of gain-of-function
        mutant
      fluorescence-activated cell sorting,
          211–213
      herpes simplex virus entry assay, 211
    gain-of-function mutant preparation
      cell culture, 208
      principles, 208
      retroviral transduction, 209–210
      transfection, 208–209
    loss-of-function mutants
      gene copy number determination,
          214–217, 219
      generation, 213–214
      types, 206–207
  glycosylation overview, 160
  glycosylphosphatidylinositol anchor
      mutants
    characterization
      enzymatic treatment of labeled
          lipids, 194
      labeling of intermediates *in vitro*,
          193–194
      labeling of intermediates *in vivo*, 193
      thin-layer chromatography, 191–193
    expression cloning of mutant gene,
        194–196
    mutagenesis and selection, 188

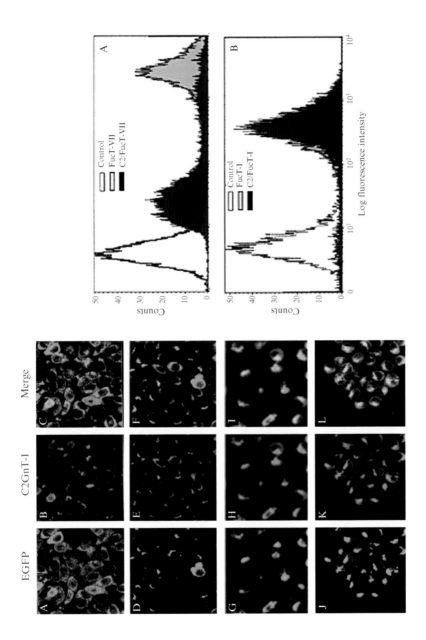

## A C2GnTI-EGFP versus the $\beta$1,3GnTIII-DsRed

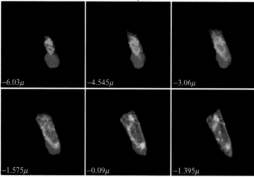

## B C2GnTI-EGFP versus the ST3GalI-DsRed2

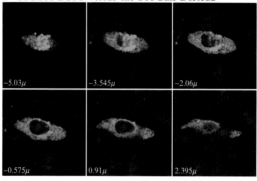

EL-BATTARI, CHAPTER 7, FIG. 5. Dual-color confocal microscopy for simultaneous comparison of intracellular distribution of enhanced green fluorescent protein (EGFP)- and DsRed2-tagged proteins. Cells stably expressing EGFP-conjugated C2GnT-I were transfected with either $\beta$1,3GnT-III or ST3Gal-I, both tagged with DsRed2, and live cells were examined by confocal microscopy. Shown are a series of z-axis slices of the intracellular distribution of C2GnTI-EGFP versus the $\beta$1,3GnTIII-DsRed2 (A) and C2GnTI-EGFP versus the ST3GalI-DsRed2 (B). Note that, in contrast to C2GnTI-EGFP, which displays a typical Golgi distribution, both DsRed2-fused $\beta$1,3GnT-III and ST3Gal-I proteins are concentrated within the two poles of the nucleus. Despite this feature, the ST3Gal-I is localized in the same planes as C2GnT-I, whereas the $\beta$1,3GnTIII distributes to areas relatively far from that of C2GnT-I.

---

EL-BATTARI, CHAPTER 7, FIG. 4. Confocal fluorescent micrographs showing the coexpression of C2GnT-I with enhanced green fluorescent protein (EGFP)–tagged proteins, FucT-VII (A, B, C), C2/FucTVII (D, E, F), FucT-I (G, H, I), and C2/FucT-I (J, K, L). Cells were fixed, permeabilized, and stained with polyclonal antibody for C2GnT-I (red) and examined by confocal microscopy. The merged images show overlapping red and green (EGFP) pixels in yellow. Magnification: 500×. Flow cytometric analyses of P-selectin binding and anti-*H* immunoreactivity. CHO/C2P1 cells were transfected with constructs coding for the chimeric C2/FucTI or C2/FucTVII enzymes or their normal counterparts fused to EGFP and assayed for P-selectin binding (A) or for the expression of the blood group *H* determinant (B). Note that no mislocalization or alteration in biological activity can be observed for FucT-I or its chimeric counterpart, although displacing FucT-VII to C2GnT-I compartment dramatically alters the P-selectin–binding activity of cells expressing the chimeric C2/FucTVII.

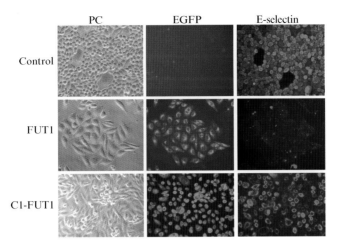

|  | PC | EGFP | E-selectin |
|---|---|---|---|
| Control | | | |
| FUT1 | | | |
| C1-FUT1 | | | |

EL-BATTARI, CHAPTER 7, FIG. 6. Altered E-selectin binding to sLe$^x$-expressing cells (CHO/F7, control) following introduction of FucT-I (FUT1) gene and the loss of inhibition in the presence of the chimeric variant fused to the CTD of $\beta$1,3GnT-III (C1-FUT1). Shown are phase contrasts (PCs) of cells and fluorescence of transfectants (enhanced green fluorescent protein [EGFP]), together with their corresponding E-selectin-IgM–binding patterns as revealed by RITC-conjugated secondary antibody (E-selectin).

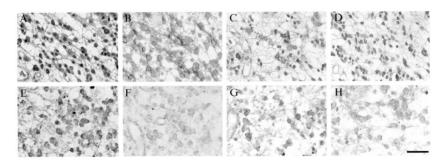

NAKAYAMA ET AL., CHAPTER 8, FIG. 1. Expression of polysialic acid (PSA), neural cell adhesion molecule (NCAM), ST8Sia II messenger RNA (mRNA), and ST8Sia IV mRNA in a diffuse astrocytoma. (A) Protoplasmic astrocytoma cells proliferating in a loose microcystic matrix are shown (hematoxylin–eosin staining). (B) NCAM was detected by the 123C3 antibody (Zymed, Carlsbad, CA). (C) PSA was detected using 5A5 antibody (University of Iowa Hybridoma Bank, Iowa City, IA), and (D) 5A5 immunoreactivity was eliminated after pretreatment with endo-N, which cleaves PSA. In situ hybridization of ST8Sia IV mRNA using (E) antisense and (F) sense control probes. In situ hybridization of ST8Sia II mRNA using (G) antisense and (H) sense control probes. Positive signals were detected in the perinuclear cytoplasm of tumor cells. Bar = 50 $\mu$m. (Reprinted, with permission, from Suzuki et al., 2005.)

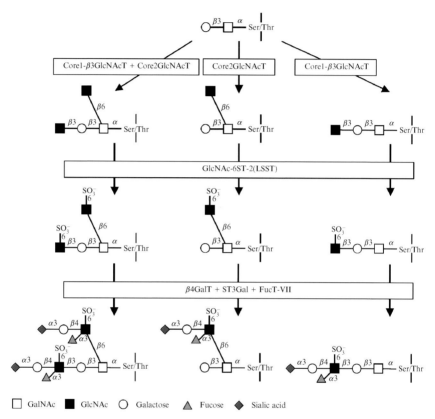

| □ GalNAc | ■ GlcNAc | ○ Galactose | ▲ Fucose | ◆ Sialic acid |
| --- | --- | --- | --- | --- |

KOBAYASHI, CHAPTER 9, FIG. 1. Structure and biosynthesis of 6-sulfo sialyl Lewis X. Core 1 *O*-glycans can be extended by Core1-β3GlcNAcT and then sulfated by GlcNAc-6ST-2 (LSST). The resultant oligosaccharide can be galactosylated, sialylated, and fucosylated to form 6-sulfo sialyl Lewis X on extended core 1 *O*-glycans. Alternatively, core 1 *O*-glycans can be branched by Core2GlcNAcT, and then sulfated by LSST, galactosylated, sialylated, and fucosylated to form 6-sulfo sialyl Lewis X on core 2 branched *O*-glycans. Core1-β3GlcNAcT can act on core 2 branched *O*-glycans, leading to bi-antennary *O*-glycans containing both core 2 branch and core 1 extension. All three *O*-glycans capped by 6-sulfo sialyl Lewis X function as L-selectin ligands (Based on Yeh *et al.*, 2001).

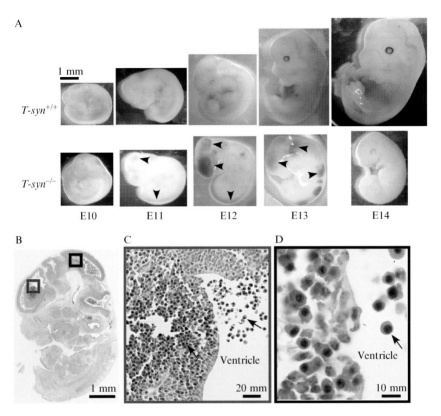

Xᴵᴬ ᴀɴᴅ McEᴠᴇʀ, Cʜᴀᴘᴛᴇʀ 21, Fɪɢ. 2. Targeted disruption of *T-syn* causes fatal embryonic hemorrhage in mice. (A) Comparison of *T-syn*⁺/⁺ and *T-syn*⁻/⁻ embryos at different developmental stages. Blood is visible in the hearts of the older *T-syn*⁺/⁺ embryos. Arrowheads indicate hemorrhage in the brain parenchyma and ventricles, spinal cord, and spinal canal of *T-syn*⁻/⁻ embryos. Less blood is detected in *T-syn*⁻/⁻ hearts because of anemia. The E14 *T-syn*⁻/⁻ embryo is dead and appears pale because blood circulation has ceased. (B) Sagittal section of the E12 *T-syn*⁻/⁻ embryo. The red and black boxes indicate hemorrhagic lesions, which are shown at higher magnification in (C) and (D). Bleeding is visible in both the brain parenchyma and ventricles. Arrows indicate erythrocytes, which are nucleated at this stage of development. Modified, with permission, from Xia *et al.*, 2004.

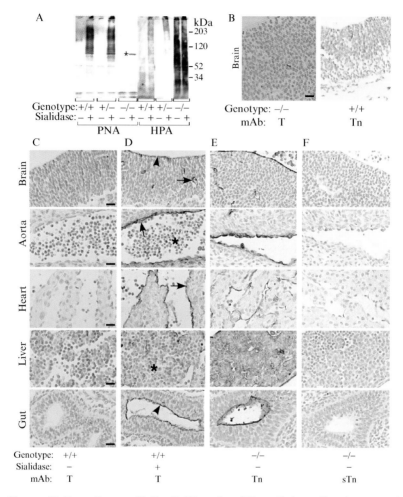

XIA AND MCEVER, CHAPTER 21, FIG. 3. Disruption of *T-syn* eliminates T-antigen expression and exposes the Tn antigen in murine embryos. (A) Blots of E12 tissue extracts probed with *Arachis hypogaea* agglutinin (PNA), which recognizes nonsialylated core 1 *O*-glycans, and with *Helix pomatia* agglutinin (HPA), which recognizes the nonsialylated Tn antigen. The extracts were incubated with or without sialidase before electrophoresis and blotting. The asterisk indicates the binding of PNA to the added sialidase. The blots are representative of three experiments. (B to F) Immunohistochemical staining of E12 tissue sections with monoclonal antibodies (mAbs) to the T antigen, Tn antigen, or sialyl-Tn (sTn) antigen. The sections in (D) were pretreated with sialidase. Brown reaction product marks sites of antibody binding. Arrows indicate endothelial cells, arrowheads indicate epithelial cells, and asterisks indicate hematopoietic cells. Reproduced, with permission, from Xia *et al.*, 2004.

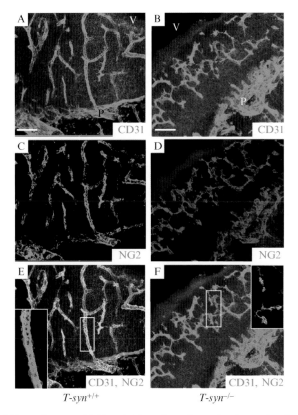

$T\text{-}syn^{+/+}$          $T\text{-}syn^{-/-}$

XIA AND MCEVER, CHAPTER 21, FIG. 5. $T\text{-}syn^{-/-}$ embryos develop a chaotic microvascular network. (A to H) Visualization of microvessels in E12 hindbrains using maximal intensity projections of z-stacked confocal images. Endothelial cells were stained with antibodies to CD31 (green), pericytes were stained with antibodies to the proteoglycan NG2 (red), and basement membrane was stained with antibodies to laminin (red). $T\text{-}syn^{+/+}$ embryos form a network of capillaries with uniform diameters and a regular branching pattern (A, C, E, and G), whereas $T\text{-}syn^{-/-}$ embryos form capillaries with heterogeneous diameters and excessive, irregular branches (B, D, F, and H). Insets in (E) and (F) are thin (5 $\mu$m) optical slices of representative vessels enlarged 2.5-fold, which illustrate the abluminal relationship of the NG2-positive pericytes to the CD31-positive endothelial cells. P, perineural vascular plexus; V, ventricle. Scale bars, 50 $\mu$m unless noted otherwise. Adapted, with permission, from Xia et al., 2004.

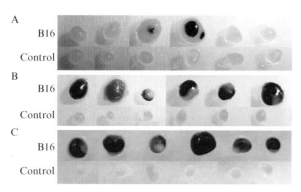

Chen and Fukuda, Chapter 24, Fig. 3. Tumor metastasis in L-selectin$^{-/-}$ and FucT-IV$^{-/-}$/VII$^{-/-}$ mice. B16-F10 melanoma cells were inoculated subcutaneously into the right hind footpads. Development of tumor foci was examined and photographed 10 days later. Tumor formation was examined in 22 C57BL/6 (A), 26 L-selectin$^{-/-}$ (B), and 14 FucT-IV$^{-/-}$/VII$^{-/-}$ (C) mice. Representative results derived from six mice from each group are shown. (Modified, with permission, from Chen *et al.*, 2005.)